Space Sciences Series of ISSI

Volume 94

Series Editor

International Space Science Institute, Bern, Switzerland

The Space Sciences Series of ISSI books are coherent reports of the findings, discussions, and ideas that result from international scientific workshops regularly held at the International Space Science Institute (ISSI) in Bern, Switzerland. ISSI's main task is to contribute to the achievement of a deeper understanding of the results from space-research missions, adding value to those results through multi-disciplinary research in an atmosphere of international cooperation. The books are reprints of special issues in the Space Science Reviews journal and occasionally of special issues in the Surveys in Geophysics journal.

Sophie Hebden • Michael Rast
Narelle van der Wel • Hannah Liddy
Richard Wood
Editors

Tipping Elements in the Earth's Climate System

Exploring the Use of Earth Observations to Improve Understanding and Monitoring Capability

Previously published in *Surveys in Geophysics*
Volume 46, Issue 2, April 2025

Springer

Editors
Sophie Hebden
Fermi Ave | Harwell Science
& Innovation Campus
ECSAT
Oxfordshire, UK

Michael Rast
Earth Sciences
International Space Science Institute
Bern, Switzerland

Narelle van der Wel
Science and Innovation
World Climate Research Programme
Genève, Switzerland

Hannah Liddy
Center for Climate Systems Research
Columbia University
New York, NY, USA

Richard Wood
Met Office
Exeter, UK

Spin-off from journal: "Surveys in Geophysics" Volume 46, Issue 2, April 2025

ISSN 1385-7525
Space Sciences Series of ISSI
ISBN 978-94-024-2348-8

Cover illustration: Greenland ice sheet melt, contains modified Copernicus Sentinel data (2021), processed by ESA.

This Springer imprint is published by the registered company Springer Nature B.V.
The registered company address is: Van Godewijckstraat 30, 3311 GX Dordrecht, The Netherlands

Contents

Surveys in Geophysics (2025) 46:259–264
https://doi.org/10.1007/s10712-025-09887-7

GUEST EDITORIAL

Special Issue: International Space Science Institute (ISSI) Workshop on Tipping Elements in the Earth's Climate System

Sophie Hebden[2] · Narelle van der Wel[3] · Hannah Liddy[4] · Jonathan Donges[5] · Richard Wood[6] · Michael Rast[1]

Published online: 16 May 2025

1 Introduction

Research into tipping elements of the Earth system has advanced significantly since Lenton et al. (2008) defined them as 'subsystems that can be switched under certain circumstances into a qualitatively different state by small perturbations'. In recent years, studies on their stability have accelerated, outpacing the average growth of scientific literature (see Fig. 1). Over the last two decades, the International Panel on Climate Change (IPCC) has progressively strengthened its risk assessment of nonlinear changes in climate tipping elements (see IPCC AR6 WG1, Table 4.10). Particular attention has been given to the potential slowing—or even shutdown—of the Atlantic Meridional Overturning Circulation (AMOC), irreversible ice loss from the Greenland and Antarctic ice sheets, and ecosystem regime shifts in tropical and boreal forests. Despite paleoclimate evidence of abrupt change in these Earth system components, Earth system models vary in their projections of the rate, timing and irreversibility of these tipping elements (Drijfhout et al. 2015). Our best global climate models still lack a full representation of potential tipping element processes, limiting the scientific guidance available for risk management. We are far from implementing an observing system capable of detecting and monitoring the onset of tipping points in real time.

✉ Sophie Hebden
sophie.hebden@esa.int

[1] International Space Science Institute (ISSI), Hallerstrasse 6, 3012 Bern, Switzerland

[2] ECSAT, Fermi Avenue, Harwell Campus, Didcot, Oxfordshire OX11 0FD, UK

[3] World Climate Research Programme, Av. de la Paix 7BIS, 1211 Geneva, Switzerland

[4] Center for Climate Systems Research @ NASA GISS, Columbia Climate School, Columbia University in the City of New York, New York, NY 10025, USA

[5] Potsdam Institute for Climate Impact Research (PIK), 14412 Potsdam, Germany

[6] The Met Office Hadley Centre, FitzRoy Road, Exeter EX1 3PB, UK

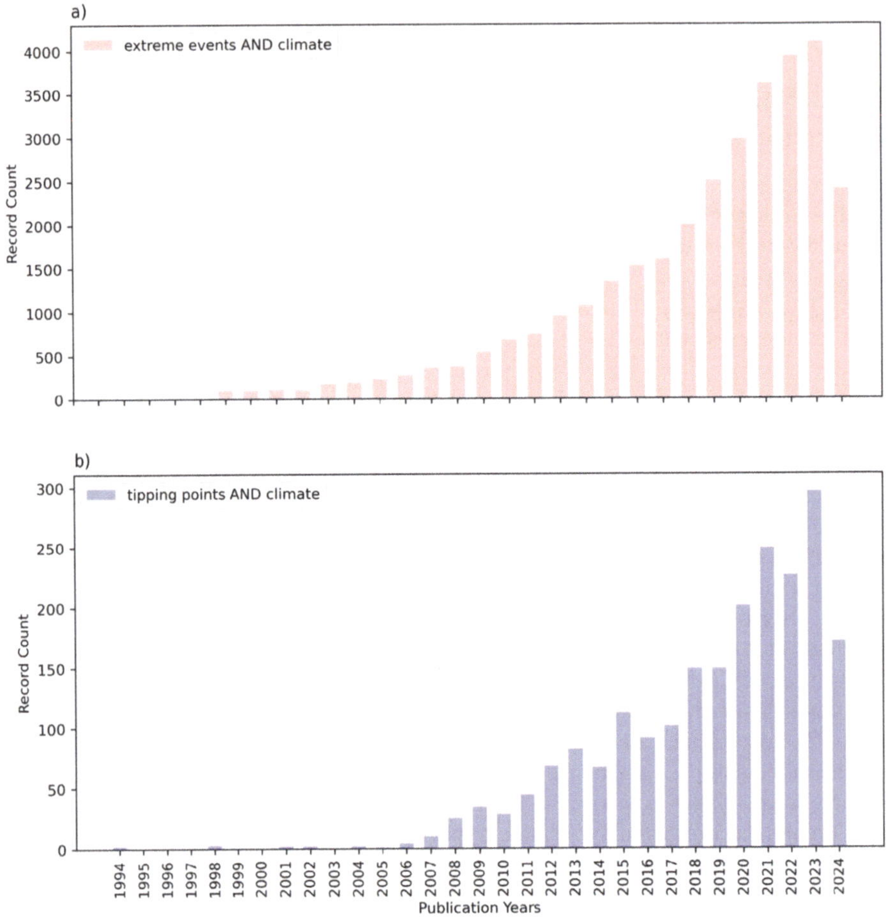

Fig. 1 **a** Number of publications in web of science core collection on the topic 'extreme events AND climate', **b** number of publications on the topic 'tipping points AND climate' (search: 23 July 2024) (Romanou et al. (2024), this issue)

2 Earth Observations for Direct Monitoring of Tipping Behaviour

The limited temporal extent of the satellite record—spanning four or five decades at best—means that direct monitoring of trends and indicators for the onset of tipping points is restricted to 'fast' tipping processes that occur over decades or less. Examples include winter sea ice in the Arctic, the subpolar gyre, the Sahel/Monsoon system, forest dieback—both tropical and boreal, freshwater ecosystems, and Arctic permafrost. The Workshop heard of the potential to develop proxy indicators of tipping behaviour and instabilities, for example, to indicate the triggering of collapse of the West Antarctic Ice Sheet.

A major strength of satellite-based Earth Observation (EO) is the ability to capture multiple temporal and spatial scales of tipping behaviour, from local to regional to global, and from daily to interannual. Across these scales, by combining the variables observed, there is potential for studying feedbacks in the Earth system that can lead to tipping points. There is opportunity to exploit observations related to the timing of events, to study extreme events and their impacts—including impacts on society, as well as cascading impacts and pace of change. There is also opportunity to exploit the fine spatial detail of satellite data, examining patterning changes in vegetation across a gradient of environmental change, for example (Lenton et al. 2024). Indicators with multiple variables that combine information from different sources can build a fuller picture of the processes occurring, for example, to study the pressures affecting vegetation. A better integration of EO data and platforms will facilitate research efforts for the detection of tipping elements.

3 Earth Observation for Modelling

EO has a key role to play in improving the ability of climate models to project possible climate futures. Climate models rely on long-term EO datasets like the Essential Climate Variables (ECVs), many of which are based on remote sensing data. ECVs developed by programmes like the ESA Climate Change Initiative are used for model development, to assess model skill (judged by how well models can reproduce past observations), and to constrain models—helping to determine certain parameters, for example, and to set initial conditions of part of the climate system at the start of a model simulation. The Global Climate Observing System (GCOS) is an international scientific body that maintains the definitions of the ECVs required for systematic observations of the Earth's climate. These definitions help underpin climate services, providing information for decision-making.

Applied to tipping points, there is huge scope to 'assimilate'—a statistical approach that brings the model's outputs closer to the observational data—EO to improve process-based models. For example, ocean colour and sea surface temperature data could be used to better constrain biogeochemical or ecosystem processes in ocean ecosystem models to lower the uncertainty of predictions of thresholds and timescales of regime shifts.

It is fair to say that climate tipping points as a source of climate risk are now a familiar issue in the public realm. The growing popularity of the concept underscores the need for the scientific community to better understand the risks that they pose: we need to understand the impacts of crossing tipping points at different levels of global warming in order to guide mitigation and adaptation efforts. World leaders at the COP27 Conference of the Parties to the United Nations Framework Convention on Climate Change held in Sharm el-Sheikh, Egypt, in November 2022, agreed on the need to understand the impact of tipping points on the cryosphere.

4 This Special Issue

Advancing knowledge about tipping elements, including their detection, prediction, irreversibility, and impacts on the Earth's climate and environment, are increasingly central to the priorities of the World Climate Research Programme (WCRP), which has recently launched a dedicated tipping points and high-impact event assessment. In addition, a newly endorsed Coupled Model Intercomparison Project (CMIP) protocol for tipping elements seeks to provide a basis for directly comparing the mechanisms, timing, and impacts of abrupt change across Earth system models.

To respond to the need for more information, the European Space Agency (ESA) sought to understand the unique contribution that Earth Observation (EO) data can make in identifying potential tipping elements, improving resilience monitoring and early warning of tipping points, and assessing their possible interactions. In January 2021, ESA partnered with the International Space Science Institute (ISSI) in Bern, Switzerland, to hold a Forum on 'Tipping points in the Earth's Climate'. One of the outcomes of the Forum was the dedicated Workshop held 10–14 October 2022, 'Tipping Points and Understanding EO data needs for a Tipping Element Model Intercomparison Project (TipMIP)'. The Workshop brought researchers together from EO, climate modelling, theoretical and process understanding, and enabled communities to understand one another's priorities.

Animated discussions covered how EO together with modelling efforts can support monitoring and process understanding of tipping points and their interactions, and the plans for joint model experiments. The EO and modelling communities have operated somewhat in silos. This Workshop illustrated how important it is to get scientists from both communities into one room to understand one another's challenges and priorities. This Special Issue contains the scientific contributions to this Workshop. It addresses both generic issues of how observations can be used to inform science and policy around climate tipping points and specific assessments of the observing needs for particular tipping elements.

The papers in this Special Issue expand on the contents of three short 'white papers' that were produced shortly after the Workshop for use in programme planning. These white papers are Bartsch et al. (2025), Wood et al. (2025), and an unpublished paper entitled 'Cryosphere Tipping Points'.

Tipping points and abrupt changes in the climate system have attracted the attention of the media and divided the scientific community. Often an overly simplistic 'doomsday' messaging is portrayed to the public, leading to an underemphasis of studies that are more conservative and nuanced. Despite progress, understanding of the risks is still insufficient. Another impediment to progress has been the variety of definitions and terms in use in the scientific literature. In this Special Issue, Stocker et al. (2024) discuss this situation, finding that the definitions outlined in IPCC AR6 WGI and WGII focus on the dynamical behaviour of the systems, but a broader terminology to emphasise the impacts could be more useful for the research community. They recommend the use of 'high impact–low likelihood events' (HILL) terminology, and call for a joint scientific assessment conducted under the auspice of the Intergovernmental Panel on Climate Change (IPCC) and Intergovernmental Science-Policy Platform on Biodiversity and Ecosystem Services (IPBES) to understand possible irreversible consequences of HILL events and their associated risk in the context of anthropogenic climate change, ecosystem degradation, and biodiversity losses.

With a focus on terrestrial and marine ecosystems, Bathiany et al. (2024) assess the role of Earth Observations in monitoring ecosystem resilience to large anthropogenic changes. They synthesise how the methods applied to modelled and observed data have been used to assess ecosystem resilience across different scientific communities. They then provide a methodological approach that can be used for Earth Observation data. This article usefully blends the perspectives of 'data users', defined as Earth system modellers and dynamical systems experts, with those of 'data producers', defined as remote sensing experts, to chart a path forward to improve resilience monitoring.

The frequency and intensity of some extreme events, particularly extreme heat events, have been increasing under the continued rise of anthropogenic climate change. Increases in the regional extent as well as the interactions between extreme events (e.g. compound extremes) may influence or interact with climate feedbacks associated with tipping elements. Romanou et al. (2024) explore mechanistic links between extreme weather and climate events and tipping elements using evidence from paleoclimate data and climate modelling experiments. They hypothesise that Earth Observations of extreme events in regions associated with tipping elements could be used as early warning indicators of instabilities in the climate system.

Several of the archetypal Earth system tipping points involve the ocean, either its circulation, its biogeochemistry, or its ecosystems. Wood et al. (2024) survey a number of such ocean tipping elements, focusing on large-scale events or local mechanisms that may operate over a large geographical domain. The emphasis is on how EO, in combination with in situ observations and models, can help deliver plausible storylines of the consequences of tipping events, monitoring and early warning of an impending event, and overall improvements in the modelling of tipping processes. Limited space–time coverage of in situ observations, and the limitations of most EO datasets to surface variables, means that there needs to be a particular emphasis on combining EO, in situ and model information if we are to build systems for monitoring, early warning, and prediction of ocean tipping points.

To place current observations of tipping elements in a wider climate monitoring perspective, Loriani et al. (2025) assess how well the current Essential Climate Variables (ECVs), as defined by the Global Climate Observing System (GCOS), meet the needs of monitoring and early warning of tipping, for a selection of tipping elements. The analysis identifies where enhancements in the ECVs would benefit the risk management of tipping points, and this information can be brought in alongside broader considerations in determining priorities for the evolution of the ECVs. The analysis framework used could be extended to include other tipping elements not considered in the paper.

Brovkin et al. (2025) review the climate feedbacks in the Arctic for threshold-type behaviour and irreversibility, outlining observing capability and gaps for the early detection of regime shifts. Climate change is significantly impacting permafrost landscapes and freshwater systems, with ecosystem shifts often triggered by disturbances and extreme events. Many of the feedbacks are happening at local scales that are neglected in Earth system models, and observational datasets struggle to characterise the changes in the permafrost thaw landscapes comprehensively. Taking permafrost lakes as a case in point, however, the authors characterise changes to Arctic lake distribution and abundance using a hydrological dataset, finding an overall trend of decreasing lake area until a hydrological tipping point is reached, leading to regional drying. Earth system modelling, meanwhile, suggests no clear global temperature threshold for these shifts in hydroclimate.

5 Research Frontiers

As the challenges of climate change become ever more apparent and immediate over the coming decades, decision-makers will call for increasingly detailed and reliable information about possible climate futures. The risks associated with tipping events present a particular challenge, as they may imply changes that are faster or in a different direction from the expected trends. The increasing volume of research in this area reflects the urgency felt both by climate scientists and policy makers to understand and predict events that may have serious impacts on society. The analyses presented here provide an important contribution to understanding how observations of the Earth system can help societies to develop science-based strategies to avoid, or build resilience to, the possibility of such 'unsurprising surprises'.

Recent wildfire and flooding events have highlighted society's fragility in the face of worsening climate extremes. Society also faces compound climate extremes whereby multiple hazards occur in the same location, or concurrent extremes occur at different locations. An open question is how extreme events might interplay with tipping points to drive even worse impacts on society. Research frontiers will push our current EO datasets and modelling tools into new territory, in combinations that enable an exploration of the interactions between tipping points in the climate and society.

Through activities like this ISSI Workshop and other coordinated efforts, such as those organised by Future Earth and WCRP, there will be more opportunities to build the tipping point research community in the future. Engagement across scientific communities for Earth Observations and mission development and Earth system modelling and climate analytics will enable society to better identify and constrain the risks posed by tipping points.

Finally, we thank the ISSI and its staff for organising this Workshop. We thank the reviewers for their hard work and the former Co-Editor-in-Chief, Michael Rycroft, for guiding this Special Issue to fruition. We also thank the staff of Springer Nature for publishing this Special Issue in a timely fashion.

References

Bartsch A, Bastos A, Brovkin V, Calamita E, Lever J, Verbesselt J (2025) Summary of biosphere discussion at the ISSI workshop 'Tipping points and understanding EO data needs for a tipping element model intercomparison project (TipMIP)', October 10–14, 2022. Zenodo. https://doi.org/10.5281/zenodo.15165491

Bathiany S, Bastiaansen R, Bastos A et al (2024) Ecosystem resilience monitoring and early warning using Earth observation data: challenges and outlook. Surv Geophys. https://doi.org/10.1007/s10712-024-09833-z

Brovkin V, Bartsch A, Hugelius G et al (2025) Permafrost and freshwater systems in the arctic as tipping elements of the climate system. Surv Geophys. https://doi.org/10.1007/s10712-025-09885-9

Drijfhout S, Bathiany S, Beaulieu C, Brovkin V, Claussen M, Huntingford C et al (2015) Catalogue of abrupt shifts in intergovernmental panel on climate change climate models. Proc Natl Acad Sci 112(43):E5777–E5786

Lenton TM, Held H, Kriegler E et al (2008) Tipping elements in the Earth's climate system. Proc Natl Acad Sci 105(6):1786–1793. https://doi.org/10.1073/pnas.0705414105

Lenton TM, Abrams JF, Bartsch A et al (2024) Remotely sensing potential climate change tipping points across scales. Nat Commun 15:343. https://doi.org/10.1038/s41467-023-44609-w

Loriani S, Bartsch A, Calamita E et al (2025) Monitoring the multiple stages of climate tipping systems from space: do the GCOS essential climate variables meet the needs? Surv Geophys. https://doi.org/10.1007/s10712-024-09866-4

Romanou A, Hegerl GC, Seneviratne SI et al (2024) Extreme events contributing to tipping elements and tipping points. Surv Geophys. https://doi.org/10.1007/s10712-024-09863-7

Stocker TF, Jones RG, Hegglin MI et al (2024) Reflecting on the science of climate tipping points to inform and assist policy making and address the risks they pose to society. Surv Geophys. https://doi.org/10.1007/s10712-024-09844-w

Wood RA, Baker JA, Beaugrand G, Boutin J, Conversi A, Donner RV, Frenger I, Goberville E, Hayashida H, Koeve W, Kvale K, Landolfi A, Maslowski W, Oschlies A, Romanou A, Somes CJ, Stocker TF, Swingedouw D (2024) Opportunities for earth observation to inform risk management for ocean tipping points. Surv Geophys. https://doi.org/10.1007/s10712-024-09859-3

Wood R, Conversi A, Donner R, Angela L, Anastasia R, Stocker T, Swingedouw D, Maslowski W (2025) Opportunities for exploitation of Earth observation data to inform risk management of ocean tipping points. Tipping points and understanding EO data needs for a tipping element model intercomparison project (TipMIP), International Space Science Institute (ISSI) in Bern, Switzerland. Zenodo. https://doi.org/10.5281/zenodo.15364154

Surveys in Geophysics (2025) 46:327–374
https://doi.org/10.1007/s10712-024-09866-4

Monitoring the Multiple Stages of Climate Tipping Systems from Space: Do the GCOS Essential Climate Variables Meet the Needs?

S. Loriani[1] · A. Bartsch[2] · E. Calamita[3] · J. F. Donges[1,4,5] · S. Hebden[6,17] · M. Hirota[7,8] · A. Landolfi[9] · T. Nagler[10] · B. Sakschewski[1] · A. Staal[11] · J. Verbesselt[12,13] · R. Winkelmann[1,14,15] · R. Wood[16] · N. Wunderling[1,5,18]

Received: 12 January 2024 / Accepted: 10 September 2024 / Published online: 18 February 2025

Abstract
Many components of the Earth system feature self-reinforcing feedback processes that can potentially scale up a small initial change to a fundamental state change of the underlying system in a sometimes abrupt or irreversible manner beyond a critical threshold. Such tipping points can be found across a wide range of spatial and temporal scales and are expressed in very different observable variables. For example, early-warning signals of approaching critical transitions may manifest in localised spatial pattern formation of vegetation within years as observed for the Amazon rainforest. In contrast, the susceptibility of ice sheets to tipping dynamics can unfold at basin to sub-continental scales, over centuries to even millennia. Accordingly, to improve the understanding of the underlying processes, to capture present-day system states and to monitor early-warning signals, tipping point science relies on diverse data products. To that end, Earth observation has proven indispensable as it provides a broad range of data products with varying spatio-temporal scales and resolutions. Here we review the observable characteristics of selected potential climate tipping systems associated with the multiple stages of a tipping process: This includes i) gaining system and process understanding, ii) detecting early-warning signals for resilience loss when approaching potential tipping points and iii) monitoring progressing tipping dynamics across scales in space and time. By assessing how well the observational requirements are met by the Essential Climate Variables (ECVs) defined by the Global Climate Observing System (GCOS), we identify gaps in the portfolio and what is needed to better characterise potential candidate tipping elements. Gaps have been identified for the Amazon forest system (vegetation water content), permafrost (ground subsidence), Atlantic Meridional Overturning Circulation, AMOC (section mass, heat and fresh water transports and freshwater input from ice sheet edges) and ice sheets (e.g. surface melt). For many of the ECVs, issues in specifications have been identified. Of main concern are spatial resolution and missing variables, calling for an update of the ECVS or a separate, dedicated catalogue of tipping variables.

Sina Loriani and Annett Bartsch have contributed equally.

Extended author information available on the last page of the article

Keywords Tipping points · Earth Observation · GCOS · Essential Climate Variables

Article Highlights

- The diversity of potential climate tipping systems brings about a variety of variables, spatial and temporal scales required to be observed
- Many complementary remote sensing products can monitor the different stages that tipping systems pass through
- The GCOS ECVs capture many tipping-relevant variables and scales, but are often not sufficiently specified and some are missing, calling for a partial update of the GCOS ECVs or a new set of Essential Climate Tipping Variables

1 Introduction

Climate tipping elements are large-scale subcomponents of the Earth system with the potential to undergo nonlinear and often irreversible state shifts under increasing anthropogenic pressures (Armstrong McKay et al. 2022; Lenton et al., 2023; Rockström, et al., 2023; Wang et al., 2023). They are associated with large risks for human societies and the Earth system overall, and are hence decisive for human development within safe and just Earth system and planetary boundaries (Rockström, et al., 2023; Richardson et al., 2023). Among potential climate tipping elements are biosphere components such as the Amazon forest system (Flores et al., 2024), circulatory patterns such as the Atlantic Meridional Overturning Circulation (AMOC, an ocean current system; Rahmstorf 2024), and cryosphere systems such as the ice sheets on Greenland and Antarctica (Gregory et al., 2020; Fox-Kemper et al. 2021). For all these systems, self-reinforcing feedback dynamics could not only amplify anthropogenic change but even drive further change in a self-sustained manner. Thereby, compared to gradual changes in the Earth system, tipping points pose significant additional risks by leading to nonlinear and sometimes abrupt and irreversible change. For example, certain feedback processes in ice sheets amplify human-caused destabilisation and could grow to an extent that irreversibly continues to drive further ice sheet disaggregation, potentially leading to large-scale sea level rise on the order of several metres possible within 2 °C of global warming (Pattyn and Morlighem 2020). Several of the climate tipping elements have been projected to be at increasing risk of destabilisation if global warming levels overshoot 1.5 °C (Bustamante et al 2023; Bauer et al., 2023; Wunderling et al. 2023a, b; Armstrong McKay et al. 2022). Moreover, there are further smaller-scale tipping or regime shift processes such as lakes tipping from a clear to a turbid water state (Scheffer et al. 2001), or dramatic decreases of ocean oxygen concentrations (Rocha et al. 2018) that are important for their feedbacks on the Earth system and as natural resources. As they obey similar system dynamics as the large-scale climate tipping elements, we here refer to both categories as *tipping systems.*

While substantial community efforts on systematic Earth system modelling and analysis of climate tipping systems and their impacts are underway (e.g. through the Tipping Points Modelling Intercomparison Project, TIPMIP), observational efforts that leverage global-scale remote sensing data are playing an increasingly important role (Lenton et al. 2024). Satellite-based data products have been used to characterise tipping systems at risk, for example by detecting the accelerating loss of Antarctic Peninsula ice shelves and ice sheets, helping to identify the destabilisation of the West Antarctic ice sheet (Mouginot et al 2014), and have been used for monitoring biosphere tipping systems' resilience over space and time (Verbesselt et al 2016; Boulton et al 2022; Dakos et al. 2024). Lenton et al. (2024) call for a smarter and combined use of remote sensing data in a tipping point sensing system, to detect and forewarn of tipping risk and support policymaking and environmental management across scales. This requires, among other things, an improvement in the analysis-ready data available for Earth system monitoring and reporting of the changes. There are existing efforts to harmonise the characterisation of Earth's climate via so-called Essential Climate Variables (ECVs), which are physical, chemical or biological variables (or a group of linked variables) established by the Global Climate Observing System (GCOS) (WMO, 2022a). They are quantified by measurable parameters called ECV products. ECVs are selected within the GCOS programme of the World Meteorological Organization (WMO) which collects and documents the data needs for monitoring the climate system and assessing the impacts of climate variability and change (WMO 2022b). ECVs have been also previously reviewed for specific environments, e.g. mountain regions (Thornton et al. 2021) for which more than 20 variables have been found not considered in GCOS.In this contribution, we compare the currently observed ECVs with the requirements to more stringently monitor Earth systems with potential tipping points.

1.1 The Stages of a Tipping System

Tipping systems are characterised by positive feedbacks, which become self-propelling beyond a critical threshold of forcing (tipping point). This means that, beyond the tipping point, the dynamics of a tipping system are dominated by self-amplifying feedbacks, which drive the state of the system from one stable state to another. More precisely, at the tipping point a small change in a critical system-dependent forcing variable (e.g. temperature, precipitation or their combination) is sufficient to push the tipping system across its critical threshold leading to significant state change in the affected system (Armstrong McKay et al. 2022; Lenton et al. 2008; Scheffer, 2001).

We can broadly divide the journey of a tipping system throughout its state change into four stages of tipping, giving rise to various Earth observation opportunities along these stages (see Fig. 1, top). As an illustrative and simplified example of these four stages, a tipping system (grey marble in Fig. 1) can be imagined to reside in a potential landscape (ball-and-cup diagram) with two minima (Scheffer and Stephen 2003).

1. In the first stage, the tipping system is located in its *stable state* and is relatively far away from its tipping point.
2. In the second stage, the tipping point is approached through increased forcing (yellow triangle in Fig. 1), *destabilising* the currently assumed stable state. This stage can under certain conditions be characterised by measurable temporal and spatial early-warning signals (EWS) as resilience decreases and return times to the original stable equilibrium following stochastic shocks increase (Scheffer et al. 2009). In the illustrative potential landscape, this is because the potential valley where the tipping system still resides becomes shallower as the forcing increases.

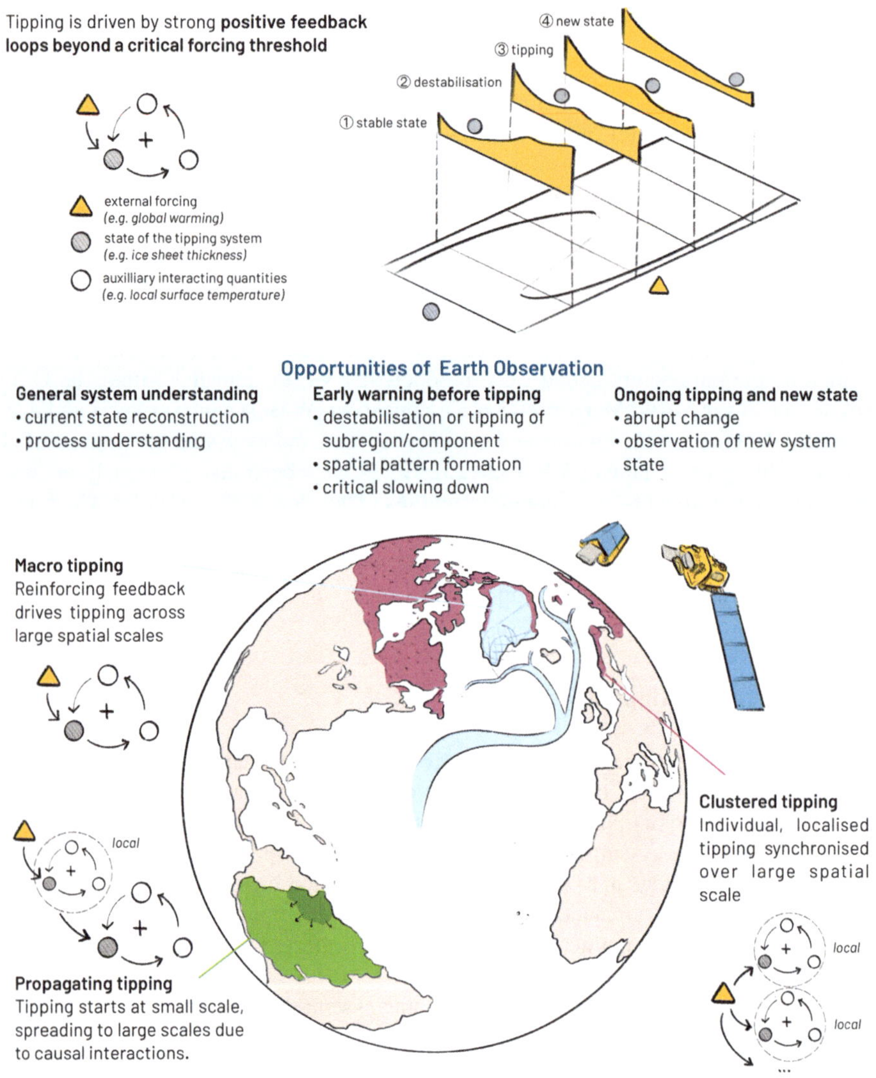

Fig. 1 Earth observation opportunities to monitor tipping across scales. Tipping occurs in systems where positive feedback loops maintain change beyond a critical forcing threshold. These dynamics can play out on different spatial scales depending on the range of the feedbacks – tipping on large scales can be the consequence of either a feedback spanning a wide (macro) range, a regional feedback that propagates to larger scales, or local feedbacks that act independently but are subject to the same external conditions and thereby synchronise over large spatial extents (see Lenton et al. 2024)

3. The third stage describes a tipping system that is in the *ongoing process of transitioning* from one stable state to another. In this stage, the formerly stable potential minimum becomes unstable and the grey marble rolls over to a tipped / alternative state.
4. In the final stage four, the tipping system is characterised by the *completed tipping process* and resides in the new state.

While we focus on so-called bifurcation-induced tipping driven by external forcing or internal slow change of a critical variable or parameter (slow in the sense of time-scale separation), other types of tipping are conceivable. For example, for noise- and rate-induced tipping (Ashwin et al. 2012) the state change is triggered by stochastic variability or rapid external forcing. While this may have implications for observability of early-warning signals (Swingedouw et al 2020; Dakos et al. 2024), the underlying positive feedbacks leading to tipping are analogous.

In earlier literature, large tipping systems have often been viewed as simple bistable systems (e.g. the Greenland Ice Sheet as a system fully intact vs collapsed, or the Amazon forest system as intact forest vs savanna state) governed by a single equation (Scheffer et al. 2009; Brummitt et al. 2015; Wunderling et al. 2021). However, reality is more intricate and multiple stable states may exist, not least due to the development of spatial patterns and regionally different properties (Robinson et al. 2012; Rietkerk et al. 2021; Lenton et al., 2023). The mechanisms underlying tipping systems can be broadly divided into three different types because of their varying feedback structure (Lenton et al. 2024) as shown in Fig. 1.

Macro tipping: The system has a feedback that plays out on a system-wide scale and leads to large-scale tipping. Examples of macro tipping are the large-scale instability of the ice sheets on Greenland and West Antarctica due to the melt-elevation feedback (see Sect. 2.1).

Propagating tipping: The system has coupled components, such that destabilisation of some parts propagates to the larger scale in a cascading manner. This means local failures can spread system-wide. An example for propagating tipping is the Amazon rainforest, where localised dieback in the northeast may propagate southwest due to a reduction in moisture recycling feedback (see Sect. 2.3).

Clustered tipping: Localised, individual and disconnected tipping events are triggered over large spatial extents by an external forcing. If the external forcing is synchronised over large spatial extents, large parts of a tipping system disintegrate at the same time even without any large feedback that connects these disjoint parts (contrary to macro tipping). An example for clustered tipping is the abrupt thaw of Arctic permafrost (see Sect. 2.4) or dieback of tropical coral reefs, with highly localised dynamics on the level of individual melt ponds or corals.

Lastly, since each tipping system is embedded in the interacting Earth system, they may influence and further destabilise each other, and interact with non-tipping systems (which can also help stabilise them) (Wunderling/von der Heydt et al., 2024).

In summary, given this diversity, tipping unfolds in a huge variety of spatial and temporal scales in the Earth system. At the same time, the stage and type of the tipping system are decisive for its dynamics. In particular, even within one tipping system, dynamics related to tipping (initial state, early-warning signals, the actual tipping transition and the established new state, see Fig. 1) play out in many different dimensions and variables and on various spatial as well as temporal scales. Therefore, rigorous Earth observation efforts are necessary in order to swiftly measure impending changes within Earth system components relevant for planetary stability, to inform policy making and environmental management, in particular for climate mitigation and adaptation.

1.2 Earth Observation for Climate Change Monitoring

Climate data records (CDR) characterise the evolution of global climate and comprise data from in situ instruments, paleoclimate, airborne and satellite sensors. Crucially, they need to be sufficient in time span, consistency and continuity to detect climate trends—typically several decades are necessary (Yang et al. 2016). The Global Climate Observing System (GCOS) is tasked with defining the requirements for CDRs that are essential for monitoring global climate change, the 'Essential Climate Variables' (ECVs). With the advances in satellite observations from passive and active sensors, on both geostationary and polar orbiting missions, satellites make an indispensable contribution to global climate monitoring efforts (Lenton et al. 2024), and support about 60% of the GCOS ECVs (WMO, 2022a). ECV requirements are determined by community consultation and expressed in terms of spatial and temporal resolution, measurement uncertainty, stability and timeliness (a newly introduced aspect which identifies how fast data records should be made available after measurement). Recently (end of 2022), GCOS proposed new ECVs and their requirements (WMO, 2022b), alongside major practical actions to be undertaken in the next 5–10 years within the GCOS Implementation Plan (WMO, 2022a). The European Organisation for the Exploitation of Meteorological Satellites (EUMETSAT) hosts a satellite-derived ECV inventory on behalf of the Committee on Earth observation satellites (CEOS) and Coordination Group for Meteorological Satellites (CGMS), which currently lists more than 800 existing and 300 planned data records (ECVInventory v4.1).

Here we discuss these new GCOS ECVs requirements from the perspective of tipping point science. To this end, we review how remote sensing provides insights on the status and key positive feedback processes of selected potential tipping systems, and discuss how satellite observations may serve to monitor systems losing in resilience and approaching criticality (early warning of tipping). In this work, we build on first steps done on evaluating the remote sensing potential for tipping points monitoring (Lenton et al. 2024) by synthesising observational needs involved with the different tipping stages, and checking the ECV requirements set by GCOS (WMO, 2022b) and the EUMETSAT inventory (ECVInventory v4.1) to identify gaps. We constrain the in-depth analysis on a selection of global core tipping elements (Armstrong McKay et al. 2022), including the ice sheets, the Amazon forest system, ocean circulations and high-latitude permafrost. Finally we complement our analysis with two selected tipping systems (lake stratification and ocean oxygen). They are exemplary for the wide class of additional potential tipping systems beyond the global core tipping elements, which may undergo large and irreversible changes in response to ongoing climate change (Rockstrom et al., 2009, Lenton et al. 2008, Heinze et al. 2021) but are currently regarded as largely uncertain due to limited evidence (Armstrong McKay et al. 2022). Overall, our selection of systems provides study cases out of many different Earth system realms—land-based and marine biosphere, ocean circulations and freshwater systems, and cryospheric systems. This shortlist may be extended in future work.

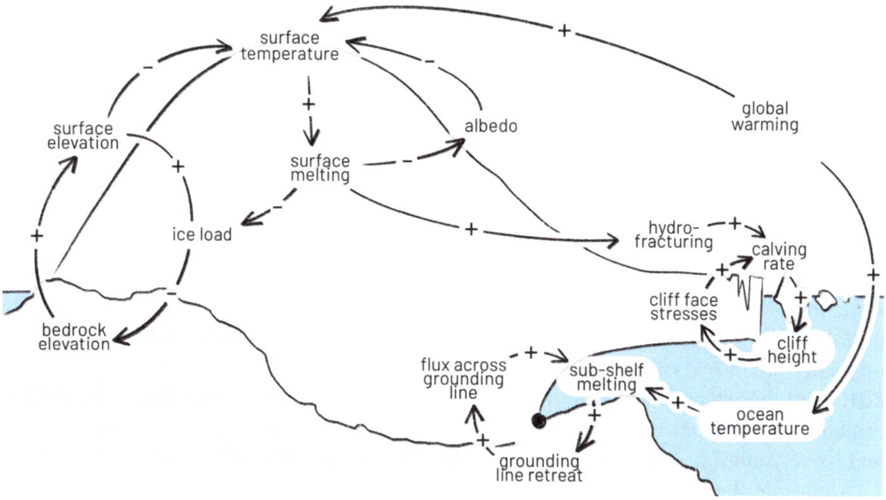

Fig. 2 Dominant feedback loops in ice sheets. Based on Zeitz (2022) and Lenton et al. (2023)

2 Selected Biophysical Tipping systems

2.1 Ice Sheets

2.1.1 Tipping System and Feedback Mechanisms

The ice sheets over Greenland and Antarctica cover extensive terrestrial regions, holding the majority of the Earth's freshwater. On the one hand, these systems exhibit high inertia, such that the bulk of induced change may unfold only in the order of centuries. On the other hand, they are strongly interacting with the atmosphere, oceans and biosphere, and are significantly susceptible to global warming. Continued, accelerating melt and disintegration of the ice sheets is contributing to sea level rise and alteration of regional climate (Fox-Kemper et al. 2021). Based on modelling evidence and reconstructions of paleo data, the ice sheets have been identified as tipping elements in previous assessments (Armstrong McKay et al. 2022; Lenton et al. 2023; Wang et al. 2023).

A series of feedbacks determine the long-term dynamics of the ice sheets, with positive ones potentially fuelling self-sustained melt beyond critical thresholds. As the ice sheet surface gets to lower altitudes due to melting, it is exposed to higher surface temperatures, which in turn lead to an increased melting and so on. Modelling and evidence from paleo data show that this melt-elevation feedback could lead to self-sustained ice sheet disintegration, and might have driven Greenland ice sheet collapse in the past (Robinson et al. 2012; Levermann and Winkelmann 2016). On time scales of millenia, this is partially compensated by glacial isostatic adjustment, in which the compressed bedrock rises vertically with decreasing ice load, acting as a negative feedback (Whitehouse et al. 2019). Surface melt can be amplified by an albedo feedback, in which the melting ice exposes the underlying ground material with lower albedo, leading to increased solar radiation absorption ensuing higher ambient surface temperature and melt (Box et al. 2012). Concerning marine-terminating glaciers, a potential mechanism accelerating melt is the marine ice sheet instability (MISI): Warming oceans increase sub-shelf melting in the shelf regions, leading to the retreat of the grounding lines (the boundary where an ice sheet transitions from being grounded on bedrock to floating on the ocean). On retrograde grounds (i.e. the bedrock descends land-inwards, see Fig. 2), the consequence is a larger ice sheet cross section at the grounding line, leading to an enhanced outflow. The ensuing further grounding line retreat closes the loop and may self-sustain the sub-shelf melt until pro-grade conditions (Schoof 2007; Feldmann and Levermann 2015; Garbe et al. 2020). A more speculative feedback, however with potentially significant consequences, is marine ice cliff instability (MICI). It is posited that the collapse of marine-terminating ice sheets may lead to high, unstable cliffs at the edge with increased iceberg calving rate, which in turn maintains the creation of these unstable cliffs (Bassis and Walker, 2011; DeConto and Pollard 2016). Mechanisms opposing the retreat of ice sheets include buttressing by ice shelves (Gudmundsson et al. 2012) or other states of ice (e.g. mélange of sea ice and icebergs, see Schlemm and Levermann 2021; Robel 2017), in which the three-dimensional geometry of the ice sheet and its surrounding (e.g. narrow fjords) can provide mechanical stability.

Alongside these major feedback loops, there are other processes concerning the feedback of the ice sheet on the ambient atmosphere and ocean, which in turn may affect the melting. For example, changes in ice sheet topography will modify the orographic precipitation around the margins, likely resulting in less precipitation at the margins but in an increase of precipitation towards the interior of the ice sheet. The net effect of this precipitation change is estimated to be negative for the Greenland Ice sheet (Hakuba et al. 2012). The surface mass balance in Antarctica is mostly determined by advection of the surrounding ocean, such that changes in the precipitation over the Southern Ocean have a large impact on the ice sheet dynamics. However, different satellite and reanalysis products diverge on the assessment of that precipitation (Bromwich et al. 2011). Similarly, the freshwater inflow into the ocean increases stratification, with potentially far-reaching consequences. The resulting increase in stratification could, for example, weaken the AMOC, which in turn would lead to cooling over Greenland and a decreased melt, hence potentially constituting a negative feedback (Wunderling/von der Heydt et al., 2023). Similarly, the retreat in Antarctic ice shelf fronts may lead to modulation in regional ocean dynamics and heat transport, presumably acting as a negative feedback (Yoon et al 2022). In summary, there needs to be a close monitoring of not only the ice sheet itself but also key ambient system variables, e.g. sea surface salinity and surface wind vectors. This will improve the understanding of how potential ice sheet tipping may emerge from the interplay of the ice sheet and ambient systems.

2.1.2 Earth Observation of Potential Tipping

2.1.2.1 Introduction: Monitoring the Ice Sheets from Space Satellite-based Earth observation, applying various measurement principles, has been indispensable in the study of ice sheets. Laser altimetry applies a ranging technique to measure the distance between satellite and ice surface with high accuracy. Repeated observations are used to retrieve changes in surface elevation of ice sheets at high spatial resolution (e.g. 100 m nominal resolution using ICESat 1 and 2). Radar altimetry exploits a similar measurement principle but uses microwave pulses, operating at lower resolution, but a key benefit is that microwave frequencies can penetrate clouds and provide all-time monitoring. It is used at different frequencies (Ku-band or S-band) and modes (e.g. on ERS-2, Envisat, CryoSat-2 and Copernicus

Sentinel-3). Synthetic aperture radar (SAR) uses a radar antenna to generate high-resolution images of the ice surface by exploiting the Doppler effect of the reflected radar signals to synthesise a large antenna, thus improving the spatial resolution that can be obtained. SAR images can be used to measure ice sheet velocity by tracking features on the ice surface over time (feature tracking) or by comparing the phase of the radar signals from two consecutive images (interferometry) to measure surface displacement. SAR images can also be used to measure the grounding zone by detecting the tidal flexure zone where the ice sheet transitions from grounded to floating (through differential interferometry). Since 1991 several SAR missions including ERS 1/2, ENVISAT ASAR, RADARSAT-1/2, TerraSAR-X and TanDEM-X, ALOS PALSAR and Copernicus Sentinel-1 have been launched, allowing continuous observation of the ice sheets at different radar frequencies (L-band, C-band, or X-band) and modes (single, dual, or quad polarisation) with high spatial resolution and with increasing temporal frequency. For example, since the launch of the Copernicus Sentinel-1 satellites in 2014 and 2015 the margins of Antarctica and Greenland have been continuously observed with weekly to 12 days repeat observations, enabling detection of seasonal and even monthly variations of ice dynamics and calving events. Active (SAR, Scatterometer systems) and passive microwave sensors enable also to estimate surface melt spatial extent based on the significant reduction of backscatter in the presence of liquid water in the snow layer. Scatterometer backscatter maps and passive microwave data enable monitoring of the temporal dynamics of surface melt at daily or even sub-daily time steps but with coarse resolution only (several km to tens of km). SAR complements this information with less frequent (weekly to biweekly) but spatially detailed information suitable for outlet glaciers in complex mountainous regions (e.g. Antarctic Peninsula). The trade-offs between these approaches are summarised by de Roda Husman et al. (2024) who merge the multiple satellite records (MetOp ASCAT C-band (5.255 GHz) radar and 19 GHz passive microwave records from SSMIS) using machine learning to produce a prototype, Antarctic-wide surface melt extent record.

A very different method for monitoring ice sheets is gravity recovery which involves the motion of satellites influenced by the present (ice) masses. GRACE and GRACE-FO use a K-band microwave ranging system and a GPS receiver to measure the inter-satellite distance and position which allows retrieval of mass changes on Earth after correcting for non-gravitational drag effects (measured on-board by accelerometers). The upcoming ESA-NASA MAGIC mission aims to monitor Earth's gravitational field at a resolution of 100 km every 3 days (Haagmans and Tsaoussi 2020).

In essence, based on these measurement principles, there are three methods to derive the total mass balance of the ice sheets from remote sensing products: (i) gravimetry, (ii) altimetry, (iii) input/output method exploiting measurements of the ice surface velocities (via feature tracking) and surface mass balance estimates (derived from altimetry and ice sheet density models) (IMBIE, 2020). Recent assessments confirmed continued mass loss in both ice sheets. On Greenland, enhanced surface melting and iceberg calving are major drivers of mass loss (King et al. 2020; IMBIE, 2020). Several outlet glaciers have experienced significant retreats and acceleration in the past two decades. For instance, Jakobshavn Isbrae, Helheim Glacier, and Kangerdlugssuaq Glacier have shown a dynamic response to oceanic and atmospheric forcing, as detected by SAR interferometry and feature tracking using data from ERS-1/2, Envisat, and Sentinel-1 satellites (Joughin et al., 2004; King et al. 2020; IMBIE, 2020; Smith et al. 2020). In Antarctica, most melting is observed on the Antarctic Peninsula, and is dominated by ocean-induced sub-shelf melting. The collapse of the Larsen A (1995) and B (2002) ice shelves on the Antarctic Peninsula were captured by ERS-1/2 and RADARSAT-1 SAR images and revealed a

Table 1 Parameters relevant to ice sheet tipping, and corresponding GCOS-ECVs

Variable related to tipping dynamics	GCOS-ECV parameters		Comment/gap(s)	Suitable?
	Variables	Specs (resolution)		
General ice sheet state				
Ice mass*	Ice volume change	Threshold annual at 50 km, goal monthly	Goal spatial resolution undefined	
Ice velocity and Ice velocity change*	Ice velocity	Threshold annual at 1 km, goal monthly at 50 m	No specifications available for velocity change	
Particularly for melt-elevation feedback (incl glacial isostatic adjustment)				
Surface elevation	Surface Elevation Change	Threshold annual at 100 m, goal monthly	Goal spatial resolution undefined	
Surface air temperature	Air temperature (near surface)	Threshold 500 km and 3 h, goal 10 km hourly		
Surface melt extent/ melt duration	–	–	Not included; 10 km, daily (IGOS Cryosphere Theme Report 2007, page 86/87)	M
Vertical land motion	–	–	Not included	M
Particularly for melt-albedo feedback				
Surface albedo	Spectral and Broadband (Visible, Near Infrared and Shortwave) DHR and BHR6 with Associated Spectral Bidirectional Reflectance Distribution Function (BRDF) Parameters	Threshold 5 days at 250 m, goal daily at 10 m		
Grounding line (zone) location*	Grounding Line (zone) Location and Thickness	Threshold annual at 1 km, goal 100 m	Goal temporal resolutionun defined	
Mass Flux across the grounding line	–	–	Not included	M
Sub-shelf topography	–	–	Not included	M
Particularly for marine ice cliff instability				
Calving rate / calving front position*	–	–	Not included	M
Areas of hydrofracturing	–	–	Not included	M

Table 1 (continued)

Variable related to tipping dynamics	GCOS-ECV parameters		Comment/gap(s)	Suitable?
	Variables	Specs (resolution)		
Sea ice properties*	Sea ice thickness	Threshold 50 km and monthly, goal 1 km and daily		
	Sea ice concentration	Threshold 50 km and monthly, goal 1 km and better than daily		

A star (*) indicates what is directly addressed in ESA CCI baseline projects. The last column indicates where the GCOS-ECV parameters do not match the specific needs of tipping science either by insufficient or undefined specs (X) or entirely missing ECV (M)

rapid disintegration of the ice shelves into thousands of icebergs (Rott et al. 1996; Scambos et al., 2004). In the Amundsen Sea sector of West Antarctica, the Pine Island Glacier and the Thwaites Glacier are also undergoing rapid changes due to enhanced sub-shelf melt (Pritchard et al., 2012). These processes have been monitored by radar altimetry and gravity recovery using data from ERS-1/2, Envisat, CryoSat-2, and GRACE satellites (Rignot et al., 2019; IMBIE, 2020; Scambos et al., 2004) and by laser altimetry with the ICESat satellites. Future missions such as the Copernicus Polar Ice and Snow Topography Altimeter (CRISTAL) mission and the NASA-ISRO SAR (NISAR) mission will provide more accurate and high-resolution measurements of ice sheet elevation and mass changes, which are essential for improving ice sheet models and projections (Markus et al. 2017; Flechtner et al. 2016) (Table 1).

2.1.2.2 Insights Into key Feedback Loops Relevant for Tipping For marine-terminating ice sheets, the migration of the grounding line (zone) is a sensitive measure of ice sheet stability and potential onset of the MISI feedback. It can be inferred from observational data (Hogg et al., 2018) by determining the vertical displacement of floating ice at tidal frequencies through InSAR (Rignot et al. 2011; Scheuchl et al. 2016), altimetry (e.g. Brunt et al., 2011), or by analysing shadows in optical satellite imagery (Scambos et al. 2007). A recent study based on InSAR data suggests that the rapid retreat of Pope, Smith and Kohler glaciers in the Amundsen bay observed in the past decades happened on retrograde bed slopes (Milillo et al., 2022). This serves as a strong study case for MISI, and is therefore especially relevant for the neighbouring larger Pine and Thwaites glaciers. These glaciers share similar characteristics in terms of underlying physical processes but hold more than a metre of global sea level rise potential (Fox-Kemper et al. 2021; Milillo et al., 2022).

Inherent to the nature of the feedback, it is challenging to directly observe MICI (Needle and Holschuh, 2023). High temporal sampling and a complementary suite of remote sensing products are needed to monitor the rapid breakup of ice sheet fronts and the variety of factors that determine their geometric (in-) stability. In particular, it is necessary to not only assess the calving front position (see e.g. Greene et al., 2022), calving rate and the ice velocity but also surface properties (such as areas of hydrofracture) along with the regional conditions on ground (bed topography) and sea (presence of sea ice or mélange). Individual study cases such as Crance glacier's retreat following the collapse of the Larsen B ice shelf can help to constrain model parameterisations of critical ice cliff heights (Needle and Holschuh, 2023).

Variables involved in the feedbacks associated with elevation and albedo are in principle more accessible via remote sensing. The ice sheet surface elevation is a critical parameter for the melt-elevation feedback, and is made readily available from radar altimetry (from ERS1/2, Envisat, and CryoSat-2) within the ESA CCI programme (Sandberg Sørensenet al. 2018; IMBIE, 2020). For mass balance estimates, changes in surface elevation need to be corrected for density and the isostatic adjustment, i.e. requiring independent assessments of vertical ground motion. Finally, surface air temperatures and melt rates require close monitoring to draw conclusions on the onset of this positive feedback. Characterisation of the melt-albedo feedback needs measurements of surface melt extent, (near-) surface air temperature and surface albedo.

As outlined in the Introduction, bistable tipping systems are expected to have increasingly long recovery times from external perturbations when approaching their tipping point. This is expressed as increasing auto-correlation in spatial or temporal data. However, the characteristic time scales for ice sheet-related processes are naturally orders of magnitude longer than for e.g. living systems, such that a complete disintegration would take centuries to millennia and fluctuations play out on annual or decadal time scales. This makes it challenging to observe early-warning signals—in fact, all studies of critical slowing down in ice sheets have been based on either station data (Boers and Rypdal 2021) or modelling exercises (Rosier et al., 2021).

However, remote sensing is invaluable to make sense of the past (and ongoing) collapse and rapid retreat events, leading to better modelling and predictive capacities. Several studies analysed in hindsight the observational evidence leading up to the collapse of the Larsen ice shelves, attributing the rapidness to e.g. hydrofracture networks (Banwell et al. 2013). Similarly, substantial thinning of the glacier terminus before the onset of rapid retreat could be identified in all regions of the Greenland Ice sheet (King et al. 2020). These analyses were underpinned by significant progress in system understanding and modelling capabilities, pointing towards a promising direction for early warning for ice sheet collapse. Indeed, improved assimilation of remote sensing products with modelling could serve as a transient forecast, similar to recent advances in sea ice extent forecasting (Navarro et al. 2022).

2.1.2.3 Discussion: Gaps in GCOS and Recommendations The different remote sensing methods for the ice sheets each have their respective strengths and drawbacks. To infer ice mass from altimetry, accurate models of snow accumulation and compaction are required to infer firn layer thicknesses. This is informed by surface mass balance models, which in turn require accurate capturing of regional climate, thereby heavily relying on atmospheric and surface ECVs. The input/output method additionally requires estimates of frontal ablation, which is limited by the outlined challenges in estimating hard-to-observe processes such as sub-shelf melting and rapid calving. As a direct measurement of mass, gravimetry is relieved from some of these difficulties, but it is particularly sensitive to glacial isostatic adjustment (GIA). To disentangle this signal from changes in the ice sheet mass, accurate modelling of GIA is needed, and/or assimilation of complementary approaches (e.g. GPS, altimetry) with a different GIA sensitivity (van der Wal et al. 2015). Closely related are the condition and dynamics of regional topography, as it has significant implications for MISI (e.g. through retrograde slopes or cavities) and ice sheet geometry potentially determining MICI. The vertical land motion as well as sub-shelf topography are therefore key observables which are currently missing from the GCOS-ECV portfolio.

Recent intercomparisons (King et al. 2020; IMBIE, 2018, 2020) show fairly consistent results between the different total mass balance estimation approaches in many regions. However, large uncertainties emerge from the poor sampling (both limited in situ data and spatial coverage) of some regions such as East Antarctica (Rignot et al. 2019) or the grounding zones of fast flowing glaciers. At the general level, it seems that such issues are increasingly recognised, such that the GCOS 2022 Implementation Plan, identifies "improved ECV satellite observations in polar regions" as an emergent need (WMO, 2022a). However, some specific associated variables remain excluded or poorly defined in the GCOS-ECV portfolio (WMO, 2022b). This is particularly true for tipping points as our assessment in shows that many of the key variables for the main positive feedback processes are missing entirely in the portfolio.

In the context of tipping points, the prime benefit of remote sensing appears to be in improved system understanding to guide model development. High-resolution (particularly in space, but also in time) monitoring—especially in the ice sheet fringe regions—is essential to inform conceptual and process-based numerical models. From this, parametrisations for whole-ice sheet models can be derived (Benn et al. 2017), providing insights into larger parts of the system. In terms of early warning, the time scales of ice sheet dynamics likely prohibit meaningful predictive power of critical slowing down indicators in the near future. However, transient forecasts and the lessons learned from the recent accelerated ice sheet retreat history allow for improved estimates of the short-term future of components of the ice sheets, such as Pine and Thwaites glaciers. To this end, the concept of propagating tipping (see introduction) is becoming increasingly central to the assessment of ice sheet instability: Given that frontal ablation effects are currently major drivers for ice sheet retreat, regional self-amplified melt through MISI or MICI can have far reaching effects propagating much further inland into the glacier (Reese et al., 2018). This lends even more importance to new/updated ECVs that capture the dynamics of the ice sheets at their fringes.

2.2 Amazon Forest System

2.2.1 Tipping System and Feedback Mechanisms

The Amazon forest is a system of high regional and global significance. It hosts a great share of global biodiversity, contributes to net global cooling via evapotranspiration and carbon storage, and plays a key role in moisture distribution in the region (Nobre et al., 2021; Flores et al., 2024). A multitude of drivers, most notably land use and anthropogenic climate change, continuously degrade the forest system (Bullock et al. 2020; Lapola et al. 2023).

Amazon rainforest dieback into a degraded or savannah-like state has been identified as a global core tipping element (Armstrong McKay et al. 2022; Lenton et al. 2023; Wang et al. 2023). Central to this dieback is the role of forests in sustaining atmospheric moisture. Remarkably, about 20% of Amazonian precipitation stems from tree transpiration, which is about two thirds of all moisture that is recycled in the Amazon (Staal et al. 2018). The ability to achieve such water recycling stems from a series of feedback mechanisms delicately co-determining the forest extent itself. Forests, particularly through their deep roots, are crucial for maintaining high moisture fluxes to the atmosphere during dry periods (Sakschewski et al. 2021; Singh et al. 2021), which in turn influences local and downstream precipitation within the Amazon (Zemp et al. 2017; Staal et al. 2018). Loss of forest coverage, e.g. through human deforestation, usually reduces evapotranspiration, thus diminishing both local and downwind precipitation. This reduction is compounded by a decrease in tree-produced volatile organic compounds (VOCs), which serve as cloud condensation nuclei, further reducing precipitation (Ehn et al. 2014; Yáñez-Serrano et al., 2021; Sanaei et al. 2023). Additionally, fire aerosol emissions from burning forests also have an effect on rain patterns, potentially inhibiting rainfall and exacerbating drought conditions. Furthermore, the forest's decreased roughness length leads to increased wind speeds, shortening the residence time of moisture in the forest ecosystem and impacting cloud formation (Barkhordarian et al. 2019; Zhou et al. 2021). These factors collectively contribute to higher temperatures and increased drought stress, exacerbated by less evaporative cooling and reduced shading from the forest canopy (Sampaio et al. 2007). This set of feedback mechanisms, collectively acting as a regional vegetation-atmosphere feedback, can start to dry out the ecosystems and help fires to take effect.

Fig. 3 Key feedbacks in the Amazon rainforest are vegetation-rainfall and vegetation-fire

The vegetation-fire feedback loop then is a major contributor for locking the respective former forest area into a degraded savanna or low-tree-cover state (Hoffmann et al. 2012; Flores and Holmgren 2021a, b; Drüke et al. 2023). Grass-fueled fires, thriving in open landscapes, help maintain a grass-dominated, often degraded ecosystem. This dynamic not only increases fire probabilities due to a more open canopy and drier understory, but also contributes to greater erosion due to less vegetative coverage and higher wind speeds, further diminishing the remaining forest resilience and possibility of recovery (Spracklen and Garcia-Carreras 2015). Conversely, a closed tree canopy suppresses grass growth by outshading and creating a humid microclimate (Flores et al., 2024).

Finally, any form of forest loss, whether through deforestation or degradation, results in significant CO_2 emissions, whose contributions to climate change are ongoing if the forest is unable to recover and act as a future carbon sink. This cascading effect of forest loss, increased CO_2 emissions, and exacerbated climate change forms a vicious cycle, perpetuating further forest degradation and loss, thereby disrupting the Amazon's critical ecological balance (Fig. 3).

2.2.2 Earth Observation of Tipping

2.2.2.1 Introduction: Monitoring the Amazon Rainforest from Space Remote sensing of Amazon tipping dynamics requires capturing variables associated with the feedback mechanisms described above. Due to their long operation times Landsat, SPOT-Vegetation, MODIS (and Sentinel-3, Sentinel-2 and Sentinel-1 in the last decade) data are commonly used to study forest cover over time. Landsat time series show that about 17% has been degraded between 1995–2017 as a result of direct human activities and of droughts (Bullock et al. 2020; Lapola et al. 2023).

Beyond forest cover, vertical profiling measurements of carbon compounds using aircraft were carried out over nine years to assess the changing carbon balance of the Amazon and, complemented by remote sensing data, its causes (Gatti et al. 2021). Remote sensing data played a crucial role in providing spatially extensive and temporally consistent information. The combination of data sources enabled a more nuanced and complete understanding of the factors contributing to the Amazon's shift from a carbon sink to a carbon source. For most areas, deforestation was found to be the primary cause. In contrast, in the southeast of the basin, climate change and especially drought was identified as the most important driver. In both cases anthropogenic fires—which are not naturally occuring in tropical forest—can strongly accelerate biomass loss (Basso et al. 2023). Signatures of the vegetation-fire feedback can be detected from space, for instance in MODIS data, where fire occurrence is strongly anti-correlated to forest cover (Staver et al. 2011) and even rises steeply above a threshold around 50% grass cover (Van Nes et al. 2018). Today, observations using MODIS VIIRS, Sentinel-3 and geostationary satellites enable improvements to near real time fire radiative power and smoke detection. But observing understory fire remains a challenge, and the resolution of sensors (Copernicus Atmosphere Monitoring Service (CAMS) products from the Global Fire Assimilation System (GFAS) available are about 11 km resolution globally) constrains early detection of small fires (Andela et al,. 2022).

Building on the findings of Amazon carbon sink loss, Staal et al. (2023) showed with a moisture tracking model that about half of all the land-derived precipitation in the net carbon source area in the Amazon is derived from moisture that evaporated from this same region. This implies a positive feedback in which reduced photosynthesis due to drying leads to further drying in the same precipitation-limited portion of the Amazon. Moisture tracking models reconstruct moisture fluxes from evaporation source to precipitation sink (e.g. Van der Ent et al. 2014; Tuinenburg and Staal 2020) hence aim at quantifying the vegetation-atmosphere feedback. More specifically, these models are forced with three-dimensional wind fluxes, atmospheric moisture content and evaporation and precipitation fluxes from ERA5 (Tuinenburg and Staal 2020) and therefore indirectly integrate remote sensing data. Recently, differently complex models were able to show effects of deforestation on precipitation reduction (Smith et al. 2023, Araújo et al., 2023).

In general the integration of diverse techniques of remote sensing and modelling are becoming more important to study the status and potential tipping of the Amazon. Sentinel-2's capacity to map plant functional diversity (Ma et al. 2020) and GEDI's lidar-based models for biomass density (Duncanson et al. 2022,) provide essential insights into forest health indicators, putting emphasis on biodiversity-ecosystem functioning relationships in the Amazon (Poorter et al. 2015; Sakschewski et al. 2016). Further, the synergy of remote sensing with dynamic forest models (Hiltner et al. 2022), and novel methodologies like radar-based forest productivity estimation (Henniger et al 2023) and lidar-derived tree size distribution (Taubert et al. 2021), enrich our understanding of forest dynamics. The significance of border effects in biomass estimation (Knapp et al. 2021) and the role of forest structure in biomass and productivity (Exbrayat et al. 2019) highlight the importance of diverse and contextual measurements, and data precision. Additionally, remotely sensed functional diversity (Pacheco-Labrador et al. 2022) enables study of forest composition and resilience over space and time. Moreover, integrating multi- and hyperspectral data from numerous in use satellites such as PRISMA, EnMAP, Sentinel-2 and upcoming missions such as NASA's Surface Biology Geology mission, to be harmonised with ESA's new CHIME mission, will provide detailed spectral information enhancing our understanding of all points above and give a new level of detail for the physiological status and stress responses of tropical forest (Qian 2021). Taking a holistic approach, leveraging lidar, radar, and multispectral data, offers

Table 2 Parameters relevant to Amazon rainforest tipping, and corresponding GCOS-ECVs

Variable related to tipping dynamics	GCOS-ECV parameters		Comment/gap(s)	Suitable?
	Variables	Specs		
General forest state				
Forest Cover Land cover fractions*	Land Cover High resolution land cover IPCC land classes	Threshold 5 years at 1 km, goal monthly at 100–300 m Threshold 5 years at 30–100 m, goal monthly at < 10 m Threshold 5 years at 1–100 km, goal monthly at 10–300 m	Weekly required Land use change	X
Soil moisture*	Surface soil moisture Root zone soil moisture	Threshold 2 days at 50 km, goal 6 h at 1 km	Surface soil moisture only from satellite records	X
Vegetation structure	–	–	Tree height (maybe height classes and understory distinction), Tree Cover, Tree density are not specified by GCOS but are noted: additional canopy structure variables would be helpful	M
Vegetation composition	–	–	Functional diversity (measureable tree traits), species composition	M
Vegetation water content	–	–	not included	M
Photosynthesis*	Leaf area index (LAI), Fraction of absorbed photosynthetically active radiation (FAPAR)	Threshold 10 days at 250 m, goal daily at 10 m		
Biomass*	Above ground biomass (AGB)	Threshold 5 years with 1 km, goal bi-annual and 10 m	Weekly required Tree height, Tree Cover, Tree density are not specified by GCOS but are noted: additional canopy structure variables would be helpful	X

Table 2 (continued)

Variable related to tipping dynamics	GCOS-ECV parameters		Comment/gap(s)	Suitable?
	Variables	Specs		
GHG emissions (CO2 dynamics)*	CO2 mole fraction CO2 column average dry air mixing ratio	Threshold bi-annual and 2000 km, breakthrough monthly, goal daily and 100 km Threshold biweekly and 10 km, goal daily and 1 km	Daily required	X
Albedo	Albedo	Threshold 5 days at 250 m, goal daily at 10 m		
Particularly for vegetation-atmosphere feedback				
Evapotranspiration	Evaporation from land (includes Sensible heat flux, Latent heat flux, Bare soil evaporation, Interception loss, Transpiration)	Threshold daily and 25 km, goal hourly and 100 m	Relates to monitoring of precipitation, newly included into 2022 ECVs	
Precipitation	Precipitation in either liquid or solid form	Threshold daily and 25 km, goal hourly and 100 m; Threshold uncertainty 5 mm, goal 1 mm	Accurate measurements are needed (uncertainties need to be met)-related metrics including dry season length and maximum climatological water deficit are key variables for monitoring Amazon tipping	X
Particularly for vegetation-fire feedback				
Fires (fire activity, Burned Area)*	Burned area Active fire Fire radiative power (FRP)	Threshold monthly at 1 km, goal daily at 10 m Threshold twice daily at 5 km, goal 5 min at 50 m		

A star (*) indicates what is directly addressed in ESA CCI baseline projects. The last column indicates where the GCOS-ECV parameters do not match the specific needs of tipping science either by insufficient or undefined specs (X) or entirely missing ECV (M)

a comprehensive framework for predicting and managing forest tipping points, informed by a multi-dimensional perspective on ecosystem health. Remote sensing in combination with different modelling techniques offers the opportunity for near real-time tipping point observation systems for the Amazon basin and beyond (Table 2).

2.2.2.2 Monitoring the Amazon Rainforest's Loss in Resilience Loss in resilience, quantified by appropriate metrics (Bathiany et al., 2024), can serve as an early warning indicator for approaching a tipping point (Dakos et al. 2024). In general, global-scale remote sensing is useful for characterising changes in disturbance regime e.g. for fire, from breakpoints in time-series (Finer and Mamani, 2022), and response via forest resilience assessment from time series of vegetation products and indices (Bathiany et al., 2024). However, remotely sensed burned area is highly spatial resolution-dependent. Of the EO products currently available, burned area, relativized burned area (for fire severity), fire radiative power, land cover, land surface temperature, precipitation, and soil moisture, could be used to distinguish old growth forests that store the most carbon. Above-ground biomass can be a proxy for vegetation structure and water content, and red-edge information from hyperspectral imaging can indicate vegetation health. It is possible to distinguish types of deforestation and attribute the causes by combining optical and SAR (Sentinel-2 and Sentinel-1) imagery: recent innovations based on deep learning classification techniques show promising results in terms of accuracy and usefulness (Slagter et al 2023).

For process understanding, multispectral EO can be used to track the impacts from compounding disturbances and feedbacks at play, i.e. the combination of impacts from droughts, wildfires, deforestation and other human-driven degradation (see for instance Lapola et al. 2023). Compounding disturbance effects are one of the major research challenges, addressing their synergies, to inform efforts to counter their impacts.

The resilience indicators developed so far for the Amazon rainforest are still too uncertain to be useful as EWS, with different conclusions drawn depending on the data product and respective bands and sensors used. For instance, based on K-band Vegetation Optical Depth (VOD), 75% of the Amazon rainforest were categorised as losing resilience since 2003, meaning that such forests could be approaching a local-scale tipping point (Boulton and Boers, 2022). However, VOD may be contaminated by soil moisture: analogous analyses using QSCAT and ASCAT active radar products (which are less sensitive to soil moisture and flooding, and able to penetrate the canopy for vegetation moisture change) have not found signals of resilience loss (Tao et al. 2023), although still actively debated (Boulton and Smith, 2023).

Complementary avenues towards monitoring the Amazon's resilience comprise tracking species diversity, in particular functional diversity (Sakschewski et al. 2016). Indicators related to functional diversity and richness show great promise for tipping EWS at local to regional-scales, but development of these products, for example from high-resolution hyperspectral imagery or active lidar, relies on the availability of large amounts of accurate in situ data to improve the data quality and uncertainty assessment, so is not currently feasible at large-scale (Jetz et al. 2016; Schneider et al. 2020).

2.2.2.3 Discussion: Gaps in GCOS and Recommendations The review above shows that a wide range of parameters need to be observed across the Amazon to capture the complexity of the potential tipping system. Relevant variables include burnt area, leaf area index (LAI), vegetation optical depth (VOD), albedo and surface roughness, normalised difference vegetation index (NDVI), cloud cover, LST for plant stress, vegetation water content, soil moisture, and canopy structure indicators such as tree height, tree cover and tree density. For carbon storage in forests biomass is needed: GCOS includes this but the temporal specifications do not yet meet requirements for this specific application. Tree height, tree cover and tree density are not considered by GCOS. Land cover maps need to become more dynamic (regular updates e.g. weekly) via cover fractions such as tree cover, while indicating the drivers of change (such as land use change drivers). Vegetation water content is indicative for forest state but is currently also not considered by GCOS. There is also an urgent need for more comprehensive measurement of stocks, fluxes and atmospheric CO_2 in humid tropical forests (Friedlingstein et al. 2022; Crisp et al. 2022) as there are large uncertainties in the carbon fluxes between atmosphere and land/ocean reservoirs.

In total, the complexity of the system and the possibility to leverage remote sensing for all stages of the tipping process necessitate the observation of a wide range of variables at high resolution. We contrast the observational requirements from the perspective of tipping points science with the specifications from GCOS and find that many key variables are either missing or insufficiently resolved.

2.3 Atlantic Ocean Circulations

2.3.1 Tipping System and Feedback Mechanisms

Two major ocean circulations in the North Atlantic are considered as potential tipping systems (Armstrong McKay et al. 2022; Lenton et al., 2023). The Atlantic Meridional Overturning Circulation (AMOC) transports warm water in an upper surface branch from the equator to the North Atlantic, where it cools, sinks, and returns southwards in a lower deepwater branch. Thereby, the circulation is playing a key role in marine ecosystems and climate, connecting remote regions of the Earth. A slowdown or collapse would reduce the northward heat transport and cause a cooling of the North Atlantic and Europe, as well as a weakening of the monsoon systems and would result in sea level rise along the eastern coast of North America. The subpolar gyre (SPG) circulation south of Greenland constitutes another potential tipping system linked to potential collapse of convection in the Labrador and Irminger seas (Sgubin et al. 2017; Swingedouw et al. 2021). Such a rapid change in overturning would entail significant regional cooling, impact European weather and shift circulations such as the jet stream and the intertropical convergence zone. The critical global warming threshold for tipping has been estimated at about 1.8 °C (with a range 1.1 to 3.8 °C), with a decadal collapse timescale (Armstrong McKay et al. 2022).

Both systems are self-sustained due to a salt-advection feedback—in the "on"-mode of the circulations, the import of salty water sustains a sufficient surface salinity for convection, thereby continuing to drive the overturning circulation. This process is however susceptible to freshwater influxes, e.g. through increased ice sheet melt, precipitation or river influxes. The induced decrease in surface water salinity reduces the sub-polar water column density and weakens the convection, thereby the overturning strength and saltwater import, again decreasing the surface water salinity. Other, potentially negative feedbacks are related to interaction with sea ice, transport of freshwater, interactions with the atmosphere have in parts stabilising effects (Jackson et al., 2017; Wood et al. 2024) (Table 3) (Fig. 4).

2.3.2 Earth Observation of Tipping

2.3.2.1 Introduction: Monitoring the Ocean Circulations from Space Wood et al. (2024) provide an in-depth overview on the opportunities and merits of Earth observation for monitoring the AMOC and SPG. In short, our best tool for understanding potential tipping in the overturning circulations are climate models of varying complexity, ranging from early box models to high-complexity coupled climate models (Stommel, 1961, Jackson et al. 2023). Direct observations of overturning strength are so far limited to in situ hydrographic arrays providing continuous measurements since 2004 in selected locations (Swingedouw et al. 2020).

However, a range of remote sensing approaches can support the monitoring (Wood et al. 2024; Swingedouw et al. 2020). Leveraging decades of satellite observations with wide spatial coverage, the hydrographic time series have been extended in time and space via altimetry of sea level (Frajka-Williams, 2015; Mercier et al. 2015) and via gravimetry of density variations in the deep ocean (Landerer et al. 2015). Complementary approaches assess proxies of the overturning strength, e.g. by determining the dynamics at the convection sites by inferring surface density from measurements of sea surface temperature (SST) and sea surface salinity (SSS; Swingedouw et al. 2020). SST and SSS are monitored by radiometers at microwave (e.g. on GMI, SLSTR, AVHRSS) and L-band frequencies (sensitive to salinity-induced changes in the di-electric constant of seawater; e.g. on SMOS or planned CIMR), respectively. The observed changes in spatial and seasonal SST (the 'warming hole' south of Greenland) have been attributed to an AMOC slowdown (Caesar et al. 2018) or SPG weakening (Sgubin et al. 2017; Swingedouw et al. 2021), although remote influences may also play a role in creating such SST patterns (e.g. Hu and Fedorov 2020).

The time span covered by hydrographic arrays or even satellite observations is, however, too short to exploit statistical properties such as temporal autocorrelation as early warning signatures (stage 2; see Lobelle et al. 2020; Boulton et al., 2014). Therefore, several

Fig. 4 The salt-advection feedback as one key positive feedback driving the Atlantic Meridional Overturning circulation and Subpolar Gyre, based on Loriani et al., (2023)

Table 3 Parameters relevant to AMOC/SPG tipping, and corresponding GCOS-ECVs

Variable related to tipping dynamics	GCOS-ECV parameters		Comment/gap(s)	Suitable?
	Variables	Specs		
Salt-advection feedback and overturning proxies				
Sea surface temperature*	Sea surface temperature	Threshold 100 km, weekly, goal 5 km, daily	HIgher resolution is needed for eddies at threshold 10 km, goal 1–3 km, at least in key boundary/choke point regions	X
Sea surface salinity*	Sea surface salinity	Threshold 50–100 km and weekly, goal daily at 10 km	Higher resolution valuable in key boundary and choke point regions	X
Sea surface height*	Global mean sea level	Threshold 100 km and monthly (weekly at regional scale), goal 10 km and daily;	Detection changes in overturning; Regions for higher sampling undefined	?
Ocean section freshwater transports	–	–	Derived mainly from fixed arrays/Argo. EO or reanalysis data for wind stress needed to obtain transport estimates	M
Other influence on overturning strength				
Sea ice concentration*	Sea ice concentration	Threshold 50 km and monthly, goal 1 km and better than daily	High resolution particularly in key boundary and choke point regions	X
Sea ice thickness*	Sea ice thickness	Threshold 50 km and monthly, goal 1 km and daily	High resolution particularly in key boundary and choke point regions	X
Sea ice motion	Sea ice drift	Threshold 50 km and monthly, goal 1 km and daily	High resolution particularly in key boundary and choke point regions	X
Wind change	Surface currents			
Precipitation over the ocean	Accumulated precipitation	Threshold 250 km, Goal 50 km. Threshold time scale yearly, Breakthrough Monthly	Monthly time resolution allows resolution of seasonal convection in subpolar gyre. Need consistency with evaporation—precipitation is the more challenging of the two	X
Evaporation over the ocean	Latent heat flux	Threshold 100 km, Goal 10 km		

Table 3 (continued)

Variable related to tipping dynamics	GCOS-ECV parameters		Comment/gap(s)	Suitable?
	Variables	Specs		
Freshwater input—rivers	River discharge	Annually to daily	Gauging station specific. Consistent basin scale estimates needed to construct fresh water budgets. Accuracy needs to be consistent with precipitation minus evaporation, and basin scale fresh water content	X
Freshwater input—ice edge melting rate	–	–	Not defined	M

A star (*) indicates what is directly addressed in ESA CCI baseline projects. The last column indicates where the GCOS-ECV parameters do not match the specific needs of tipping science either by insufficient or undefined specs (X) or entirely missing ECV (M)

studies rely on long-time paleo proxies or historical instrumental observations for temperature and salinity to infer AMOC weakening (Moffa-Sánchez et al. 2019; Caesar et al. 2021) or approaching criticality (Boers 2021; Ditlevsen and Ditlevsen 2023; Michel et al. 2022). There are however ongoing debates about potential false warnings (Michel et al. 2023) and the feasibility of these proxies for AMOC tipping in general (Kilbourne et al. 2022; Jackson and Wood 2020). One main difficulty is that the assessment of potential bistability of both AMOC and SPG and potential proximity to a critical threshold depends on good representation of the non-linear feedbacks in the models (Jackson et al. 2023). This is potentially constrained by biases in stratification (Swingedouw et al. 2021), salinity (Liu et al. 2017; Mecking et al. 2022) or limited resolution of eddies (Heuze, 2017; Swingedouw et al., 2021) and lies at the focus of current and upcoming modelling initiatives like NAHosMIP (Jackson et al. 2023) and TIPMIP.

Concerning the salt-advection feedback, the influx of fresh water plays a key role. Therefore, a significant leverage of remote sensing data for the Atlantic ocean circulations lies in the monitoring of the mass balance of the Greenland Ice sheet (IMBIE 2020, Dukhovskoy et al. 2021), discharges from Arctic rivers (Druckenmiller et al., 2021) and connected basins such as the Beaufort Gyre (Zhang and Thomas 2021; Lin et al. 2023) as well as atmospheric circulation changes (Holliday et al. 2020) and sea ice (Drijfhout et al. 2015; Sgubin et al. 2017), and through combination with hydrographic and fixed array data, oceanic fresh water transports (McDonagh et al. 2015) (Table 3).

2.3.2.2 Discussion: Gaps in GCOS and recommendations Despite the outlined challenges and uncertainties about EO use for direct monitoring of AMOC and SPG, it serves as an indispensable component at the interplay of models, in situ measurements and evidence from paleo records. Based on the key variables in the salt-advection feedback and current approaches to reconstructing overturning strength (including proxy-based methods), several direct surface quantities need to be monitored, that is sea surface salinity, temperature and elevation. All of them are included in the GCOS ECVs with largely sufficient specifications with the exception of spatial resolution for eddy-resolving modelling, and coverage needs to be improved at high latitudes (Estella-Perez et al. 2020). Sea surface salinity can be also indirectly estimated through freshwater influx (precipitation—evaporation, North Atlantic river runoff, melting of glaciers and sea ice). This calls for improved resolution analyses near coasts and for estuaries, with special attention to key regions such as the Fram and Davis straits and subpolar basins (Wood et al. 2024). Sea surface height analyses using satellite altimetry and density change through gravimetric observations have provided added value (Swingedouw et al. 2020). The recent discussions around the suitability of a range of proxies show that the main value of EO for AMOC/SPG tipping lies in the wide spatial coverage of multiple variables, offering potential in resolving tipping fingerprints in multivariate analyses (Michel et al. 2023). The GCOS ECVs should continue to be scrutinised with respect to potential additional variables useful for tipping research. GCOS currently considers the majority of so far identified parameters of relevance for AMOC monitoring. Density change and freshwater input are, however, not addressed. A potential area for further development would be to monitor integrated regional ocean heat and fresh water budgets in key regions such as the subpolar gyre. This will need a combination of in situ (Argo, fixed section arrays) and EO data, and a focus on specifying *combinations* of ECVs to achieve consistent accuracy for budget calculation of such integral quantities (e.g. regional heat content and section-integrated ocean transports) (Forster et al. 2021).

2.4 Permafrost

2.4.1 Tipping System and Feedback Mechanisms

Permafrost is defined based on ground temperature (Matthews et al. 2021). Values below zero degree Celsius need to occur for at least two consecutive years. This excludes the upper soil (active layer) which thaws and refreezes seasonally. Permafrost is thus a subsurface phenomenon. The discussion in the context of climate tipping elements centres around the release of soil carbon which is currently still frozen (Swingedouw et al. 2020).

If ground thaws, subground carbon becomes available for decomposition and release as CO_2 or CH_4, depending on soil saturation, resulting in temperature increase. This leads to enhanced thaw and subsequent carbon release and is expected to translate into a range of 0.13–0.27 °C additional global warming by 2100 (Schuur et al. 2015). Permafrost thaw has only recently been discussed as a tipping system under abrupt thaw conditions. Otherwise greenhouse gases are expected to be gradually released with permafrost thaw with quasi-linear behaviour (Lenton et al. 2008; Schuur et al. 2015; Nitzbon et al. 2024). It has been suggested that microbial properties are also important, as when activated, microbial organisms are expected to cause additional warming and increased thaw (Bathiany et al. 2016; Steffen et al. 2018). This effect is expected to become relevant at a threshold rate of local warming (Luke and Cox, 2010). The hydrological feedback between permafrost and climate has also been suggested to be relevant to tipping science recently (Brovkin et al. in review, this issue). Moisture in the ground and seasonal storage in overlying snow influences heat transfer. The feedback magnitude is less well studied than the biogeochemical feedback, but is essential also considering teleconnections.

Localised positive feedback mechanisms have been identified related to ground thaw (Armstrong McKay et al. 2022). Abrupt thaw which is often triggered by unusually warm summers results in terrain change. This includes ground subsidence (immediate loss of ice in ground) and thaw slumps (permafrost specific landslides) (Brovkin et al. in review, this issue). Localised tipping occurs in clusters synchronised across large areas (Lenton et al. 2024). This has been specifically investigated for permafrost lakes for the last two decades (Nitze et al. 2018).

2.4.2 Earth Observation of Tipping

2.4.2.1 Introduction: Monitoring Permafrost from Space Subground temperature is essential for monitoring permafrost state. Monitoring requires in situ monitoring with boreholes, which are of limited availability and representativeness (Biskaborn et al. 2015). Satellite observations are used to complement the in situ measurements through modelling and land surface proxies (Trofaier et al. 2017; Bartsch et al. 2023a, b). Current approaches rely on models that utilise satellite-derived land surface temperature from thermal sensors which are impacted by cloud cover (e.g. Obu et al. 2019). Records are therefore gap-filled based on reanalysis data. Insulation/heat transfer properties are considered mechanisms in the modelling, specifically properties of snow (snow water equivalent) and soils (Westermann et al. 2015, 2023). Land cover is used as a proxy, specifically to separate areas with peat and wetlands. Common soil moisture detection schemes using microwave satellite data fail due to land cover heterogeneity (Högström and Bartsch 2017; Högström et al. 2018; Wrona et al. 2017).

For actual assessment of tipping with respect to release of soil carbon, GHG emissions need to be quantified. Advanced satellite missions as well as improved retrieval, adjusted to the Arctic, are needed to facilitate monitoring at the required detail (Miner et al. 2022). The added value of remote sensing in this context is currently comparably low (Swingedouw et al. 2020; Lenton et al. 2024). The potential lies largely in identification of localised tipping and abrupt thaw detection.

Land surface characteristics including lake extent change and subsidence are commonly used as indicators for local tipping (abrupt thaw features) and permafrost thaw in general. Regional clusters of ground thaw can be identified through land cover change reliably for up to 20 years back in time (Nitze et al. 2018; Runge et al. 2022). The investigation of abrupt thaw features which results in terrain change (thaw slumps) requires availability of terrain information at high spatial resolution. The ground also subsides where ice is lost in the ground due to thaw: this can be be quantified with synthetic aperture radar (SAR) data if long-term continuous records exist (e.g. Strozzi et al. 2018). However, soil organic carbon (SOC) content needs to be known to assess the potential magnitude of feedbacks. Interpolated maps from in situ records are currently the main source for SOC. They lack detail and consistency across the Arctic (Bartsch et al. 2016a, b). Land cover and land surface properties, as identified from space using multispectral as well as SAR data, have been to date only partially used to fill the gaps (Hugelius et al. 2013; Bartsch et al. 2016a) (Table 4).

2.4.2.2 Discussion: Gaps in GCOS and Recommendations Specifications dedicated to permafrost monitoring using satellite data are lacking for GCOS (see Table 4). Only in situ monitoring is addressed. Specifications under the more generic ECV soil temperature are not applicable in the current form. In case of lake extent, land cover and snow water equivalent, goal requirements need to be met to be applicable for permafrost. Ground subsidence is regionally feasible and can be potentially fully implemented on for the entire permafrost domain but is not considered as an ECV for land (only for glaciers).

2.5 Other potential tipping systems

2.5.1 Lakes

Although lakes and freshwaters in general are not considered tipping elements in the climate system (Lenton et al. 2008; Armstrong McKay et al. 2022), lake shifts provide important direct or indirect feedbacks to the Earth's climate system. Among others, abrupt lake water level or extent changes directly affect water availability with direct consequence for vegetation and thus respiration (Richardson et al. 2013). Changes in the capacity of lakes to store carbon in their sediments could affect the carbon budget of lakes and biogeochemical alterations in lakes could lead to changes in the exchange of greenhouse gases between lakes and the atmosphere (Bartosiewicz et al., 2021). More studies should explore the climate–hydrology–biogeochemistry linkages that lead to lake ecosystem shifts and potential feedbacks to the climate system (Lehnherr et al. 2018).

Climate pressure can push lakes towards tipping points with important consequences. One of the typical lake shift analyses is the transition from phytoplankton- to macrophytes-dominated state of lakes (Scheffer 1991, 1993, 2001; Loverde-Oliveira et al. 2009; Kosten et al., 2012). However, more and more studies are nowadays focusing on other lake shifts

Table 4 Parameters relevant to permafrost tipping, and corresponding GCOS-ECVs

Variables related to tipping dynamics	GCOS-ECV parameters		Comment/gap(s)	Suitable?
	Variables	Specs		
Ground thaw—carbon release feedback				
Ground temperature*	Thermal state of permafrost Active layer thickness	Sufficient sites to characterise each bio-climate zone, in situ specific specifications	Specifications not applicable for satellite based retrievals	X
	Land surface temperature	Threshold 1 km (at 6 h), Goal < 1 km	Goal spatial resolution insufficiently defined	X
	Soil temperature	Down to 180 cm, Threshold 139– 278 km daily, Goal 5 km at 3 h	Vertical and spatial resolution requirements do not address permafrost needs	
Soil organic carbon content	Soil organic carbon content % of organic carbon in the topmost 30 cm and sub-soil 30–100 cm	20 to 1000 km (20 to 1000 m for peatlands + plus mineral soil bulk density)	Peatland specifications applicable Currently satellite-derived products meeting the requirements are only locally available	X
GHG flux*	CO_2 mole fraction/column average CH_4 mole fraction/column average	100–2000 km for mole fraction 1–10 km for column	Limited in situ availability for permafrost, constraints for satellite retrievals in high latitudes	X
Misc: linked to permafrost state				
Thermokarst lake water extent	Lake water extent	Threshold monthly at 1 km, goal 5 days at 10 m	Lake drainage as indicator of thaw, Threshold spatial requirements insufficient, goal requirements applicable	X
Surface elevation change	–	–	Long-term subsidence required as indicator of thaw High spatial detail needed for abrupt thaw feature monitoring	M
Snow water equivalent*	Snow water equivalent	Threshold every 2nd day at 25 km, goal 6 h at 0.5 km	Insulation effect, Threshold spatial requirements insufficient	X

Table 4 (continued)

Variables related to tipping dynamics	GCOS-ECV parameters		Comment/gap(s)	Suitable?
	Variables	Specs		
Land cover*	Land cover high resolution land cover	Threshold 5 years at 1 km, goal monthly at 100–300 m Threshold 5 years at 30–100 m, goal monthly at < 10 m	Abrupt thaw indicator Only high resolution land cover applicable	X
Soil moisture	Surface soil moisture Root zone soil moisture	Threshold 2 days at 50 km, goal 6 h at 1 km	Too coarse to represent Arctic landcover heterogeneity; Surface soil moisture only from satellite records, and operational schemes fail at high latitudes	X

A star (*) indicates what is directly addressed in ESA CCI baseline projects. The last column indicates where the GCOS-ECV parameters do not match the specific needs of tipping science either by insufficient or undefined specs (X) or entirely missing ECV (M)

such as shifts in lake mixing regime (Woolway and Merchant 2019) and the cascading effects on oxygen regime (Schwefel et al 2016). Surface warming increases the stability of the water column of lakes resulting in longer periods of stratification. Longer stratification means reduced mixing and thus less exchange between epilimnion and hypolimnion. As a result, the nutrient as well as the oxygen transport along the water column are reduced with consequences for the lake biota as well.

As for other tipping systems, feedback mechanisms can be identified in lakes. A recent review on the physical trends related to increased duration and strength of density stratification, and the related chemical and biological consequences (Mesman et al. 2021) found that the internal feedbacks can reinforce (positive feedbacks) or slow down (negative feedbacks) shifts between mixing regimes in lakes that stratify at least during one season. Lake internal feedback can be different depending on the mixing regime of lakes (and within these classes the morphometry of the lake can also differentiate the feedback strength). Lake internal feedback could stabilise and even determine the mixing regime, especially in situations where morphometry and climate can support multiple mixing regimes. For example recent studies predicted shifts from dimictic to monomictic regime, where ice cover and inverse stratification in winter are disappearing; and from holomictic to oligomictic or meromictic regime in non-ice covered lakes (Shatwell et al. 2019).

The main source of data to analyse lake shifts are of four different types: in situ measurements, paleolimnological records, models or satellite remote sensing (Calamita et al. 2024); all with different spatial and temporal resolution and different timespan. EO data has been mainly used for studies about two types of lake shifts: lake water extent (Buma et al. 2018; Liang and Li 2019; Nitze et al. 2018; Bai et al. 2021; Zhang et al. 2021), and ice cover (Sun et al. 2018; Wang et al. 2018; Liu et al., 2020). Both are considered ECVs by GCOS and requirements are specified. They differ between the two parameters regarding goal specifications for horizontal resolution, 10 m versus 50 m, but threshold is 1 km for both. However, the scope of usage of EO in the field of lake shifts has broadened in recent years (Calamita et al. 2024). EO has been recently used also to detect and analyse ecological shifts (Bergamino et al. 2007; Duan et al. 2014; Free et al. 2021) and shifts in mixing regime (Woolway and Merchant 2017, 2018; Fichot et al. 2019). This includes observations of chlorophyll-a concentrations (Bergamino et al. 2007; Free et al. 2021). The ECV ‘lake water-leaving reflectance’ is expected to support retrieval. An additional parameter is surface concentration of particulate organic carbon (POC; Duan et al. 2014) which is currently not addressed through GCOS. Monitoring of mixing shifts with remote sensing is addressed through analyses of lake surface water temperature. Goal and threshold requirements for temporal and spatial scale have been defined for this ECV but are not well justified.

2.5.2 Ocean Deoxygenation

The dissolved oxygen ocean content has decreased over the last 50 years by about 2% (Schmidtko et al. 2017), reflecting climate-driven changes in ocean circulation and ventilation and changes in biological oxygen demand. Ocean deoxygenation is a severe threat to life in the ocean, impairing vital functions, causing mass mortality and habitat compression (Breitburg et al. 2018) and affecting the microbially-mediated cycling of key elements, mostly carbon (C), nitrogen (N), phosphorus (P) and iron (Fe). Under low oxygen conditions the process of denitrification can lead to a net loss of fixed nitrogen and reduced ocean fertility, with potential impact on ocean productivity and biologically-driven carbon storage. Further, ocean deoxygenation affects the oceanic production of a potent greenhouse gas N_2O which is strongly enhanced under low-oxygen conditions, potentially providing a positive feedback to global warming.

Ocean deoxygenation is currently regarded as an uncertain potential tipping system (Armstrong McKay et al. 2022), due to the large uncertainties over drivers, mechanisms, time-scale and thresholds for triggering a widespread oxygen loss in the ocean (for more detail see Wood et al. 2024). Given the profound implications that oxygen changes have on marine ecosystems and their services, O_2 is identified as an ECV by GCOS (WMO, 2022a), and it is of high priority to reliably monitor and assess the likelihood of widespread ocean deoxygenation under current global warming.

Observations from multiple observing platforms (ship-based, fixed-point moorings, autonomous like BGC-Argo, gliders, AUV) key for identifying ongoing subsurface ocean deoxygenation trends and for assessing intraseasonal, seasonal and interannual oxygen variability. Although the oxygen-observing network is expanding (Grégoire et al. 2021), and machine learning (ML) is increasingly used as a powerful method to fill the observational gap (eg. Sharp et al. 2023), further improved observational coverage, spatial and temporal resolution is needed to meet the GCOS requirements to better monitor and quantify oxygen changes (WMO, 2022a). Satellite products are becoming increasingly used for assessing climate change as well as for the near-real time monitoring of ocean health (e.g. Groom et al. 2019; Vinogradova et al. 2019). Yet, their use in support of monitoring ocean deoxygenation and identifying areas susceptible to hypoxia is still limited. This is due to the lack of a direct EO-observed proxy for ocean oxygen and limited vertical coverage of EO to the surface ocean. However, machine learning (ML) algorithms have been developed to reconstruct 3D field of physical (e.g. temperature and salinity; Buongiorno Nardelli 2020) and biological variables (e.g. chlorophyll; Sammartino et al. 2020) from 2D remote sensing imageries. ML approaches have also been used to reconstruct biogeochemical variables at global scale from Biogeochemical-Argo (BGC-Argo) data combined with satellite measurements (Sauzède et al. 2015, 2016, 2017). Hence, the use of artificial intelligence (AI) in conjunction with Earth observed variables (associated with O_2 variability) and O_2 observations appears as a promising avenue towards the reconstruction of O_2 fields over the satellite era, for monitoring short time scale deoxygenation events (eutrophication) and climate trends. Chlorophyll-a, recognized by GCOS as an ECV, is a proxy of phytoplankton biomass and is used for assessing biological productivity space–time dynamics. The increase in the number of spectral bands, from 4 for Coastal Zone Colour Scanner to 21 for Sentinel 3, has improved the quality of OC products with a better differentiation between chlorophyll-a (Chl-a) and non-organic materials, yet its GCOS requirements are insufficiently defined. Estimation of sea surface salinity (SSS) from space in combination with sea surface temperature (SST) and sea surface height (SSH), offer the possibility to follow the evolution of the sea surface density which can be related to the intensity of the ventilation process, circulation, plume tracking, oxygen saturation level and, in combination with in situ data, stratification dynamics. Although these parameters are included in the GCOS portfolio, their requirements may need to be revised for their suitability to support early warning for ocean deoxygenation. Further potential may derive from indirect proxies of ocean deoxygenation such the oceanic emissions of N_2O. This GHG has proven difficult to estimate by satellite because of its weak spectral signature in the infra-red band and its low variability in the troposphere (ppb), although retrieval of tropospheric N_2O concentration profiles may be possible from high spectral resolution nadir measurements (Thermal Infrared 1200–1300 cm-1; AIRS and IASI sensors). However, the low spatial resolution, contamination with other gases (H_2O, CH_4), and low sensitivity to tropospheric concentration currently limit the exploitation of this satellite-derived N_2O product (Garcia et al., 2018). Next generation instruments (IASI-NG) could improve the observation capabilities in this direction (Ricaud et al. 2021).

3 Summary of Gaps in GCOS Portfolio and Outlook

3.1 Do the GCOS ECVs Meet the Needs of Tipping Points Science?

A wide range of parameters which are of relevance for tipping processes in the Earth's climate system are considered for monitoring in the GCOS specifications. The utility of remote sensing is often limited for direct observation of ongoing tipping dynamics or early warning signals of approaching tipping points (Lenton et al. 2024). The main merit rather lies in provision of means for improvement of numerical models of climate tipping systems (with or without data assimilation), thereby allowing for improved understanding of the processes and system state (stage 1). Spatial coverage through satellite data is in most cases an asset. Stage 2 (EWS) observation is mostly constrained because of the relatively short record length of the satellite record (about five decades). Integrated use of multiple sensors and platforms (satellites, aircrafts, drones, ground stations etc.) is therefore crucial in conjunction with modelling and paleoclimate evidence.

The majority of variables needed to monitor key tipping systems are considered as GCOS ECVs. Gaps for key elements have been identified for the Amazon rainforest (vegetation water content), permafrost (ground subsidence), AMOC (section mass, heat and fresh water transports and freshwater input from ice sheet edges) and specifically ice sheets. In the latter case, recognition as an ECV is missing for vertical land motion for isostatic effects, melt area, mass flux across the grounding line, calving rate and areas of hydrofracturing. Vertical motion and melt area are required for monitoring the melt-elevation feedback. Glacier melt area has been also previously reported as required but missing in the context of mountain ECVs (Thornton et al. 2021). In addition, lake properties such as particulate organic carbon and chlorophyll-a are lacking. Most parameters not considered under GCOS for the tipping systems discussed here are also not addressed in the framework of the ESA Climate Change Initiative (CCI) program,showcasing the important implications of whether or not variables are recognised as ECVs. Further on, all considered systems require monitoring of parameters which are (in parts) listed in GCOS but so far not addressed by the CCI program.

For many of the ECVs, issues in specifications (from threshold to goal) have been identified. Of main concern is spatial resolution, but also the means of measurements. This applies specifically for cases where the ECVs refer to subsurface properties which cannot be directly measured from space, but through use in models. For example, GCOS ground temperature specifications for permafrost target only in situ point measurements while the majority of other ECVs, including e.g. soil organic carbon content and sea ice thickness, considers spatially continuous observations through satellite observations. In many cases, the threshold spatial resolution target is insufficient to address tipping, only meeting the goal requirements would ensure applicability. Consistency of specification across several related variables is also an important issue (e.g. to define regional ocean heat/water budgets). Several ECVs are related to lake shifts, including lake water extent and lake ice cover, chlorophyll-a concentrations and lake water-leaving reflectance and lake surface water temperature (LSWT). Goal and threshold requirements for temporal and spatial scale have been defined for LSWT but are not well justified. Surface concentration of particulate organic carbon is missing.

Another issue concerns ECVs that are recognised by GCOS but where the requirements are not well justified for remote sensing or defined (or only in parts). This includes lake surface temperature, permafrost ground temperature, lake and ocean chlorophyll-a concentration, surface elevation of ice sheets and grounding line location. Goal requirements for land cover and biomass do not meet the requirements of Amazon rain forest monitoring.

Monitoring of proxies for actual ECVs is of high relevance in many cases, especially when subground phenomena are investigated. For example in case of permafrost abrupt thaw relevant parameters (thaw lake change, terrain change/ground subsidence) and soil properties are needed to assess the vulnerability of tipping (presence of carbon in the ground). The potential for reaching full spatial coverage with satellite data is limited in these cases since comparably high spatial and temporal resolution is needed. A major constraint for early warning regarding permafrost is the limited availability of CO_2 and CH_4 observation in high latitudes. Similarly, the state of AMOC and SPG is challenging to assess from space directly, and reconstructions rely on combinations of models and proxy data such as sea surface temperature and salinity to infer the buoyancy conditions that determine the overturning. Remote sensing shines here by offering a wide range of variables and broad spatial coverage, the utility of which is however determined by the capacity of the models to capture the key tipping dynamics.

Regarding ocean deoxygenation, O_2 is identified by GCOS as ECV, with various different approaches underway to improve observational coverage to meet the requirements (through AI reconstruction, indirect proxies like N_2O and SSS, SST, SSH). Revisioning of the requirements of the proxies is needed to support EWS of ocean deoxygenation.

Several parameters are identified as of relevance across several of the discussed elements. The different requirements are not reflected in the GCOS portfolio. For example lake water extent is of relevance for shift in lakes as well as proxy for permafrost (thaw lakes). The spatial resolution needs to be higher than currently defined as goal in both cases. Similarly, monitoring the ice loss at the fringes of the Greenland ice sheet does not only capture the ice sheet state but is also of high importance for stability assessments of the AMOC and SPG.

3.2 Recommendations

In this analysis, we have made first steps towards specifying actual observation needs from a tipping points perspective for future ECV considerations. The discussed tipping systems only present a selection and, looking forward, merit more in-depth studies from the individual domain communities. In light of the very active tipping points research, requirements specifications in the GCOS portfolio need to be continuously revised with respect to emerging tipping systems. However, in an update of essential climate variables it is critical to consider a balance between parsimony and specificity, taking into account the resources required to meet the needs. While our work set out to identify shortcomings in the GCOS-ECV portfolio from the perspective of tipping points science, our findings don't necessarily warrant closing all these gaps in this portfolio. Rather, subsequent work should identify which of the observational requirements should find their way into the ECVs, and analyse the benefit of a complementary set of essential tipping points variables. In line with similar efforts on (climate) variables for ocean (Miloslavich et al. 2018), biodiversity (Pereira et al., 2013) and mountains (Thornton et al. 2021), the tipping points community could define its needs in an independent portfolio centred around the monitoring of critical, positive feedback loops.

Our analysis needs to be extended to further systems and different angles (numerical and conceptual modelling, paleo reconstruction, in situ observation, etc.) considered to provide a full assessment where monitoring of tipping is limited in terms of observational evidence. The rapid advances in computational capabilities (including high-resolution modelling, big data and machine learning approaches) and continued availability of (open access) data hold the potential for ever-increasing utility of EO for tipping research. We encourage the tipping community to revise what their requirements are in terms of monitored variables, spatial and temporal resolution. This may guide mission and even sensor development for better informed risk assessments with respect to emerging climate tipping dynamics under increasing human pressures in the future.

Acknowledgements This paper is an outcome of the Workshop "Tipping Points and Understanding EO data needs for a Tipping Element Model Intercomparison Project" held at the International Space Science Institute (ISSI) in Bern, Switzerland (10-14 October 2022). The authors would like to thank the three anonymous reviewers for their critical assessment, which significantly improved the readability and content of the manuscript. S.L. acknowledges financial support from the Earth Commission. The work of the Earth Commission, a program of Future Earth, is made possible through the support of the Global Commons Alliance, a sponsored project of Rockefeller Philanthropy Advisors. S.L. J.F.D and N.W. acknowledge financial support from the European Research Council Advanced Grant project ERA (Earth Resilience in the Anthropocene, grant ERC-2016-ADG-743080). J.F.D. is grateful for support by the German Federal Ministry for Education and Research (BMBF, project PIK Change, grant 01LS2001A). S.H. acknowledges support from the European Space Agency through the ESA-Future Earth Joint Program. A.B. was supported by the European Space Agency through Permafrost_cci (4000123681/18/INB) and AMPAC-Net (4000137912/22/I-DT), H2020 OptimESM (101081193) and the European Research Council Synergy project No. 951288 (Q-Arctic). A. L. acknowledges support from European Space Agency MiTHO project (ESA ESRIN/Contract No. 4000142100/23/I-DT). A. S. is supported by the Dutch Research Council (NWO) Talent Program Grant VI.Veni.202.170. M.H. acknowledges support from the Serrapilheira Institute (grant number Serra-1709-18983). T. N. acknowledges support through Greenland Ice Sheet CCI (ESA Contract No. 4000126023) and Antarctic Ice Sheet CCI (Contract No. 4000126813). E.C. has been funded by the European Space Agency (ESA) as part of the Climate Change Initiative (CCI) fellowship (ESA ESRIN/Contract No. 4000133621/20/I/NB). RW was supported by the Met Office Hadley Centre Climate Programme funded by BEIS.

Author contributions Authors Bartsch, Calamita, Donges, Hebden, Landolfi, Loriani, Verbesselt and Winkelmann contributed to the conceptualisation of the paper, initiated at the ISSI Workshop on "Tipping Points and Understanding EO data needs for a Tipping Element Model Intercomparison Project" in October 2022. All authors contributed to the writing and preparation of the manuscript.

Funding Open Access funding enabled and organized by Projekt DEAL.

Declarations

Conflict of interest The authors declare no competing interests.

References

Andela N, Morton DC, Schroeder W, Chen Y, Brando PM, Randerson JT (2022) Tracking and classifying Amazon fire events in near real time. Sci adv 8(30):eabd2713

Andela N, Morton DC, Schroeder W, Chen Y, Brando PM, Randerson JT (2022) Tracking and classifying Amazon fire events in near real time. Sci Adv. https://doi.org/10.1126/sciadv.abd2713

Araujo R, Assunção J, Hirota M, Scheinkman JA (2023) Estimating the spatial amplification of damage caused by degradation in the Amazon. Proc Natl Acad Sci 120:e2312451120. https://doi.org/10.1073/pnas.2312451120

Armstrong McKay DI, Staal A, Abrams JF, Winkelmann R, Sakschewski B, Loriani S, Fetzer I, Cornell SE, Rockström J, Lenton TM (2022) Exceeding 1.5°C global warming could trigger multiple climate tipping points. Science. https://doi.org/10.1126/science.abn7950

Ashwin P, Wieczorek S, Vitolo R, Cox P (2012) Tipping points in open systems: bifurcation, noise-induced and rate-dependent examples in the climate system. Phil Trans R Soc A 370:1166–1184. https://doi.org/10.1098/rsta.2011.0306

Bai M, Mo X, Liu S, Hu S (2021) Detection and attribution of lake water loss in the semi-arid Mongolian Plateau—A case study in the Lake Dalinor. Ecohydrology 14:e2251. https://doi.org/10.1002/eco.2251

Banwell AF, MacAyeal DR, Sergienko OV (2013) Breakup of the Larsen B Ice Shelf triggered by chain reaction drainage of supraglacial lakes. Geophys Res Lett 40:5872–5876. https://doi.org/10.1002/2013GL057694

Barkhordarian A, Saatchi SS, Behrangi A, Loikith PC, Mechoso CR (2019) A Recent Systematic Increase in Vapor Pressure Deficit over Tropical South America. Sci Rep 9:15331. https://doi.org/10.1038/s41598-019-51857-8

Bartosiewicz M, Maranger R, Przytulska A, Laurion I (2021) Effects of phytoplankton blooms on fluxes and emissions of greenhouse gases in a eutrophic lake. Water Res 196:116985. https://doi.org/10.1016/j.watres.2021.116985

Bartsch A, Widhalm B, Kuhry P, Hugelius G, Palmtag J, Siewert MB (2016a) Can C-band synthetic aperture radar be used to estimate soil organic carbon storage in tundra? Biogeosciences 13:5453–5470. https://doi.org/10.5194/bg-13-5453-2016

Bartsch A, Höfler A, Kroisleitner C, Trofaier AM (2016b) Land cover mapping in northern high latitude permafrost regions with satellite data: achievements and remaining challenges. Remote Sensing 8:979. https://doi.org/10.3390/rs8120979

Bartsch A, Strozzi T, Nitze I (2023a) Permafrost Monitoring from Space. Surv Geophys 44:1579–1613. https://doi.org/10.1007/s10712-023-09770-3

Bartsch A, Bergstedt H, Pointner G, Muri X, Rautiainen K, Leppänen L, Joly K, Sokolov A, Orekhov P, Ehrich D, Soininen EM (2023b) Towards long-term records of rain-on-snow events across the Arctic from satellite data. Cryosphere 17:889–915. https://doi.org/10.5194/tc-17-889-2023

Bassis JN, Walker CC (2011) Upper and lower limits on the stability of calving glaciers from the yield strength envelope of ice. Proc Royal Soc: Mathem, Phys Eng Sci 468:913–931. https://doi.org/10.1098/rspa.2011.0422

Basso LS, Wilson C, Chipperfield MP, Tejada G, Cassol HLG, Arai E, Williams M, Smallman TL, Peters W, Naus S, Miller JB, Gloor M (2023) Atmospheric CO2 inversion reveals the Amazon as a minor carbon source caused by fire emissions, with forest uptake offsetting about half of these emissions. Atmos Chem Phys 23:9685–9723. https://doi.org/10.5194/acp-23-9685-2023

Bathiany S, Dijkstra H, Crucifix M, Dakos V, Brovkin V, Williamson MS, Lenton TM, Scheffer M (2016) Beyond bifurcation: using complex models to understand and predict abrupt climate change. Dynam Statist Climate Syst. https://doi.org/10.1093/climsys/dzw004

Bathiany S, Bastiaansen R, Bastos A, Blaschke L, Lever J, Loriani S, Keersmaecker W, Dorigo W, Milenković M, Cornelius S, Taylor S, Jan V, Niklas B (2024) Ecosystem resilience monitoring and early warning using earth observation data: Challenges and outlook. Surv Geophys. https://doi.org/10.1007/s10712-024-09833-z

Bauer N, Keller DP, Garbe J, Karstens K, Piontek F, von Bloh W, Thiery W, Zeitz M, Mengel M, Strefler J, Thonicke K, Winkelmann R (2023) Exploring risks and benefits of overshooting a 1 5 °C carbon budget over space and time. Environ. Res. Lett. 18:054015. https://doi.org/10.1088/1748-9326/accd83

Benn DI, Cowton T, Todd J, Luckman A (2017) Glacier calving in greenland. Curr Clim Change Rep 3:282–290. https://doi.org/10.1007/s40641-017-0070-1

Berenguer E, Lennox GD, Ferreira J, Malhi Y, Aragão LEOC, Barreto JR, Del Bon Espírito-Santo F, Figueiredo AES, França F, Gardner TA, Joly CA, Palmeira AF, Quesada CA, Rossi LC, de Seixas MMM, Smith CC, Withey K, Barlow J (2021) Tracking the impacts of El Niño drought and fire in human-modified Amazonian forests. Procee Natio Acad Sci. https://doi.org/10.1073/pnas.2019377118

Bergamino N, Loiselle SA, Cózar A, Dattilo AM, Bracchini L, Rossi C (2007) Examining the dynamics of phytoplankton biomass in Lake Tanganyika using Empirical Orthogonal Functions. Ecol Model 204:156–162. https://doi.org/10.1016/j.ecolmodel.2006.12.031
Biskaborn BK, Lanckman J-P, Lantuit H, Elger K, Streletskiy DA, Cable WL, Romanovsky VE (2015) The new database of the Global Terrestrial Network for Permafrost (GTN-P). Earth System Science Data 7:245–259. https://doi.org/10.5194/essd-7-245-2015
Boehrer B, Fukuyama R, Chikita K (2008) Stratification of very deep, thermally stratified lakes. Geophys Res Lett. https://doi.org/10.1029/2008gl034519
Boers N (2021) Observation-based early-warning signals for a collapse of the Atlantic Meridional Overturning Circulation. Nat Clim Chang 11:680–688. https://doi.org/10.1038/s41558-021-01097-4
Boers N, Rypdal M (2021) Critical slowing down suggests that the western Greenland Ice Sheet is close to a tipping point. Proc Natl Acad Sci 118:e2024192118. https://doi.org/10.1073/pnas.2024192118
Boulton CA, Allison LC, Lenton TM (2014) Early warning signals of Atlantic Meridional Overturning Circulation collapse in a fully coupled climate model. Nat Commun 5:5752. https://doi.org/10.1038/ncomms6752
Boulton CA, Lenton TM, Boers N (2022) Pronounced loss of Amazon rainforest resilience since the early 2000s. Nat Clim Chang 12:271–278. https://doi.org/10.1038/s41558-022-01287-8
Box JE, Fettweis X, Stroeve JC, Tedesco M, Hall DK, Steffen K (2012) Greenland ice sheet albedo feedback: thermodynamics and atmospheric drivers. Cryosphere 6:821–839. https://doi.org/10.5194/tc-6-821-2012
Breitburg D, Levin LA, Oschlies A, Grégoire M, Chavez FP, Conley DJ, Garçon V, Gilbert D, Gutiérrez D, Isensee K, Jacinto GS, Limburg KE, Montes I, Naqvi SWA, Pitcher GC, Rabalais NN, Roman MR, Rose KA, Seibel BA, Telszewski M, Yasuhara M, Zhang J (2018) Declining oxygen in the global ocean and coastal waters. Science. https://doi.org/10.1126/science.aam7240
Bromwich DH, Nicolas JP, Monaghan AJ (2011) An Assessment of Precipitation Changes over Antarctica and the Southern Ocean since 1989 in Contemporary Global Reanalyses. J Clim 24:4189–4209. https://doi.org/10.1175/2011JCLI4074.1
Brothers S, Köhler J, Attermeyer K, Grossart HP, Mehner T, Meyer N, Scharnweber K, Hilt S (2014) A feedback loop links brownification and anoxia in a temperate, shallow lake. Limnol Oceanogr 59:1388–1398. https://doi.org/10.4319/lo.2014.59.4.1388
Brummitt CD, Barnett G, D'Souza RM (2015) Coupled catastrophes: sudden shifts cascade and hop among interdependent systems. J R Soc Interface 12:20150712. https://doi.org/10.1098/rsif.2015.0712
Brunt KM, Fricker HA, Padman L (2011) Analysis of ice plains of the Filchner-Ronne Ice Shelf, Antarctica, using ICESat laser altimetry. J Glaciol 57:965–975. https://doi.org/10.3189/002214311798043753
Bullock EL, Woodcock CE, Souza C, Olofsson P (2020) Satellite-based estimates reveal widespread forest degradation in the Amazon. Glob Change Biol 26:2956–2969. https://doi.org/10.1111/gcb.15029
Buma WG, Lee S-I, Seo JY (2018) Recent Surface Water Extent of Lake Chad from Multispectral Sensors and GRACE. Sensors 18:2082. https://doi.org/10.3390/s18072082
Buongiorno Nardelli B (2020) A multi-year time series of observation-based 3D horizontal and vertical quasi-geostrophic global ocean currents. Earth Syst Sci Data 12:1711–1723. https://doi.org/10.5194/essd-12-1711-2020
Bustamante M, Roy J, Ospina D, Achakulwisut P, Aggarwal A, Bastos A, Broadgate W, Canadell JG, Carr ER, Chen D, Cleugh HA, Ebi KL, Edwards C, Farbotko C, Fernández-Martínez M, Frölicher TL, Fuss S, Geden O, Gruber N, Harrington LJ, Hauck J, Hausfather Z, Hebden S, Hebinck A, Huq S, Huss M, Jamero MLP, Juhola S, Kumarasinghe N, Lwasa S, Mallick B, Martin M, McGreevy S, Mirazo P, Mukherji A, Muttitt G, Nemet GF, Obura D, Okereke C, Oliver T, Orlove B, Ouedraogo NS, Patra PK, Pelling M, Pereira LM, Persson Å, Pongratz J, Prakash A, Rammig A, Raymond C, Redman A, Reveco C, Rockström J, Rodrigues R, Rounce DR, Schipper ELF, Schlosser P, Selomane O, Semieniuk G, Shin Y-J, Siddiqui TA, Singh V, Sioen GB, Sokona Y, Stammer D, Steinert NJ, Suk S, Sutton R, Thalheimer L, Thompson V, Trencher G, van der Geest K, Werners SE, Wübbelmann T, Wunderling N, Yin J, Zickfeld K, Zscheischler J (2023) Ten New Insights in Climate Science 2023/2024. Global Sustainability. https://doi.org/10.1017/sus.2023.25
Caesar L, Rahmstorf S, Robinson A, Feulner G, Saba V (2018) Observed fingerprint of a weakening Atlantic Ocean overturning circulation. Nature 556:191–196. https://doi.org/10.1038/s41586-018-0006-5
Caesar L, McCarthy GD, Thornalley DJR, Cahill N, Rahmstorf S (2021) Current Atlantic Meridional Overturning Circulation weakest in last millennium. Nat Geosci 14:118–120. https://doi.org/10.1038/s41561-021-00699-z
Calamita E, Lever JJ, Albergel C, Woolway RI, Odermatt D (2024) Detecting climate-related shifts in lakes: a review of the use of satellite Earth Observation. Limnology and Oceanography. 69(4):723–41

Calamita E, Lever JJ, Albergel C, Woolway RI, Odermatt D (2024) Detecting climate-related shifts in lakes: A review of the use of satellite Earth Observation. Limnol Oceanogr 69:723–741. https://doi.org/10.1002/lno.12498
Ciemer C, Boers N, Hirota M, Kurths J, Müller-Hansen F, Oliveira RS, Winkelmann R (2019) Higher resilience to climatic disturbances in tropical vegetation exposed to more variable rainfall. Nat Geosci 12:174–179. https://doi.org/10.1038/s41561-019-0312-z
Crisp D, Dolman H, Tanhua T, McKinley GA, Hauck J, Bastos A, Sitch S, Eggleston S, Aich V (2022) How Well Do We Understand the Land-Ocean-Atmosphere Carbon Cycle? Reviews of Geophysics. https://doi.org/10.1029/2021RG000736
Dakos V, Boulton CA, Buxton JE, Abrams JF, Arellano-Nava B, Armstrong McKay DI, Bathiany S, Blaschke L, Boers N, Dylewsky D, López-Martínez C, Parry I, Ritchie P, van der Bolt B, van der Laan L, Weinans E, Kéfi S (2024) Tipping point detection and early warnings in climate, ecological, and human systems. Earth System Dynamics 15:1117–1135. https://doi.org/10.5194/esd-15-1117-2024
de Roda Husman S, Lhermitte S, Bolibar J, Izeboud M, Hu Z, Shukla S, van der Meer M, Long D, Wouters B (2024) A high-resolution record of surface melt on Antarctic ice shelves using multi-source remote sensing data and deep learning. Remote Sens Environ 301:113950. https://doi.org/10.1016/j.rse.2023.113950
DeConto RM, Pollard D (2016) Contribution of Antarctica to past and future sea-level rise. Nature 531:591–597. https://doi.org/10.1038/nature17145
Ditlevsen P, Ditlevsen S (2023) Warning of a forthcoming collapse of the Atlantic meridional overturning circulation. Nat Commun 14:4254. https://doi.org/10.1038/s41467-023-39810-w
Drijfhout S, Bathiany S, Beaulieu C, Brovkin V, Claussen M, Huntingford C, Scheffer M, Sgubin G, Swingedouw D (2015) Catalogue of abrupt shifts in Intergovernmental Panel on Climate Change climate models. Proc Natl Acad Sci USA 112:E5777–E5786. https://doi.org/10.1073/pnas.1511451112
Druckenmiller ML, Moon T, Thoman R (eds) (2021) The Arctic, in the Arctic bulletin of the American meteorological society 102(8):S263–S316. https://doi.org/10.1175/BAMS-D-21-0086.1
Drüke M, Sakschewski B, von Bloh W, Billing M, Lucht W, Thonicke K (2023) Fire may prevent future Amazon forest recovery after large-scale deforestation. Commun Earth Environ 4:1–10. https://doi.org/10.1038/s43247-023-00911-5
Duan H, Feng L, Ma R, Zhang Y, Arthur Loiselle S (2014) Variability of particulate organic carbon in inland waters observed from MODIS Aqua imagery. Environ Res Lett 9:084011. https://doi.org/10.1088/1748-9326/9/8/084011
Dukhovskoy DS, Yashayaev I, Chassignet EP, Myers PG, Platov G, Proshutinsky A (2021) Time scales of the Greenland Freshwater Anomaly in the Subpolar North Atlantic. Journal of Climate. https://doi.org/10.1175/JCLI-D-20-0610.1
Duncanson L et al (2022) Aboveground biomass density models for NASA's Global Ecosystem Dynamics Investigation (GEDI) lidar mission. Remote Sensing Environ. https://doi.org/10.1016/j.rse.2021.112845
Ehn M et al (2014) A large source of low-volatility secondary organic aerosol. Nature 506:476–479. https://doi.org/10.1038/nature13032
Estella-Perez V, Mignot J, Guilyardi E, Swingedouw D, Reverdin G (2020) Advances in reconstructing the AMOC using sea surface observations of salinity. Clim Dyn 55:975–992. https://doi.org/10.1007/s00382-020-05304-4
Exbrayat J-F, Bloom AA, Carvalhais N, Fischer R, Huth A, MacBean N, Williams M (2019) Understanding the Land Carbon Cycle with Space Data: Current Status and Prospects. Surv Geophys 40:735–755. https://doi.org/10.1007/s10712-019-09506-2
Feldmann J, Levermann A (2015) Collapse of the West Antarctic Ice Sheet after local destabilization of the Amundsen Basin. Proc Natl Acad Sci 112:14191–14196. https://doi.org/10.1073/pnas.1512482112
Fichot CG, Matsumoto K, Holt B, Gierach MM, Tokos KS (2019) Assessing change in the overturning behavior of the Laurentian Great Lakes using remotely sensed lake surface water temperatures. Remote Sens Environ 235:111427. https://doi.org/10.1016/j.rse.2019.111427
Finer M, Mamani N (2022) Amazon Deforestation Hotspots 2021. MAAP: 153
Flechtner, F., Neumayer, K.-H., Dahle, C., Dobslaw, H., Fagiolini, E., Raimondo, J.-C., and Güntner, A. 2016 What Can be Expected from the GRACE-FO Laser Ranging Interferometer for Earth Science Applications?, in: Remote Sensing and Water Resources, edited by: Cazenave, A., Champollion, N., Benveniste, J., and Chen, J., Springer International Publishing, Cham
Flores BM, Holmgren M (2021a) White-Sand Savannas Expand at the Core of the Amazon After Forest Wildfires. Ecosystems 24:1624–1637. https://doi.org/10.1007/s10021-021-00607-x

Flores BM, Holmgren M (2021b) Why forest fails to recover after repeated wildfires in Amazonian floodplains? Experimental evidence on tree recruitment limitation. J Ecol 109:3473–3486. https://doi.org/10.1111/1365-2745.13769
Forster, P., Storelvmo, T., Armour, K., Collins, W., Dufresne, J.-L., Frame, D., Lunt, D. J., Mauritsen, T., Palmer, M. D., Watanabe, M., Wild, M., and Zhang, H.: Chapter 7: The Earth's Energy Budget, Climate Feedbacks and Climate Sensitivity, edited by: Masson-Delmotte, V., Zhai, P., Pirani, A., Connors, S. L., Péan, C., Berger, S., Caud, N., Chen, Y., Goldfarb, L., Gomis, M. I., Huang, M., Leitzell, K., Lonnoy, E., Matthews, J. B. R., Maycock, T. K., Waterfield, T., Yelekçi, O., Yu, R., and Zhou, B., Cambridge University Press, 2021.
Forzieri G, Dakos V, McDowell NG, Ramdane A, Cescatti A (2022) Emerging signals of declining forest resilience under climate change. Nature 608:534–539. https://doi.org/10.1038/s41586-022-04959-9
Fox-Kemper, B., Hewitt, H. T., Xiao, C., Aðalgeirsdóttir, G., Drijfhout, S. S., Edwards, T. L., Golledge, N. R., Hemer, M., Kopp, R. E., Krinner, G., Mix, A., Notz, D., Nowicki, S., Nurhati, I. S., Ruiz, L., Sallée, J.-B., Slangen, A. B. A., and Yu, Y.: Chapter 9: Ocean, Cryosphere and Sea Level Change, in: Climate Change 2021: The Physical Science Basis. Contribution of Working Group I to the Sixth Assessment Report of the Intergovernmental Panel on Climate Change, edited by: Masson-Delmotte, V., Zhai, P., Pirani, A., Connors, S. L., Péan, C., Berger, S., Caud, N., Chen, Y., Goldfarb, L., Gomis, M. I., Huang, M., Leitzell, K., Lonnoy, E., Matthews, J. B. R., Maycock, T. K., Waterfield, T., Yelekçi, O., Yu, R., and Zhou, B., Cambridge University Press, 2021.
Frajka-Williams E (2015) Estimating the Atlantic overturning at 26°N using satellite altimetry and cable measurements. Geophys Res Lett 42:3458–3464. https://doi.org/10.1002/2015GL063220
Frajka-Williams E, Ansorge IJ, Baehr J, Bryden HL, Chidichimo MP, Cunningham SA, Danabasoglu G, Dong S, Donohue KA, Elipot S, Heimbach P, Holliday NP, Hummels R, Jackson LC, Karstensen J, Lankhorst M, Le Bras IA, Lozier MS, McDonagh EL, Meinen CS, Mercier H, Moat BI, Perez RC, Piecuch CG, Rhein M, Srokosz MA, Trenberth KE, Bacon S, Forget G, Goni G, Kieke D, Koelling J, Lamont T, McCarthy GD, Mertens C, Send U, Smeed DA, Speich S, Van Den Berg M, Volkov D, Wilson C (2019) Atlantic Meridional Overturning Circulation: Observed Transport and Variability. Front Mar Sci 6:260. https://doi.org/10.3389/fmars.2019.00260
Free G, Bresciani M, Pinardi M, Ghirardi N, Luciani G, Caroni R, Giardino C (2021) Detecting Climate Driven Changes in Chlorophyll-a in Deep Subalpine Lakes Using Long Term Satellite Data. Water 13:866. https://doi.org/10.3390/w13060866
Friedlingstein P et al (2022) Global Carbon Budget 2022. Earth System Science Data 14:4811–4900. https://doi.org/10.5194/essd-14-4811-2022
Garbe J, Albrecht T, Levermann A, Donges JF, Winkelmann R (2020) The hysteresis of the Antarctic Ice Sheet. Nature 585:538–544. https://doi.org/10.1038/s41586-020-2727-5
García OE, Schneider M, Ertl B, Sepúlveda E, Borger C, Diekmann C, Wiegele A, Hase F, Barthlott S, Blumenstock T, Raffalski U, Gómez-Peláez A, Steinbacher M, Ries L, De Frutos AM (2018) The MUSICA IASI CH₄ and N₂O products and their comparison to HIPPO, GAW and NDACC FTIR references. Atmos Meas Tech 11:4171–4215. https://doi.org/10.5194/amt-11-4171-2018
Gatti LV, Basso LS, Miller JB, Gloor M, Gatti Domingues L, Cassol HLG, Tejada G, Aragão LEOC, Nobre C, Peters W, Marani L, Arai E, Sanches AH, Corrêa SM, Anderson L, Von Randow C, Correia CSC, Crispim SP, Neves RAL (2021) Amazonia as a carbon source linked to deforestation and climate change. Nature 595:388–393. https://doi.org/10.1038/s41586-021-03629-6
Greene CA, Gardner AS, Schlegel N-J, Fraser AD (2022) Antarctic calving loss rivals ice-shelf thinning. Nature 609:948–953. https://doi.org/10.1038/s41586-022-05037-w
Gregory et al. (2020) Large and irreversible future decline of the Greenland ice sheet. The Cryosph 14(12):4299–4322. https://doi.org/10.5194/tc-14-4299-2020
Grégoire M et al (2021) A Global Ocean Oxygen Database and Atlas for Assessing and Predicting Deoxygenation and Ocean Health in the Open and Coastal Ocean. Front Mar Sci 8:724913. https://doi.org/10.3389/fmars.2021.724913
Groom S, Sathyendranath S, Ban Y, Bernard S, Brewin R, Brotas V, Brockmann C, Chauhan P, Choi J, Chuprin A, Ciavatta S, Cipollini P, Donlon C, Franz B, He X, Hirata T, Jackson T, Kampel M, Krasemann H, Lavender S, Pardo-Martinez S, Mélin F, Platt T, Santoleri R, Skakala J, Schaeffer B, Smith M, Steinmetz F, Valente A, Wang M (2019) Satellite Ocean Colour: Current Status and Future Perspective. Frontiers in Marine Science. https://doi.org/10.3389/fmars.2019.00485
Gudmundsson GH, Krug J, Durand G, Favier L, Gagliardini O (2012) The stability of grounding lines on retrograde slopes. Cryosphere 6:1497–1505. https://doi.org/10.5194/tc-6-1497-2012
Haagmans R, Tsaoussi L (eds) (2020) Next generation gravity mission as a mass-change and geosciences international constellation (MAGIC) mission requirements document, earth and mission science

division, European Space Agency. NASA Earth Sci Div. https://doi.org/10.5270/esa.nasa.magic-mrd.2020
Haagmans R, Siemes C, Massotti L, Carraz O, Silvestrin P (2020) ESA's next-generation gravity mission concepts. Rend Fis Acc Lincei 31:15–25. https://doi.org/10.1007/s12210-020-00875-0
Hakuba MZ, Folini D, Wild M, Schär C (2012) Impact of Greenland's topographic height on precipitation and snow accumulation in idealized simulations. Journal of Geophysical Research: Atmospheres. https://doi.org/10.1029/2011JD017052
Heinze C, Blenckner T, Martins H, Rusiecka D, Döscher R, Gehlen M, Gruber N, Holland E, Hov Ø, Joos F, Matthews JBR, Rødven R, Wilson S (2021) The quiet crossing of ocean tipping points. Proc Nat Academy Sci 118:e2008478118. https://doi.org/10.1073/pnas.2008478118
Henniger H, Huth A, Bohn FJ (2023) A new approach to derive productivity of tropical forests using radar remote sensing measurements. Royal Society Open Science 10:231186. https://doi.org/10.1098/rsos.231186
Heuzé C (2017) North Atlantic deep water formation and AMOC in CMIP5 models. Ocean Sci 13:609–622. https://doi.org/10.5194/os-13-609-2017
Hiltner U, Huth A, Fischer R (2022) Importance of the forest state in estimating biomass losses from tropical forests: combining dynamic forest models and remote sensing. Biogeosciences 19:1891–1911. https://doi.org/10.5194/bg-19-1891-2022
Hirota M, Holmgren M, Van Nes EH, Scheffer M (2011) Global Resilience of Tropical Forest and Savanna to Critical Transitions. Science 334:232–235. https://doi.org/10.1126/science.1210657
Hoffmann WA, Geiger EL, Gotsch SG, Rossatto DR, Silva LC, Lau OL, Haridasan M, Franco AC (2012) Ecological thresholds at the savanna-forest boundary: how plant traits, resources and fire govern the distribution of tropical biomes. Ecology letters. 15(7):759–68
Hogg AE, Shepherd A, Gilbert L, Muir A, Drinkwater MR (2018) Mapping ice sheet grounding lines with CryoSat-2. Adv Space Res 62:1191–1202. https://doi.org/10.1016/j.asr.2017.03.008
Högström E, Heim B, Bartsch A, Bergstedt H, Pointner G (2018) Evaluation of a MetOp ASCAT-Derived Surface Soil Moisture Product in Tundra Environments. JGR Earth Surface 123:3190–3205. https://doi.org/10.1029/2018JF004658
Hogstrom E, Bartsch A (2017) Impact of Backscatter Variations Over Water Bodies on Coarse-Scale Radar Retrieved Soil Moisture and the Potential of Correcting With Meteorological Data. IEEE Trans Geosci Remote Sensing 55:3–13. https://doi.org/10.1109/TGRS.2016.2530845
Holliday NP, Bersch M, Berx B, Chafik L, Cunningham S, Florindo-López C, Hátún H, Johns W, Josey SA, Larsen KMH, Mulet S, Oltmanns M, Reverdin G, Rossby T, Thierry V, Valdimarsson H, Yashayaev I (2020) Ocean circulation causes the largest freshening event for 120 years in eastern subpolar North Atlantic. Nat Commun 11:585. https://doi.org/10.1038/s41467-020-14474-y
Hu S, Fedorov AV (2020) Indian Ocean warming as a driver of the North Atlantic warming hole. Nat Commun 11:4785. https://doi.org/10.1038/s41467-020-18522-5
Hugelius G, Tarnocai C, Broll G, Canadell JG, Kuhry P, Swanson DK (2013) The Northern Circumpolar Soil Carbon Database: spatially distributed datasets of soil coverage and soil carbon storage in the northern permafrost regions. Earth System Science Data 5:3–13. https://doi.org/10.5194/essd-5-3-2013
Hutchinson GE, Löffler H (1956) THE THERMAL CLASSIFICATION OF LAKES. Proc Natl Acad Sci 42:84–86. https://doi.org/10.1073/pnas.42.2.84
Jackson LC, Wood RA (2018) Timescales of AMOC decline in response to fresh water forcing. Clim Dyn 51:1333–1350. https://doi.org/10.1007/s00382-017-3957-6
Jackson LC, Wood RA (2020) Fingerprints for Early Detection of Changes in the AMOC. J Clim 33:7027–7044. https://doi.org/10.1175/JCLI-D-20-0034.1
Jackson LC, Alastrué De Asenjo E, Bellomo K, Danabasoglu G, Haak H, Hu A, Jungclaus J, Lee W, Meccia VL, Saenko O, Shao A, Swingedouw D (2023) Understanding AMOC stability: the North Atlantic Hosing Model Intercomparison Project. Geosci. Model Dev. 16:1975–1995
Jackson LC, Smith RS, Wood RA (2017) Ocean and atmosphere feedbacks affecting AMOC hysteresis in a GCM. Clim Dyn 49:173–191. https://doi.org/10.1007/s00382-016-3336-8
Jenkyns HC (2010) Geochemistry of oceanic anoxic events. Geochemistry, Geophysics, Geosystems. https://doi.org/10.1029/2009gc002788
Jetz W, Cavender-Bares J, Pavlick R, Schimel D, Davis FW, Asner GP, Guralnick R, Kattge J, Latimer AM, Moorcroft P, Schaepman ME, Schildhauer MP, Schneider FD, Schrodt F, Stahl U, Ustin SL (2016) Monitoring plant functional diversity from space. Nature Plants 2:1–5. https://doi.org/10.1038/nplants.2016.24
Joughin I, Abdalati W, Fahnestock M (2004) Large fluctuations in speed on Greenland's Jakobshavn Isbræ glacier. Nature 432:608–610. https://doi.org/10.1038/nature03130

Karspeck AR, Stammer D, Köhl A, Danabasoglu G, Balmaseda M, Smith DM, Fujii Y, Zhang S, Giese B, Tsujino H, Rosati A (2017) Comparison of the Atlantic meridional overturning circulation between 1960 and 2007 in six ocean reanalysis products. Clim Dyn 49:957–982. https://doi.org/10.1007/s00382-015-2787-7

Kilbourne KH, Wanamaker AD, Moffa-Sanchez P, Reynolds DJ, Amrhein DE, Butler PG, Gebbie G, Goes M, Jansen MF, Little CM, Mette M, Moreno-Chamarro E, Ortega P, Otto-Bliesner BL, Rossby T, Scourse J, Whitney NM (2022) Atlantic circulation change still uncertain. Nat Geosci 15:165–167. https://doi.org/10.1038/s41561-022-00896-4

King MD, Howat IM, Candela SG, Noh MJ, Jeong S, Noël BPY, van den Broeke MR, Wouters B, Negrete A (2020) Author Correction: Dynamic ice loss from the Greenland Ice Sheet driven by sustained glacier retreat. Commun Earth Environ 1:1–1. https://doi.org/10.1038/s43247-020-00019-0

Knapp N, Huth A, Fischer R (2021) Tree Crowns Cause Border Effects in Area-Based Biomass Estimations from Remote Sensing. Remote Sensing 13:1592. https://doi.org/10.3390/rs13081592

Kosten S, Huszar VLM, Bécares E, Costa LS, van Donk E, Hansson L-A, Jeppesen E, Kruk C, Lacerot G, Mazzeo N, De Meester L, Moss B, Lürling M, Nõges T, Romo S, Scheffer M (2012) Warmer climates boost cyanobacterial dominance in shallow lakes. Glob Change Biol 18:118–126. https://doi.org/10.1111/j.1365-2486.2011.02488.x

Kraft NJB, Comita LS, Chase JM, Sanders NJ, Swenson NG, Crist TO, Stegen JC, Vellend M, Boyle B, Anderson MJ, Cornell HV, Davies KF, Freestone AL, Inouye BD, Harrison SP, Myers JA (2011) Disentangling the Drivers of β Diversity Along Latitudinal and Elevational Gradients. Science 333:1755–1758. https://doi.org/10.1126/science.1208584

Landerer FW, Wiese DN, Bentel K, Boening C, Watkins MM (2015) North Atlantic meridional overturning circulation variations from GRACE ocean bottom pressure anomalies. Geophys Res Lett 42:8114–8121. https://doi.org/10.1002/2015GL065730

Lapola DM et al (2023) The drivers and impacts of Amazon forest degradation. Science. https://doi.org/10.1126/science.abp8622

Lehnherr I et al (2018) The world's largest High Arctic lake responds rapidly to climate warming. Nature Communications. https://doi.org/10.1038/s41467-018-03685-z

Lenton TM, Held H, Kriegler E, Hall JW, Lucht W, Rahmstorf S, Schellnhuber HJ (2008) Tipping elements in the Earth's climate system. Proc Natl Acad Sci 105:1786–1793. https://doi.org/10.1073/pnas.0705414105

Lenton T, Armstrong McKay DI, Loriani S, Abrams JF, Lade SJ, Donges JF, Milkoreit M, Powell T, Smith SR, Zimm C, Buxton JE, Bailey E, Laybourn L, Ghadiali A, Dyke JG (eds) (2023) The Global Tipping Points Report 2023. University of Exeter, Exeter, UK

Lenton TM, Abrams JF, Bartsch A, Bathiany S, Boulton CA, Buxton JE, Conversi A, Cunliffe AM, Hebden S, Lavergne T, Poulter B, Shepherd A, Smith T, Swingedouw D, Winkelmann R, Boers N (2024) Remotely sensing potential climate change tipping points across scales. Nat Commun 15:343. https://doi.org/10.1038/s41467-023-44609-w

Levermann A, Winkelmann R (2016) A simple equation for the melt elevation feedback of ice sheets. Cryosphere 10:1799–1807. https://doi.org/10.5194/tc-10-1799-2016

Lewis WM Jr (1983) A Revised Classification of Lakes Based on Mixing. Can J Fish Aquat Sci 40:1779–1787. https://doi.org/10.1139/f83-207

Liang K, Li Y (2019) Changes in Lake Area in Response to Climatic Forcing in the Endorheic Hongjian Lake Basin. China, Remote Sensing 11:3046. https://doi.org/10.3390/rs11243046

Lin P, Pickart RS, Heorton H, Tsamados M, Itoh M, Kikuchi T (2023) Recent state transition of the Arctic Ocean's Beaufort Gyre. Nat Geosci 16:485–491. https://doi.org/10.1038/s41561-023-01184-5

Liu W, Xie S-P, Liu Z, Zhu J (2017) Overlooked possibility of a collapsed Atlantic Meridional Overturning Circulation in warming climate. Sci Adv 3:e1601666. https://doi.org/10.1126/sciadv.1601666

Liu Y, Kumar M, Katul GG, Porporato A (2019) Reduced resilience as an early warning signal of forest mortality. Nat Clim Chang 9:880–885. https://doi.org/10.1038/s41558-019-0583-9

Liu Y, Chen H, Li H, Wang H (2020) The impact of preceding spring Antarctic oscillation on the variations of lake ice phenology over the Tibetan Plateau ABSTRACT. J Climate 33(2):639–656. https://doi.org/10.1175/JCLI-D-19-0111.1

Lobelle D, Beaulieu C, Livina V, Sévellec F, Frajka-Williams E (2020) Detectability of an AMOC Decline in Current and Projected Climate Changes. Geophysical Research Letters. https://doi.org/10.1029/2020GL089974

Lobelle, D., Sévellec, F., Beaulieu, C., Livina, V., and Frajka-Williams, E. 2022 AMOC Trends From 1850 to 2100 At Interannual To Multi-Decadal Time Scales Corroborated By Changes In Salinity Budget. Review. https://doi.org/10.21203/rs.3.rs-1231837/v1

Loriani S, Aksenov Y, Armstrong McKay D, Bala G, Born A, Chiessi CM, Dijkstra H, Donges JF, Drijfhout S, England MH, Fedorov AV, Jackson L, Kornhuber K, Messori G, Pausata F, Rynders S, Salée J-B, Sinha B, Sherwood S, Swingedouw D, Tharammal T (2023) Tipping points in ocean and atmosphere circulations, EGUsphere, 1–62. https://doi.org/10.5194/egusphere-2023-2589

Loverde-Oliveira SM, Huszar VLM, Mazzeo N, Scheffer M (2009) Hydrology-Driven Regime Shifts in a Shallow Tropical Lake. Ecosystems 12:807–819. https://doi.org/10.1007/s10021-009-9258-0

Luke CM, Cox PM (2010) Soil carbon and climate change: from the Jenkinson effect to the compost-bomb instability. Eur J Soil Sci 62:5–12. https://doi.org/10.1111/j.1365-2389.2010.01312.x

Ma X, Migliavacca M, Wirth C, Bohn FJ, Huth A, Richter R, Mahecha MD (2020) Monitoring Plant Functional Diversity Using the Reflectance and Echo from Space. Remote Sensing 12:1248. https://doi.org/10.3390/rs12081248

Markus T, Neumann T, Martino A, Abdalati W, Brunt K, Csatho B, Farrell S, Fricker H, Gardner A, Harding D, Jasinski M, Kwok R, Magruder L, Lubin D, Luthcke S, Morison J, Nelson R, Neuenschwander A, Palm S, Popescu S, Shum C, Schutz BE, Smith B, Yang Y, Zwally J (2017) The Ice, Cloud, and land Elevation Satellite-2 (ICESat-2): Science requirements, concept, and implementation. Remote Sens Environ 190:260–273. https://doi.org/10.1016/j.rse.2016.12.029

Martinka K (1999) European Organisation for the Exploitation of Meteorological Satellites (EUMETSAT): ECV Inventory. Meteorologicky Casopis. 4:3–4

Masolele RN, De Sy V, Herold M, Marcos D, Verbesselt J, Gieseke F, Mullissa AG, Martius C (2021) Spatial and temporal deep learning methods for deriving land-use following deforestation: A pan-tropical case study using Landsat time series. Remote Sens Environ 264:112600. https://doi.org/10.1016/j.rse.2021.112600

Matthews, J. B. R., Möller, V., van Diemen, R., Fuglestvedt, J. S., Masson-Delmotte, V., Méndez, C., Semenov, S., and Reisinger, A.: IPCC, 2021: Annex VII: Glossary, edited by: Masson-Delmotte, V., Zhai, P., Pirani, A., Connors, S. L., Péan, C., Berger, S., Caud, N., Chen, Y., Goldfarb, L., Gomis, M. I., Huang, M., Leitzell, K., Lonnoy, E., Matthews, J. B. R., Maycock, T. K., Waterfield, T., Yelekçi, O., Yu, R., and Zhou, B., Cambridge University Press, 2021.

McDonagh EL, King BA, Bryden HL, Courtois P, Szuts Z, Baringer M, Cunningham SA, Atkinson C, McCarthy G (2015) Continuous Estimate of Atlantic Oceanic Freshwater Flux at 26.5°N. Journal of Climate 28:8888–8906. https://doi.org/10.1175/JCLI-D-14-00519.1

Mecking SS, Drijfhout LC, Jackson MB, Andrews (2022) The effect of model bias on Atlantic freshwater transport and implications for AMOC bi-stability. Tellus A: Dynamic Meteorol Oceanogr 69(1):1299910. https://doi.org/10.1080/16000870.2017.1299910

Mercier H, Lherminier P, Sarafanov A, Gaillard F, Daniault N, Desbruyères D, Falina A, Ferron B, Gourcuff C, Huck T, Thierry V (2015) Variability of the meridional overturning circulation at the Greenland-Portugal OVIDE section from 1993 to 2010. Prog Oceanogr 132:250–261. https://doi.org/10.1016/j.pocean.2013.11.001

Mesman JP, Stelzer JAA, Dakos V, Goyette S, Jones ID, Kasparian J, McGinnis DF, Ibelings BW (2021) The role of internal feedbacks in shifting deep lake mixing regimes under a warming climate. Freshw Biol 66:1021–1035. https://doi.org/10.1111/fwb.13704

Michel SLL, Swingedouw D, Ortega P, Gastineau G, Mignot J, McCarthy G, Khodri M (2022) Early warning signal for a tipping point suggested by a millennial Atlantic Multidecadal Variability reconstruction. Nat Commun 13:5176. https://doi.org/10.1038/s41467-022-32704-3

Michel S, Dijkstra H, Guardamagna F, Jaques-Dumas V, van Westen R, von der Heydt A (2023) Deep learning reconstruction of Atlantic meridional overturning circulation strength validates ongoing twenty-first century decline. https://www.researchsquare.com/article/rs-3377545/v1

Milillo P, Rignot E, Rizzoli P, Scheuchl B, Mouginot J, Bueso-Bello JL, Prats-Iraola P, Dini L (2022) Rapid glacier retreat rates observed in West Antarctica. Nature Geoscience 15(1):48–53. https://doi.org/10.1038/s41561-021-00877-z

Miloslavich P, Bax NJ, Simmons SE, Klein E, Appeltans W, Aburto-Oropeza O, Andersen Garcia M, Batten SD, Benedetti-Cecchi L, Checkley DM Jr, Chiba S, Duffy JE, Dunn DC, Fischer A, Gunn J, Kudela R, Marsac F, Muller-Karger FE, Obura D, Shin Y-J (2018) Essential ocean variables for global sustained observations of biodiversity and ecosystem changes. Global Change Biology 24:2416–2433. https://doi.org/10.1111/gcb.14108

Miner KR, Turetsky MR, Malina E, Bartsch A, Tamminen J, McGuire AD, Fix A, Sweeney C, Elder CD, Miller CE (2022) Permafrost carbon emissions in a changing Arctic. Nature Reviews Earth & Environment 3:55–67. https://doi.org/10.1038/s43017-021-00230-3

Moffa-Sánchez P, Moreno-Chamarro E, Reynolds DJ, Ortega P, Cunningham L, Swingedouw D, Amrhein DE, Halfar J, Jonkers L, Jungclaus JH, Perner K, Wanamaker A, Yeager S (2019) Variability in the

Northern North Atlantic and Arctic Oceans Across the Last Two Millennia: A Review. Paleoceanog and Paleoclimatol 34:1399–1436. https://doi.org/10.1029/2018PA003508
Moreno MV, Conedera M, Chuvieco E, Pezzatti GB (2014) Fire regime changes and major driving forces in Spain from 1968 to 2010. Environ Sci Policy 37:11–22. https://doi.org/10.1016/j.envsci.2013.08.005
Morrow R, Fu L-L, Ardhuin F, Benkiran M, Chapron B, Cosme E, d'Ovidio F, Farrar JT, Gille ST, Lapeyre G, Le Traon P-Y, Pascual A, Ponte A, Qiu B, Rascle N, Ubelmann C, Wang J, Zaron ED (2019) Global Observations of Fine-Scale Ocean Surface Topography With the Surface Water and Ocean Topography (SWOT) Mission. Frontiers in Marine Science. https://doi.org/10.3389/fmars.2019.00232
Mouginot J, Rignot E, Scheuchl B (2014) Sustained increase in ice discharge from the Amundsen Sea Embayment, West Antarctica, from 1973 to 2013. Geophys Res Lett 41:1576–1584. https://doi.org/10.1002/2013GL059069
Navarro JCA, García-Serrano J, Lapin V, Ortega P (2022) Added value of assimilating springtime Arctic sea ice concentration in summer-fall climate predictions. Environ Res Lett 17:064008. https://doi.org/10.1088/1748-9326/ac6c9b
Needell C, Holschuh N (2023) Evaluating the Retreat, Arrest, and Regrowth of Crane Glacier Against Marine Ice Cliff Process Models. Geophysical Research Letters. https://doi.org/10.1029/2022GL102400
Nitzbon J, Schneider von Deimling T, Aliyeva M, Chadburn SE, Grosse G, Laboor S, Lee H, Lohmann G, Steinert NJ, Stuenzi SM, Werner M, Westermann S, Langer M (2024) No respite from permafrost-thaw impacts in the absence of a global tipping point. Nat. Clim. Chang. 14:573–585. https://doi.org/10.1038/s41558-024-02011-4
Nitze I, Grosse G, Jones BM, Romanovsky VE, Boike J (2018) Remote sensing quantifies widespread abundance of permafrost region disturbances across the Arctic and Subarctic. Nature Communications. https://doi.org/10.1038/s41467-018-07663-3
O'Reilly CM et al (2015) Rapid and highly variable warming of lake surface waters around the globe. Geophysical Research Letters. https://doi.org/10.1002/2015gl066235
Obu J, Westermann S, Bartsch A, Berdnikov N, Christiansen HH, Dashtseren A, Delaloye R, Elberling B, Etzelmüller B, Kholodov A, Khomutov A, Kääb A, Leibman MO, Lewkowicz AG, Panda SK, Romanovsky V, Way RG, Westergaard-Nielsen A, Wu T, Yamkhin J, Zou D (2019) Northern Hemisphere permafrost map based on TTOP modelling for 2000–2016 at 1?km2 scale. Earth Sci Rev 193:299–316. https://doi.org/10.1016/j.earscirev.2019.04.023
Oschlies A, Brandt P, Stramma L, Schmidtko S (2018) Drivers and mechanisms of ocean deoxygenation. Nat Geosci 11:467–473. https://doi.org/10.1038/s41561-018-0152-2
Pacheco-Labrador J, Migliavacca M, Ma X, Mahecha MD, Carvalhais N, Weber U, Benavides R, Bouriaud O, Barnoaiea I, Coomes DA, Bohn FJ, Kraemer G, Heiden U, Huth A, Wirth C (2022) Challenging the link between functional and spectral diversity with radiative transfer modeling and data. Remote Sens Environ 280:113170. https://doi.org/10.1016/j.rse.2022.113170
Pattyn F, Morlighem M (2020) The uncertain future of the Antarctic Ice Sheet. Science 367:1331–1335. https://doi.org/10.1126/science.aaz5487
Pereira HM, Ferrier S, Walters M, Geller GN, Jongman RH, Scholes RJ, Bruford MW, Brummitt N, Butchart SH, Cardoso AC, Coops NC (2013) Essential biodiversity variables. Science. 339(6117):277–8
Pisano A, Marullo S, Artale V, Falcini F, Yang C, Leonelli FE, Santoleri R, Buongiorno Nardelli B (2020) New Evidence of Mediterranean Climate Change and Variability from Sea Surface Temperature Observations. Remote Sensing 12:132. https://doi.org/10.3390/rs12010132
Poorter L, van der Sande MT, Thompson J, Arets EJMM, Alarcón A, Álvarez-Sánchez J, Ascarrunz N, Balvanera P, Barajas-Guzmán G, Boit A, Bongers F, Carvalho FA, Casanoves F, Cornejo-Tenorio G, Costa FRC, de Castilho CV, Duivenvoorden JF, Dutrieux LP, Enquist BJ, Fernández-Méndez F, Finegan B, Gormley LHL, Healey JR, Hoosbeek MR, Ibarra-Manríquez G, Junqueira AB, Levis C, Licona JC, Lisboa LS, Magnusson WE, Martínez-Ramos M, Martínez-Yrizar A, Martorano LG, Maskell LC, Mazzei L, Meave JA, Mora F, Muñoz R, Nytch C, Pansonato MP, Parr TW, Paz H, Pérez-García EA, Rentería LY, Rodríguez-Velazquez J, Rozendaal DMA, Ruschel AR, Sakschewski B, Salgado-Negret B, Schietti J, Simões M, Sinclair FL, Souza PF, Souza FC, Stropp J, ter Steege H, Swenson NG, Thonicke K, Toledo M, Uriarte M, van der Hout P, Walker P, Zamora N, Peña-Claros M (2015) Diversity enhances carbon storage in tropical forests. Glob Ecol Biogeogr 24:1314–1328. https://doi.org/10.1111/geb.12364
Pritchard HD, Ligtenberg SRM, Fricker HA, Vaughan DG, van den Broeke MR, Padman L (2012) Antarctic ice-sheet loss driven by basal melting of ice shelves. Nature 484:502–505. https://doi.org/10.1038/nature10968

Qian S-E (2021) Hyperspectral Satellites, Evolution, and Development History. IEEE Journal of Selected Topics in Applied Earth Observations and Remote Sensing 14:7032–7056. https://doi.org/10.1109/JSTARS.2021.3090256
Rahmstorf S (2024) Is the Atlantic Overturning Circulation Approaching a Tipping Point? Oceanography 37:1. https://doi.org/10.5670/oceanog.2024.501
Reese R, Gudmundsson GH, Levermann A, Winkelmann R (2018) The far reach of ice-shelf thinning in Antarctica. Nature Clim Change 8:53–57. https://doi.org/10.1038/s41558-017-0020-x
Ricaud P, Attié J-L, Chalinel R, Pasternak F, Léonard J, Pison I, Pattey E, Thompson RL, Zelinger Z, Lelieveld J, Sciare J, Saitoh N, Warner J, Fortems-Cheiney A, Reynal H, Vidot J, Brooker L, Berdeu L, Saint-Pé O, Patra PK, Dostál M, Suchánek J, Nevrlý V, Zwaaftink CG (2021) The Monitoring Nitrous Oxide Sources (MIN2OS) satellite project. Remote Sens Environ 266:112688. https://doi.org/10.1016/j.rse.2021.112688
Richardson AD, Keenan TF, Migliavacca M, Ryu Y, Sonnentag O, Toomey M (2013) Climate change, phenology, and phenological control of vegetation feedbacks to the climate system. Agric for Meteorol 169:156–173. https://doi.org/10.1016/j.agrformet.2012.09.012
Rietkerk M, Bastiaansen R, Banerjee S, van de Koppel J, Baudena M, Doelman A (2021) Evasion of tipping in complex systems through spatial pattern formation. Science. https://doi.org/10.1126/science.abj0359
Rignot E, Mouginot J, Scheuchl B (2011) Antarctic grounding line mapping from differential satellite radar interferometry. Geophysical Research Letters. https://doi.org/10.1029/2011GL047109
Rignot E, Mouginot J, Scheuchl B, van den Broeke M, van Wessem MJ, Morlighem M (2019) Four decades of Antarctic Ice Sheet mass balance from 1979–2017. Proc Natl Acad Sci 116:1095–1103. https://doi.org/10.1073/pnas.1812883116
Robel AA (2017) Thinning sea ice weakens buttressing force of iceberg mélange and promotes calving. Nat Commun 8:14596. https://doi.org/10.1038/ncomms14596
Robinson A, Calov R, Ganopolski A (2012) Multistability and critical thresholds of the Greenland ice sheet. Nature Clim Change 2:429–432. https://doi.org/10.1038/nclimate1449
Rocha JC, Peterson G, Bodin Ö, Levin S (2018) Cascading regime shifts within and across scales. Science 362:1379–1383. https://doi.org/10.1126/science.aat7850
Rockström J, Gupta J, Qin D, Lade SJ, Abrams JF, Andersen LS, Armstrong McKay DI, Bai X, Bala G, Bunn SE, Ciobanu D, DeClerck F, Ebi K, Gifford L, Gordon C, Hasan S, Kanie N, Lenton TM, Loriani S, Liverman DM, Mohamed A, Nakicenovic N, Obura D, Ospina D, Prodani K, Rammelt C, Sakschewski B, Scholtens J, Stewart-Koster B, Tharammal T, van Vuuren D, Verburg PH, Winkelmann R, Zimm C, Bennett EM, Bringezu S, Broadgate W, Green PA, Huang L, Jacobson L, Ndehedehe C, Pedde S, Rocha J, Scheffer M, Schulte-Uebbing L, de Vries W, Xiao C, Xu C, Xu X, Zafra-Calvo N, Zhang X (2023) Safe and just Earth system boundaries. Nature 619:102–111. https://doi.org/10.1038/s41586-023-06083-8
Rockström J, Steffen W, Noone K, Persson , Chapin FS, Lambin EF, Lenton TM, Scheffer M, Folke C, Schellnhuber HJ, Nykvist B, de Wit CA, Hughes T, van der Leeuw S, Rodhe H, Sörlin S, Snyder PK, Costanza R, Svedin U, Falkenmark M, Karlberg L, Corell RW, Fabry VJ, Hansen J, Walker B, Liverman D, Richardson K, Crutzen P, Fole JA (2009) A safe operating space for humanity. Nature 461:472–475. https://doi.org/10.1038/461472as
Rosier SHR, Reese R, Donges JF, De Rydt J, Gudmundsson GH, Winkelmann R (2021) The tipping points and early warning indicators for Pine Island Glacier. West Antarctica, the Cryosphere 15:1501–1516. https://doi.org/10.5194/tc-15-1501-2021
Rott H, Skvarca P, Nagler T (1996) Rapid Collapse of Northern Larsen Ice Shelf. Antarctica, Science 271:788–792. https://doi.org/10.1126/science.271.5250.788
Runge A, Nitze I, Grosse G (2022) Remote sensing annual dynamics of rapid permafrost thaw disturbances with LandTrendr. Remote Sens Environ 268:112752. https://doi.org/10.1016/j.rse.2021.112752
Sakschewski B, von Bloh W, Boit A, Poorter L, Peña-Claros M, Heinke J, Joshi J, Thonicke K (2016) Resilience of Amazon forests emerges from plant trait diversity. Nature Clim Change 6:1032–1036. https://doi.org/10.1038/nclimate3109
Sakschewski B, von Bloh W, Drüke M, Sörensson AA, Ruscica R, Langerwisch F, Billing M, Bereswill S, Hirota M, Oliveira RS, Heinke J, Thonicke K (2021) Variable tree rooting strategies are key for modelling the distribution, productivity and evapotranspiration of tropical evergreen forests. Biogeosciences 18:4091–4116. https://doi.org/10.5194/bg-18-4091-2021
Sammartino M, Buongiorno Nardelli B, Marullo S, Santoleri R (2020) An Artificial Neural Network to Infer the Mediterranean 3D Chlorophyll-a and Temperature Fields from Remote Sensing Observations. Remote Sensing 12:4123. https://doi.org/10.3390/rs12244123

Sampaio G, Nobre C, Costa MH, Satyamurty P, Soares-Filho BS, Cardoso M (2007) Regional climate change over eastern Amazonia caused by pasture and soybean cropland expansion. Geophysical Research Letters. https://doi.org/10.1029/2007GL030612

Sanaei A, Herrmann H, Alshaabi L, Beck J, Ferlian O, Fomba KW, Haferkorn S, van Pinxteren M, Quaas J, Quosh J, Rabe R, Wirth C, Eisenhauer N, Weigelt A (2023) Changes in biodiversity impact atmospheric chemistry and climate through plant volatiles and particles. Commun Earth Environ 4:1–12. https://doi.org/10.1038/s43247-023-01113-9

Sandberg Sørensen L, Simonsen SB, Forsberg R, Khvorostovsky K, Meister R, Engdahl ME (2018) 25 years of elevation changes of the Greenland Ice Sheet from ERS, Envisat, and CryoSat-2 radar altimetry. Earth Planet Sci Lett 495:234–241. https://doi.org/10.1016/j.epsl.2018.05.015

Sauzède R, Bittig H. Cetal (2017) Estimates of Water-Column Nutrient Concentrations and Carbonate System Parameters in the Global Ocean: A Novel Approach Based on Neural Networks. Mar. Sci, Front. https://doi.org/10.3389/fmars.2017.00128

Sauzède R, Claustre H, Jamet C, Uitz J, Ras J, Mignot A, D'Ortenzio F (2015) Retrieving the vertical distribution of chlorophyll a concentration and phytoplankton community composition from in situ fluorescence profiles: A method based on a neural network with potential for global-scale applications. Journal of Geophysical Research: Oceans 120:451–470. https://doi.org/10.1002/2014JC010355

Sauzède R et al (2016) A neural network-based method for merging ocean color and Argo data to extend surface bio-optical properties to depth: Retrieval of the particulate backscattering coefficient. JGR Oceans 121:2552–2571. https://doi.org/10.1002/2015JC011408

Scambos TA, Bohlander JA, Shuman CA, Skvarca P (2004) Glacier acceleration and thinning after ice shelf collapse in the Larsen B embayment. Antarctica, Geophysical Research Letters. https://doi.org/10.1029/2004GL020670

Scambos TA, Haran TM, Fahnestock MA, Painter TH, Bohlander J (2007) MODIS-based Mosaic of Antarctica (MOA) data sets: Continent-wide surface morphology and snow grain size. Remote Sens Environ 111:242–257. https://doi.org/10.1016/j.rse.2006.12.020

Scheffer M (1991) Should we expect strange attractors behind plankton dynamics – and if so, should we bother? J Plankton Res 13:1291–1305. https://doi.org/10.1093/plankt/13.6.1291

Scheffer M (2001) Alternative Attractors of Shallow Lakes. Scientific World Journal 1:254–263. https://doi.org/10.1100/tsw.2001.62

Scheffer M, Stephen RC (2003) Catastrophic regime shifts in ecosystems: Linking theory to observation. Trends EcolEvol 18(12):648–656. https://doi.org/10.1016/j.tree.2003.09.002

Scheffer M, Hosper SH, Meijer M-L, Moss B, Jeppesen E (1993) Alternative equilibria in shallow lakes. Trends Ecol Evol 8:275–279. https://doi.org/10.1016/0169-5347(93)90254-M

Scheffer M, Carpenter S, Foley JA, Folke C, Walker B (2001) Catastrophic shifts in ecosystems. Nature 413:591–596. https://doi.org/10.1038/35098000

Scheffer M, Bascompte J, Brock WA, Brovkin V, Carpenter SR, Dakos V, Held H, van Nes EH, Rietkerk M, Sugihara G (2009) Early-warning signals for critical transitions. Nature 461:53–59. https://doi.org/10.1038/nature08227

Scheuchl B, Mouginot J, Rignot E, Morlighem M, Khazendar A (2016) Grounding line retreat of Pope, Smith, and Kohler Glaciers, West Antarctica, measured with Sentinel-1a radar interferometry data. Geophys Res Lett 43:8572–8579. https://doi.org/10.1002/2016GL069287

Schlemm T, Levermann A (2021) A simple parametrization of mélange buttressing for calving glaciers. Cryosphere 15:531–545. https://doi.org/10.5194/tc-15-531-2021

Schmidtko S, Stramma L, Visbeck M (2017) Decline in global oceanic oxygen content during the past five decades. Nature 542:335–339. https://doi.org/10.1038/nature21399

Schmittner A, Galbraith ED (2008) Glacial greenhouse-gas fluctuations controlled by ocean circulation changes. Nature 456:373–376. https://doi.org/10.1038/nature07531

Schneider FD, Ferraz A, Hancock S, Duncanson LI, Dubayah RO, Pavlick RP, Schimel DS (2020) Towards mapping the diversity of canopy structure from space with GEDI. Environ Res Lett 15:115006. https://doi.org/10.1088/1748-9326/ab9e99

Schoof C (2007) Ice sheet grounding line dynamics: Steady states, stability, and hysteresis. Journal of Geophysical Research: Earth Surface. https://doi.org/10.1029/2006JF000664

Schuur EAG, McGuire AD, Schädel C, Grosse G, Harden JW, Hayes DJ, Hugelius G, Koven CD, Kuhry P, Lawrence DM, Natali SM, Olefeldt D, Romanovsky VE, Schaefer K, Turetsky MR, Treat CC, Vonk JE (2015) Climate change and the permafrost carbon feedback. Nature 520:171–179. https://doi.org/10.1038/nature14338

Schwefel R, Gaudard A, Wüest A, Bouffard D (2016) Effects of climate change on deepwater oxygen and winter mixing in a deep lake (Lake Geneva): Comparing observational findings and modeling. Water Resour Res 52:8811–8826. https://doi.org/10.1002/2016WR019194

Sgubin G, Swingedouw D, Drijfhout S, Mary Y, Bennabi A (2017) Abrupt cooling over the North Atlantic in modern climate models. Nat Commun 8:14375. https://doi.org/10.1038/ncomms14375

Sharp JD, Fassbender AJ, Carter BR, Johnson GC, Schultz C, Dunne JP (2023) GOBAI-O_2: temporally and spatially resolved fields of ocean interior dissolved oxygen over nearly 2 decades. Earth System Science Data 15:4481–4518. https://doi.org/10.5194/essd-15-4481-2023

Shatwell T, Thiery W, Kirillin G (2019) Future projections of temperature and mixing regime of European temperate lakes. Hydrol Earth Syst Sci 23:1533–1551. https://doi.org/10.5194/hess-23-1533-2019

Shepherd, A., Ivins, E., Rignot, E., Smith, B., van den Broeke, M., Velicogna, I., Whitehouse, P., Briggs, K., Joughin, I., Krinner, G., Nowicki, S., Payne, T., Scambos, T., Schlegel, N., A, G., Agosta, C., Ahlstrøm, A., Babonis, G., Barletta, V. R., Bjørk, A. A., Blazquez, A., Bonin, J., Colgan, W., Csatho, B., Cullather, R., Engdahl, M. E., Felikson, D., Fettweis, X., Forsberg, R., Hogg, A. E., Gallee, H., Gardner, A., Gilbert, L., Gourmelen, N., Groh, A., Gunter, B., Hanna, E., Harig, C., Helm, V., Horvath, A., Horwath, M., Khan, S., Kjeldsen, K. K., Konrad, H., Langen, P. L., Lecavalier, B., Loomis, B., Luthcke, S., McMillan, M., Melini, D., Mernild, S., Mohajerani, Y., Moore, P., Mottram, R., Mouginot, J., Moyano, G., Muir, A., Nagler, T., Nield, G., Nilsson, J., Noël, B., Otosaka, I., Pattle, M. E., Peltier, W. R., Pie, N., Rietbroek, R., Rott, H., Sandberg Sørensen, L., Sasgen, I., Save, H., Scheuchl, B., Schrama, E., Schröder, L., Seo, K.-W., Simonsen, S. B., Slater, T., Spada, G., Sutterley, T., Talpe, M., Tarasov, L., van de Berg, W. J., van der Wal, W., van Wessem, M., Vishwakarma, B. D., Wiese, D., Wilton, D., Wagner, T., Wouters, B., Wuite, J., and The IMBIE Team: Mass balance of the Greenland Ice Sheet from (1992) to 2018. Nature 579(233–239):2020. https://doi.org/10.1038/s41586-019-1855-2

Shepherd, A., Ivins, E., Rignot, E., Smith, B., van den Broeke, M., Velicogna, I., Whitehouse, P., Briggs, K., Joughin, I., Krinner, G., Nowicki, S., Payne, T., Scambos, T., Schlegel, N., A, G., Agosta, C., Ahlstrøm, A., Babonis, G., Barletta, V., Blazquez, A., Bonin, J., Csatho, B., Cullather, R., Felikson, D., Fettweis, X., Forsberg, R., Gallee, H., Gardner, A., Gilbert, L., Groh, A., Gunter, B., Hanna, E., Harig, C., Helm, V., Horvath, A., Horwath, M., Khan, S., Kjeldsen, K. K., Konrad, H., Langen, P., Lecavalier, B., Loomis, B., Luthcke, S., McMillan, M., Melini, D., Mernild, S., Mohajerani, Y., Moore, P., Mouginot, J., Moyano, G., Muir, A., Nagler, T., Nield, G., Nilsson, J., Noel, B., Otosaka, I., Pattle, M. E., Peltier, W. R., Pie, N., Rietbroek, R., Rott, H., Sandberg-Sørensen, L., Sasgen, I., Save, H., Scheuchl, B., Schrama, E., Schröder, L., Seo, K.-W., Simonsen, S., Slater, T., Spada, G., Sutterley, T., Talpe, M., Tarasov, L., van de Berg, W. J., van der Wal, W., van Wessem, M., Vishwakarma, B. D., Wiese, D., Wouters, B., and The IMBIE team: Mass balance of the Antarctic Ice Sheet from (1992) to 2017. Nature 558(219–222):2018. https://doi.org/10.1038/s41586-018-0179-y

Singh C, van der Ent R, Wang-Erlandsson L, Fetzer I (2021) Hydroclimatic adaptation critical to the resilience of tropical forests. Global Change Biology. https://doi.org/10.1111/gcb.16115

Slagter B, Reiche J, Marcos D, Mullissa A, Lossou E, Peña-Claros M, Herold M (2023) Monitoring direct drivers of small-scale tropical forest disturbance in near real-time with Sentinel-1 and -2 data. Remote Sens Environ 295:113655. https://doi.org/10.1016/j.rse.2023.113655

Smith B, Fricker HA, Gardner AS, Medley B, Nilsson J, Paolo FS, Holschuh N, Adusumilli S, Brunt K, Csatho B, Harbeck K, Markus T, Neumann T, Siegfried MR, Zwally HJ (2020) Pervasive ice sheet mass loss reflects competing ocean and atmosphere processes. Science 368:1239–1242. https://doi.org/10.1126/science.aaz5845

Smith C, Baker JCA, Spracklen DV (2023) Tropical deforestation causes large reductions in observed precipitation. Nature 615:270–275. https://doi.org/10.1038/s41586-022-05690-1

Souza CM et al (2020) Reconstructing three decades of land use and land cover changes in brazilian biomes with landsat archive and earth engine. Remote Sensing 12:2735. https://doi.org/10.3390/rs12172735

Spracklen DV, Garcia-Carreras L (2015) The impact of Amazonian deforestation on Amazon basin rainfall. Geophys Res Lett 42:9546–9552. https://doi.org/10.1002/2015GL066063

Staal A, Tuinenburg OA, Bosmans JHC, Holmgren M, van Nes EH, Scheffer M, Zemp DC, Dekker SC (2018) Forest-rainfall cascades buffer against drought across the Amazon. Nat Clim Chang 8:539–543. https://doi.org/10.1038/s41558-018-0177-y

Staal A, Fetzer I, Wang-Erlandsson L, Bosmans JHC, Dekker SC, van Nes EH, Rockström J, Tuinenburg OA (2020) Hysteresis of tropical forests in the 21st century. Nature Communications. https://doi.org/10.1038/s41467-020-18728-7

Staal A, Koren G, Tejada G, Gatti LV (2023) Moisture origins of the Amazon carbon source region. Environ Res Lett 18:044027. https://doi.org/10.1088/1748-9326/acc676

Staver AC, Archibald S, Levin SA (2011) The Global Extent and Determinants of Savanna and Forest as Alternative Biome States. Science 334:230–232. https://doi.org/10.1126/science.1210465

Steffen W, Rockström J, Richardson K, Lenton TM, Folke C, Liverman D, Summerhayes CP, Barnosky AD, Cornell SE, Crucifix M, Donges JF, Fetzer I, Lade SJ, Scheffer M, Winkelmann R, Schellnhuber HJ (2018) Trajectories of the Earth System in the Anthropocene. Proc Natl Acad Sci USA 115:8252–8259. https://doi.org/10.1073/pnas.1810141115

Stommel H (1961) Thermohaline Convection with Two Stable Regimes of Flow. Tellus 13:224–230. https://doi.org/10.1111/j.2153-3490.1961.tb00079.x

Strozzi T, Antonova S, Günther F, Mätzler E, Vieira G, Wegmüller U, Westermann S, Bartsch A (2018) Sentinel-1 SAR Interferometry for Surface Deformation Monitoring in Low-Land Permafrost Areas. Remote Sensing 10:1360. https://doi.org/10.3390/rs10091360

Sun Q, Miao C, Duan Q, Ashouri H, Sorooshian S, Hsu K-L (2018) A Review of Global Precipitation Data Sets: Data Sources. Estimation, and Intercomparisons, Reviews of Geophysics 56:79–107. https://doi.org/10.1002/2017RG000574

Swingedouw D, Ifejika Speranza C, Bartsch A, Durand G, Jamet C, Beaugrand G, Conversi A (2020) Early Warning from Space for a Few Key Tipping Points in Physical. Biological, and Social-Ecological Systems, Surv Geophys 41:1237–1284. https://doi.org/10.1007/s10712-020-09604-6

Swingedouw D, Bily A, Esquerdo C, Borchert LF, Sgubin G, Mignot J, Menary M (2021) On the risk of abrupt changes in the North Atlantic subpolar gyre in CMIP6 models. Ann N Y Acad Sci 1504:187–201. https://doi.org/10.1111/nyas.14659

Tao S, Wigneron J-P, Chave J, Tang Z, Wang Z, Zhu J, Guo Q, Liu YY, Ciais P (2023) Little evidence that Amazonian rainforests are approaching a tipping point. Nat Clim Chang 13:1317–1320. https://doi.org/10.1038/s41558-023-01853-8

Taubert F, Fischer R, Knapp N, Huth A (2021) Deriving Tree Size Distributions of Tropical Forests from Lidar. Remote Sensing 13:131. https://doi.org/10.3390/rs13010131

Thornton JM, Palazzi E, Pepin NC, Cristofanelli P, Essery R, Kotlarski S, Giuliani G, Guigoz Y, Kulonen A, Pritchard D, Li X, Fowler HJ, Randin CF, Shahgedanova M, Steinbacher M, Zebisch M, Adler C (2021) Toward a definition of Essential Mountain Climate Variables. One Earth 4:805–827. https://doi.org/10.1016/j.oneear.2021.05.005

Trofaier AM, Westermann S, Bartsch A (2017) Progress in space-borne studies of permafrost for climate science: Towards a multi-ECV approach. Remote Sens Environ 203:55–70. https://doi.org/10.1016/j.rse.2017.05.021

Tuinenburg OA, Staal A (2020) Tracking the global flows of atmospheric moisture and associated uncertainties. Hydrol Earth Syst Sci 24:2419–2435. https://doi.org/10.5194/hess-24-2419-2020

van der Ent RJ, Wang-Erlandsson L, Keys PW, Savenije HHG (2014) Contrasting roles of interception and transpiration in the hydrological cycle – Part 2: Moisture recycling. Earth System Dynamics 5:471–489. https://doi.org/10.5194/esd-5-471-2014

van der Wal W, Whitehouse PL, Schrama EJO (2015) Effect of GIA models with 3D composite mantle viscosity on GRACE mass balance estimates for Antarctica. Earth Planet Sci Lett 414:134–143. https://doi.org/10.1016/j.epsl.2015.01.001

van Nes EH, Staal A, Hantson S, Holmgren M, Pueyo S, Bernardi RE, Flores BM, Xu C, Scheffer M (2018) Fire forbids fifty-fifty forest. PLoS ONE 13:e0191027. https://doi.org/10.1371/journal.pone.0191027

Verbesselt J, Umlauf N, Hirota M, Holmgren M, Van Nes EH, Herold M, Zeileis A, Scheffer M (2016) Remotely sensed resilience of tropical forests. Nat Clim Chang 6:1028–1031. https://doi.org/10.1038/nclimate3108

Vinogradova N, Lee T, Boutin J, Drushka K, Fournier S, Sabia R, Stammer D, Bayler E, Reul N, Gordon A, Melnichenko O, Li L, Hackert E, Martin M, Kolodziejczyk N, Hasson A, Brown S, Misra S, Lindstrom E (2019) Satellite Salinity Observing System: Recent Discoveries and the Way Forward. Frontiers in Marine Science. https://doi.org/10.3389/fmars.2019.00243

Wallmann K (2010) Phosphorus imbalance in the global ocean? Global Biogeochemical Cycles. https://doi.org/10.1029/2009gb003643

Wallmann K, Flögel S, Scholz F, Dale AW, Kemena TP, Steinig S, Kuhnt W (2019) Periodic changes in the Cretaceous ocean and climate caused by marine redox see-saw. Nat Geosci 12:456–461. https://doi.org/10.1038/s41561-019-0359-x

Wang J, Kessler J, Bai X, Clites A, Lofgren B, Assuncao A, Bratton J, Chu P, Leshkevich G (2018) Decadal Variability of Great Lakes Ice Cover in Response to AMO and PDO, 1963–2017. J Clim 31:7249–7268. https://doi.org/10.1175/JCLI-D-17-0283.1

Wang S, Foster A, Lenz EA, Kessler JD, Stroeve JC, Anderson LO, Turetsky M, Betts R, Zou S, Liu W, Boos WR, Hausfather Z (2023) Mechanisms and impacts of Earth system tipping elements. Reviews of Geophysics. https://doi.org/10.1029/2021RG000757

Westermann S, Østby TI, Gisnås K, Schuler TV, Etzelmüller B (2015) A ground temperature map of the North Atlantic permafrost region based on remote sensing and reanalysis data. Cryosphere 9:1303–1319. https://doi.org/10.5194/tc-9-1303-2015

Westermann S et al (2023) The CryoGrid community model (version 1.0) – a multi-physics toolbox for climate-driven simulations in the terrestrial cryosphere. Geosci. Model Dev. 16:2607–2647

Whitehouse PL, Gomez N, King MA, Wiens DA (2019) Solid Earth change and the evolution of the Antarctic Ice Sheet. Nat Commun 10:503. https://doi.org/10.1038/s41467-018-08068-y

Wood RA, Jonathan AB, Grégory B, Jacqueline B, Alessandra C, Reik VD, Ivy F et al (2024) Opportunities for Earth observation to inform risk management for ocean tipping points. Surv Geophys. https://doi.org/10.1007/s10712-024-09859-3

Woolway RI (2021) Phenological shifts in lake stratification under climate change. Nature Communications. https://doi.org/10.1038/s41467-021-22657-4

Woolway RI, Merchant CJ (2017) Amplified surface temperature response of cold, deep lakes to inter-annual air temperature variability. Scientific Reports. https://doi.org/10.1038/s41598-017-04058-0

Woolway RI, Merchant CJ (2018) Intralake Heterogeneity of Thermal Responses to Climate Change: A Study of Large Northern Hemisphere Lakes. Journal of Geophysical Research: Atmospheres 123:3087–3098. https://doi.org/10.1002/2017jd027661

Woolway RI, Merchant CJ (2019) Worldwide alteration of lake mixing regimes in response to climate change. Nat Geosci 12:271–276. https://doi.org/10.1038/s41561-019-0322-x

World Meteorological Organization (WMO): The 2022 GCOS ECVs Requirements (GCOS 245), Geneva, 2022a.

World Meteorological Organization (WMO): The 2022 GCOS Implementation Plan (GCOS-244), Geneva, 2022b.

Wrona E, Rowlandson TL, Nambiar M, Berg AA, Colliander A, Marsh P (2017) Validation of the Soil Moisture Active Passive (SMAP) satellite soil moisture retrieval in an Arctic tundra environment. Geophys Res Lett 44:4152–4158. https://doi.org/10.1002/2017GL072946

Wunderling N, Donges JF, Kurths J, Winkelmann R (2021) Interacting tipping elements increase risk of climate domino effects under global warming. Earth System Dynamics 12:601–619. https://doi.org/10.5194/esd-12-601-2021

Wunderling N, Von Der Heydt A, Aksenov Y, Barker S, Bastiaansen R, Brovkin V, Brunetti M, Couplet V, Kleinen T, Lear CH, Lohmann J, Roman-Cuesta RM, Sinet S, Swingedouw D, Winkelmann R, Anand P, Barichivich J, Bathiany S, Baudena M, Bruun JT, Chiessi CM, Coxall HK, Docquier D, Donges JF, Falkena SKJ, Klose AK, Obura D, Rocha J, Rynders S, Steinert NJ, Willeit M (2023a) Climate tipping point interactions and cascades: A review. Climate Change/other Interactions/earth System and Climate Modeling. https://doi.org/10.5194/egusphere-2023-1576

Wunderling N, Winkelmann R, Rockström J, Loriani S, Armstrong McKay DI, Ritchie PDL, Sakschewski B, Donges JF (2023b) Global warming overshoots increase risks of climate tipping cascades in a network model. Nat Clim Chang 13:75–82. https://doi.org/10.1038/s41558-022-01545-9

Wunderling N, von der Heydt AS, Aksenov Y, Barker S, Bastiaansen R, Brovkin V, Brunetti M, Couplet V, Kleinen T, Lear CH, Lohmann J, Roman-Cuesta RM, Sinet S, Swingedouw D, Winkelmann R, Anand P, Barichivich J, Bathiany S, Baudena M, Bruun JT, Chiessi CM, Coxall HK, Docquier D, Donges JF, Falkena SKJ, Klose AK, Obura D, Rocha J, Rynders S, Steinert NJ, Willeit M (2024) Climate tipping point interactions and cascades: a review. Earth System Dynamics 15:41–74. https://doi.org/10.5194/esd-15-41-2024

Xu C, Staal A, Hantson S, Holmgren M, van Nes EH, Scheffer M (2017) Remotely sensed canopy height reveals three pantropical ecosystem states: reply. Ecology 99:235–237. https://doi.org/10.1002/ecy.2077

Xu X, Jia G, Zhang X, Riley WJ, Xue Y (2020) Climate regime shift and forest loss amplify fire in Amazonian forests. Glob Change Biol 26:5874–5885. https://doi.org/10.1111/gcb.15279

Yáñez-Serrano AM, Bach A, Bartolomé-Català D, Matthaios V, Seco R, Llusià J, Filella I, Peñuelas J (2021) Dynamics of volatile organic compounds in a western Mediterranean oak forest. Atmos Environ 257:118447. https://doi.org/10.1016/j.atmosenv.2021.118447

Yang W, John V, Zhao X, Lu H, Knapp K (2016) Satellite Climate Data Records: Development, Applications, and Societal Benefits. Remote Sensing 8:331. https://doi.org/10.3390/rs8040331

Yoon S-T, Lee WS, Nam S, Lee C-K, Yun S, Heywood K, Boehme L, Zheng Y, Lee I, Choi Y, Jenkins A, Jin EK, Larter R, Wellner J, Dutrieux P, Bradley AT (2022) Ice front retreat reconfigures meltwater-driven gyres modulating ocean heat delivery to an Antarctic ice shelf. Nat Commun 13:306. https://doi.org/10.1038/s41467-022-27968-8

Zeitz, M. 2022 Modeling the future resilience of the Greenland Ice Sheet: from the flow of ice to the interplay of feedbacks, Universität Potsdam. https://doi.org/10.25932/publishup-56883

Zemp DC, Schleussner C-F, Barbosa HMJ, Hirota M, Montade V, Sampaio G, Staal A, Wang-Erlandsson L, Rammig A (2017) Self-amplified Amazon forest loss due to vegetation-atmosphere feedbacks. Nature Communications. https://doi.org/10.1038/ncomms14681
Zhang R, Thomas M (2021) Horizontal circulation across density surfaces contributes substantially to the long-term mean northern Atlantic Meridional Overturning Circulation. Commun Earth Environ 2:1–12. https://doi.org/10.1038/s43247-021-00182-y
Zhang X, Kurbaniyazov A, Kirillin G (2021) Changing Pattern of Water Level Trends in Eurasian Endorheic Lakes as a Response to the Recent Climate Variability. Remote Sensing 13:3705. https://doi.org/10.3390/rs13183705
Zhou S, Williams AP, Lintner BR, Berg AM, Zhang Y, Keenan TF, Cook BI, Hagemann S, Seneviratne SI, Gentine P (2021) Soil moisture–atmosphere feedbacks mitigate declining water availability in drylands. Nat Clim Chang 11:38–44. https://doi.org/10.1038/s41558-020-00945-z

Authors and Affiliations

S. Loriani[1] · A. Bartsch[2] · E. Calamita[3] · J. F. Donges[1,4,5] · S. Hebden[6,17] · M. Hirota[7,8] · A. Landolfi[9] · T. Nagler[10] · B. Sakschewski[1] · A. Staal[11] · J. Verbesselt[12,13] · R. Winkelmann[1,14,15] · R. Wood[16] · N. Wunderling[1,5,18]

✉ S. Loriani
sina.loriani@pik-potsdam.de

1 Earth Resilience Science Unit and Earth System Analysis, Potsdam Institute for Climate Impact Research, Member of the Leibniz Association, Telegrafenberg 31A, 14473 Potsdam, Germany

2 b.geos, Industriestrasse 1A, 2100 Korneuburg, Austria

3 Swiss Federal Institute of Aquatic Science and Technology, 8600 Dubendorf, Switzerland

4 Stockholm Resilience Centre, Stockholm University, Albanovägen 28, 106 91 Stockholm, Sweden

5 High Meadows Environmental Institute, Princeton University, Princeton, NJ 08544, USA

6 Future Earth Secretariat, 11418 Stockholm, Sweden

7 Group IpES, Department of Physics, Federal University of Santa Catarina, Florianópolis 88034-102, Brazil

8 Department of Plant Biology, University of Campinas, Campinas 13083-970, Brazil

9 National Research Council of Italy, CNR-ISMAR-Roma, 00133 Rome, Italy

10 ENVEO Environmental Earth Observation Information Technology GmbH, Fürstenweg 176, 6020 Innsbruck, Austria

11 Copernicus Institute of Sustainable Development, Utrecht University, Princetonlaan 8a, 3584 CB Utrecht, the Netherlands

12 Belgian Science Policy Office (BELSPO), Simon Bolivarlaan 30 Bus 7 Boulevard Simon Bolivar 30 Bte 7, 1000 Brussels, Belgium

13 Laboratory of Geo-Information Science and Remote Sensing, Wageningen University and Research, Droevendaalsesteeg 4, 6708 PB Wageningen, The Netherlands

14 Institute for Physics and Astronomy, University of Potsdam, Potsdam 14476, Germany

15 Integrative Earth System Science, Max Planck Institute of Geoanthropology, Jena 07745, Germany

16 Met Office Hadley Centre, FitzRoy Road, Exeter EX1 3PB, UK

[17] ECSAT, European Space Agency, Harwell Didcot OX11 0FD, UK

[18] Center for Critical Computational Studies, Goethe University, Theodor-W.-Adorno-Platz 1, Frankfurt am Main 60629, Germany

Surveys in Geophysics (2025) 46:265–301
https://doi.org/10.1007/s10712-024-09833-z

Ecosystem Resilience Monitoring and Early Warning Using Earth Observation Data: Challenges and Outlook

Sebastian Bathiany[1,2] · Robbin Bastiaansen[3,4] · Ana Bastos[5] · Lana Blaschke[1,2] · Jelle Lever[6,7] · Sina Loriani[2] · Wanda De Keersmaecker[8] · Wouter Dorigo[9] · Milutin Milenković[10] · Cornelius Senf[11] · Taylor Smith[12] · Jan Verbesselt[13,14] · Niklas Boers[1,2,15]

Received: 19 September 2023 / Accepted: 27 February 2024 / Published online: 3 May 2024

Abstract
As the Earth system is exposed to large anthropogenic interferences, it becomes ever more important to assess the resilience of natural systems, i.e., their ability to recover from natural and human-induced perturbations. Several, often related, measures of resilience have been proposed and applied to modeled and observed data, often by different scientific communities. Focusing on terrestrial ecosystems as a key component of the Earth system, we review methods that can detect large perturbations (temporary excursions from a reference state as well as abrupt shifts to a new reference state) in spatio-temporal datasets, estimate the recovery rate after such perturbations, or assess resilience changes indirectly from stationary time series via indicators of critical slowing down. We present here a sequence of ideal methodological steps in the field of resilience science, and argue how to obtain a consistent and multi-faceted view on ecosystem or climate resilience from Earth observation (EO) data. While EO data offers unique potential to study ecosystem resilience globally at high spatial and temporal scale, we emphasize some important limitations, which are associated with the theoretical assumptions behind diagnostic methods and with the measurement process and pre-processing steps of EO data. The latter class of limitations include gaps in time series, the disparity of scales, and issues arising from aggregating time series from multiple sensors. Based on this assessment, we formulate specific recommendations to the EO community in order to improve the observational basis for ecosystem resilience research.

Keywords Resilience · Remote sensing · Earth observations · Terrestrial vegetation · Tipping points

Extended author information available on the last page of the article

Article Highlights

- Substantial differences exist in defining vegetation "resilience" and its estimators, creating sometimes unnecessary uncertainty
- Progress has been made in constructing indicators of resilience, but there are underappreciated caveats associated with the measurement process of Earth observations
- Novel developments and increasing diversity in datasets from Earth observations, in combination with theoretical insights, can help to improve resilience monitoring from space

1 Introduction

Concerns have been raised that different terrestrial ecosystems might respond to ongoing anthropogenic climate and land-use change with abrupt and potentially catastrophic regime shifts. Indeed, several vegetation ecosystems, such as the Amazon and other tropical rainforests, but also drylands or boreal forests, have been suggested to be potential Earth system tipping elements (Lenton et al. 2008; Armstrong McKay 2022; Boers et al. 2022). To reliably project the future fate of terrestrial ecosystems and to assess the risk of local- to regional-scale regime transitions it is crucial to be able to measure their resilience based on empirical spatiotemporal data.

The resilience of an ecosystem is often qualitatively defined as the ability of the system to cope with disturbances, where a disturbance is "any relatively discrete event in time that disrupts ecosystem, community, or population structure and changes resource pools, substrate availability, or the physical environment" (White and Pickett, 1985). Disturbances are usually induced by anthropogenic or natural discrete events which act to perturb a system away from its dynamic equilibrium, after which the system gradually recovers. In principle, disturbances do not necessarily have to be extreme or rare, though practically they need to be large and rare enough to be distinguished from the normal variability of the system.

We note that, in this review, "equilibrium" refers to any state of the system that does not show significant changes in its properties over time, except when external forces impose a change. A special case of an equilibrium is a fixed point x^* where the evolution operator f of an autonomous dynamical system $dx/dt=f(x)$ becomes 0, i.e., $f(x^*)=0$ (for example, see Kuznetsov 2004). Under more complicated conditions with time varying forcing, one can define a "pullback attractor" to which the system converges when the initial condition is applied at a time point shifted infinitely into the past (Chekroun et al. 2011). This perspective of equilibrium is purely mathematical and differs from the concept of thermodynamic equilibrium in physical systems, which refers to a state of maximum entropy for a given set of constraints.

From a dynamical systems perspective, a local and a global notion of resilience can be distinguished (Holling 1973; Krakovská et al 2023). The local perspective applies close to an equilibrium state and defines resilience as the recovery rate after a perturbation (e.g., Smith et al. 2022). Such a perturbation could be a single disturbance event. If the perturbation is small, and the reference state is a stable equilibrium in the sense of dynamical systems theory, this definition of resilience is the fixed point's stability, expressed by the eigenvalue of the slowest mode of recovery (Holling 1973). This equilibrium-based notion of resilience has been coined as "engineering resilience" (Pimm 1984).

In contrast, the global perspective defines resilience by how large a perturbation must be for the system to not recover from it at all. In dynamical systems terms, this non-equilibrium based definition refers to the width of a basin of attraction, or "ecological resilience" (Holling 1973) and is sometimes also called "latitude" (Walker et al. 2004). Figure 1

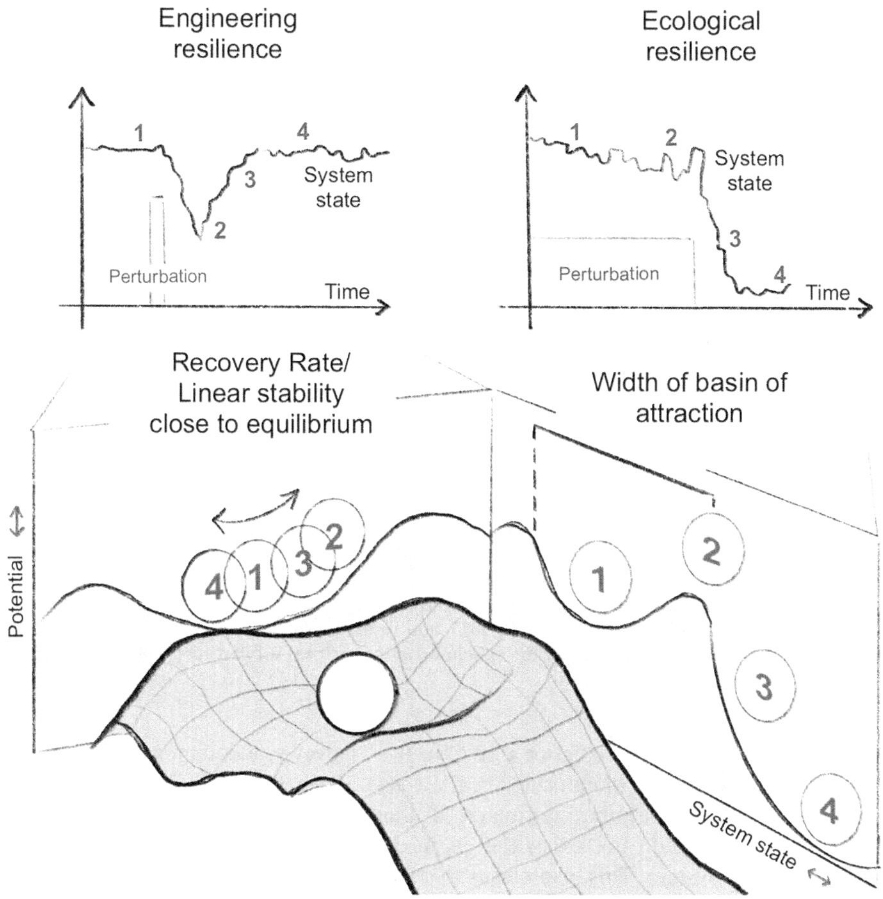

Fig. 1 Conceptual difference between engineering resilience (local linear stability, return rate) and ecological resilience (basin width). The numbers in the lower panel correspond to the numbers in the upper panel and indicate where the system is in the depicted potential landscape over time

illustrates these two perspectives of resilience with a widely used analogy, where the system's state is represented by a ball in a potential landscape.

These two views of resilience are not unrelated. A system can only display engineering resilience if it also has ecological resilience regarding a disturbance (Fig. 2, case 1). Environmental change (captured as a slowly changing parameter in models) can reduce the width of the basin and at the same time make it less steep around the equilibrium. As a consequence, the system is more vulnerable to disturbances, first in the sense that it shows a smaller recovery rate, and second in the sense that these disturbances can tip the system to an alternative state (Fig. 2, case 2). In the extreme case, the whole basin and with it the stable equilibrium disappears at a so-called bifurcation point (Fig. 2, case 3). At this point, both ecological resilience (basin width) and "engineering" resilience (recovery rate) would decrease to zero. Consequently, indicators measuring "engineering resilience", which are closely related to the concept of critical slowing down (CSD), are often referred to as "early warning signals" of abrupt shifts associated with a loss

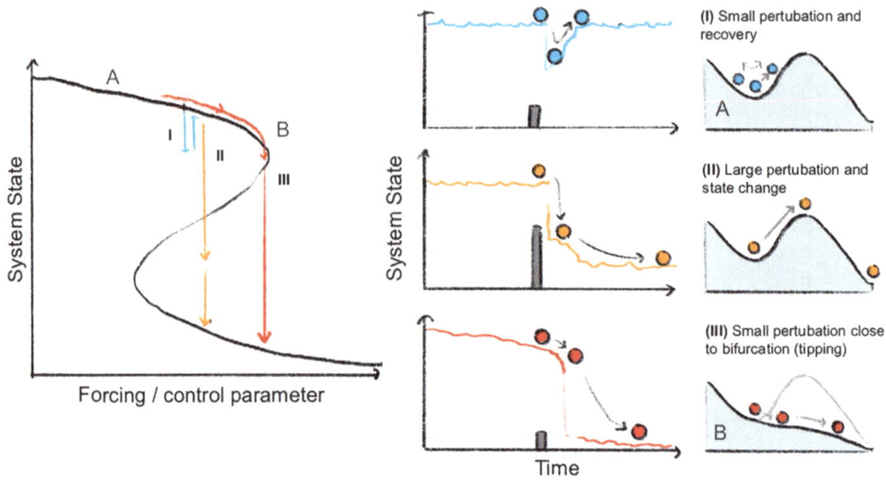

Fig. 2 Idealized cases of abrupt events and how they are related to different kinds of resilience, and resilience indicators. case I (blue): A large external state perturbation happens, and thereafter the system recovers gradually. The recovery time measures "engineering resilience". case II (yellow): A similar event occurs but this time the external shock is larger (or the resistance of the system smaller), so that the system does not show "ecological resilience", i.e.,, it does not return to the previous state. case III (red): A gradual parameter change drives rapid change in the system because it crosses a bifurcation point where the old equilibrium disappears

of ecological resilience (Scheffer et al. 2009). It is, however, a common misconception to believe that resilience monitoring (in the sense of measuring changes in engineering resilience) would only be informative if alternative equilibria existed; even when alternative equilibria do not exist, the engineering resilience measures the recovery capacities of the system. This also means that CSD indicators alone do not provide any informative value on the existence of alternative equilibria (Kéfi et al. 2013).

Another quantifier of resilience which to some degree integrates the above two views can be defined based on mean first exit times from a given basin of attraction to the basin of attraction of an alternative equilibrium (Ciemer et al. 2019; Arani et al. 2021). A given state is naturally more resilient in this sense if the expected first passage time to the alternative basin is longer. To first approximation, the resilience in terms of mean first exit times depends on the width and depth of the basin of attraction, as well as on the strength of the noise. This quantifier of resilience can be called global in distinction to the local concept of engineering resilience. For an excellent in-depth review of different concepts of ecosystem resilience we refer to Dakos and Kéfi (2022).

In this review, we focus on terrestrial ecosystems and how to measure changes in their resilience to disturbances under increasing anthropogenic pressures, as an important field of application for Earth observation (EO) systems. The motivation for this article is to work out major conceptual as well as practical caveats that need to be addressed before the value of novel and better resolved EO data can be manifested in monitoring resilience (and changes thereof) in the Earth system. While recent studies have argued for enhanced and improved resilience monitoring, highlighting specific examples (Lenton et al. 2022; Lenton et al. 2024), we here discuss in more detail regarding the conceptual, practical and computational challenges of data analysis, in light of the most recent literature.

Some of the discussed caveats are not exclusive to EO data, but all of them are important to consider in EO-based resilience monitoring. Our intention is (i) to make "data users" (e.g., Earth system modelers and dynamical systems experts) aware of these challenges and of arising opportunities, and (ii) to make "data producers" (e.g., remote sensing experts) more aware of the specific requirements that resilience monitoring implies for the data. We believe that both would contribute to more meaningful resilience assessments by future studies. The majority of work discussed in this article focuses on engineering resilience and the closely related concept of CSD; first, because there are arguably more observed examples of disturbance and recovery than examples of crashes without recovery, and second because it is less clear how to accurately define/measure ecological resilience mathematically and in practice.

We will first (Sect. 2) discuss the conceptual challenges associated with resilience quantification. After that, we focus on the practical problems related to estimating ecosystem resilience from EO data. To this end, we review how engineering resilience has been measured in practice using two different approaches.

The first approach (Sect. 3) is by computing properties of the recovery process after single large perturbations. This includes methods that detect disturbances in time series (Sect. 3.1) and methods that quantify recovery rates after such disturbances (Sect. 3.2). This can be considered the direct way to estimate the recovery rate and thus engineering resilience.

The second approach (Sect. 4) is to infer engineering resilience indirectly, from the normal variability of the system as it is permanently adjusting to many small perturbations. The fact that the recovery rate can be inferred from natural variability is suggested by theoretical insights, such as those from the theory of stochastic dynamical systems (Kuehn 2011) and the fluctuation–dissipation theorem (Kubo 1966). These theoretical relationships have only recently been empirically validated in vegetation ecosystems (Smith et al. 2022; Smith and Boers 2023a, 2023b). The major advantage of this second approach is that it allows to measure resilience everywhere and not only where large perturbations have occurred, and continuously in time. We discuss metrics to measure resilience indicators from permanent fluctuations (Sect. 4.1), and ways to cope with long-term trends and seasonality (Sect. 4.2) and spatially and mechanistically complex ecosystems (Sect. 4.3).

In Sect. 5, we discuss future opportunities and challenges for resilience monitoring using EO data. As more and more data with higher and higher resolutions from different sensors is becoming readily available, the vision to monitor vegetation resilience from EO data can be seen as a Big Data challenge. However, this requires both meaningful measures of resilience in high-dimensional systems and a way to automate their computation as well as the detection of significant changes, and there might be limitations associated with EO data. Finally, Sect. 6 provides a brief summary and our most important conclusions.

2 Challenges in Defining Resilience

2.1 Diversity of Resilience Metrics

The concept of resilience has been a central topic in the field of ecology for approximately 50 years (Yi and Jackson 2021). Defining resilience is challenging, not only because of differences in definitions and semantics, but also because of the difficulties in finding appropriate metrics to quantify resilience. While ecological and engineering resilience as outlined in Sect. 1 are the most common categories, other definitions of resilience exist, and many variants are used when quantifying resilience (Ingrisch and Bahn 2018; Yi and Jackson 2021).

The term 'resilience' is also used differently in different scientific disciplines and contexts, and even where the qualitative definitions appear to be the same, the mathematically defined metrics can differ. In principle, there are two kinds of challenges in trying to define resilience. The first is a conceptual challenge of how to precisely define resilience and distinguish it from other related properties. Often, different studies focus on different properties of disturbances and recovery processes, but nonetheless use the same terminology (Ingrisch and Bahn 2018; van Meerbeek et al. 2021), which can create unnecessary confusion, adding to the actual epistemic uncertainties. The second kind of challenge arises when such concepts are applied to observations of geoscientific or ecological systems, where practical estimation problems occur (e.g., arising from uncertainties and gaps in the measurement process or biases induced by other effects). We will deal with conceptual challenges in this section, while later sections focus on more practical problems associated with applying the definitions to EO data.

Regarding conceptual challenges, an important factor is that the recovery rate generally varies over time in systems with more than one dimension (e.g., an ecological model describing several coexisting species), even if the linearisation around the fixed point is valid. In the following, we assume that the one-dimensional simplification is justified. A more complete formal definition and comparison of resilience properties in higher dimensions is given in Arnoldi et al. (2016).

Even in a one-dimensional setting, a great deal of ambiguity arises from the fact that there are several properties that characterize and influence the recovery process (as shown in Fig. 2, case 1), and that one can measure: the magnitude of the disturbance, the ratio between this magnitude and the size of the external perturbation (resistance or robustness, also called Lloret's resilience, see Lloret et al. 2011), the total duration of the recovery process, the ratio between disturbance magnitude and duration (abruptness), the initial pulse rate, the rate of recovery (engineering resilience), the area under the pulse curve (see Jentzsch and White 2019, and Fig. 1 therein), the magnitude of the perturbation needed to change an ecosystem to a different set of processes and structures (Angeler and Allen 2016), as well as combinations of multiple properties (Mitra, Kurths and Donner, 2015). In systems more complex than the simple dynamical systems shown above, it can also happen that the system does recover to some extent, but not fully (i.e.,, an intermediate type between the blue and yellow trajectory in Fig. 2), which is referred to as the degree of recovery (e.g., see Tao et al. 2022). For a comprehensive review of resilience metrics, see Yi and Jackson (2021) and Dakos and Kefi (2022).

Among the mentioned properties of disturbance and recovery phenomena, resistance is arguably among the most popular quantifiers (Pimm 1984; Ingrisch and Bahn 2018). Resistance has been defined as the ability of individuals or ecosystems to persist or stay essentially unchanged under a disturbance (Grimm and Calabrese 2011; Nimmo et al. 2015). The higher the resistance, the smaller is the impact of a given disturbance on ecosystem state variables. Therefore, resistance is typically defined based on the concurrent impact of a disturbance, quantified either by the (relative) difference or ratio of the state variable before and following a disturbance (Yi and Jackson 2021), and is closely related to disturbance severity used as characteristics of disturbance regimes in ecology (Turner 2010).

Resistance has also been included as one component of engineering resilience, next to recovery rate (Holling 1996). However, resistance is a property that is different from the recovery rate from a perturbation (Ingrisch and Bahn 2018) (though in a simple one-dimensional dynamical system setting, both should be proportional), and it can be insightful to consider both when assessing resilience. Particular care has to be taken when controlling for one of them to estimate the other. Even when the same qualitative definition of resilience is used (e.g., recovery rate), it matters how the quantified resilience is normalized in order to obtain unitless quantities. For example, the same absolute recovery time will lead to different recovery rates depending on whether one normalizes the time series by the baseline (reference state) value, or by the impact (amplitude) of the perturbation (Ingrisch and Bahn 2018), which is determined by the system's resistance.

In the following, if not specified further, we refer to resilience as the engineering resilience in the sense of a recovery rate that is not normalized with the resistance. We will discuss some of the associated conceptual challenges in Sect. 2.2 and the practical problems of its estimation from Earth observations in Sect. 3 and 4.

2.2 Conceptual Challenges

Real ecosystems are much more complex than the idealized dynamical systems shown in Figs. 1 and 2. Hence, a number of additional caveats have to be kept in mind when assessing their resilience.

(i) Equilibrium assumption.

The notions of equilibrium (a stable fixed point in terms of dynamical systems) and out-of-equilibrium is questionable in complex ecosystems that might be better described with chaotic dynamics even under constant environmental conditions (Hastings and Wysham 2010). Another caveat is how to define a system's equilibrium in practice. Examples for a predefined state can be the empirically observed long-term average (Ingrisch and Bahn 2018), but typically covering only a few years given observational constraints, or it can be defined in absolute terms such as the FAO (Food and Agriculture Organization of the United Nations) definition of forests (>40% tree cover, 5 m height), yet how different definitions impact the outcome of resilience assessments remains largely not well quantified.

(ii) High dimensionality.

Even in case one-dimensional concepts apply mathematically (as mentioned above), there is typically a variety of variables to work with. Yi and Jackson (2021) consequently argue that it is challenging to calculate ecosystem resilience by a single universally applicable measure or method. For example, multiple state variables have been used to estimate resilience, some related with a system's functioning (e.g., tree growth, assessed by basal area increment or tree ring width), others related with a system's state (e.g., leaf area or vegetation indices related to photosynthetic activity, assessed by optical remote sensing) or a system's composition (e.g., species richness). These reflect different facets of ecological systems that do not necessarily correspond to the same measures of "resilience". As an example, when observing a deviation of the normalized Difference Vegetation Index (NDVI), the post-recovery NDVI, even if similar to pre-disturbance NDVI, may represent a different species composition or even another vegetation state (e.g., shrubs, grasslands instead of trees; Buma 2012). It is hence important to be aware of and clearly state the variables and mechanisms one aims to resolve, which also affects the sampling time (see below). For example, the generation time of a tree species may reflect the time scale of recovery for a forest after a severe dieback event, but may not matter for the recovery of leaves or biomass after less extreme events. Furthermore, measurable variables typically correspond to proxies of state variables, rather than the state variables themselves (e.g., spectral vegetation indices as proxies for photosynthetic activity/productivity). Given this complexity, it is generally of advantage to have knowledge on the drivers and main ecosystem processes involved, in order to make meaningful choices on relevant variables to assess the resilience of the system. This also applies to the right time period and sampling intervals, as we will discuss below.

(iii) Interactions between disturbance and recovery time-scales.

Ecosystems with low resilience will only slowly regain their pre-disturbance state, or may not ever recover completely. The essential metric used to assess resilience is thus 'recovery time' or 'recovery interval', which needs to be interpreted in light of the disturbance interval. As an example, a system taking 100 years to recover is likely to fully recover if disturbances only occur every say 500 years, but if same system is disturbed every 50 years, it will never reach its pre-disturbance state (see Senf and Seidl 2021a, b), and its response to later perturbations might change due to nonlinear or even irreversible legacy effects (Hansen et al. 2022; Anderegg et al. 2020).

This is an important point to consider, as changing disturbance frequencies (i.e.,, due to climate change) might change the resilience of a system even though recovery times stay similar. To judge whether a system is actually at risk of experiencing a permanent state change, we hence need to consider both changing recovery times but also changing disturbance intervals. An example of this is increasing fire frequency observed in some regions, where ecosystems will be stuck in successional stages because they reburn before late-successional stages can be reached (a so-called fire-trap; see Freeman et al. 2017 and Hoffmann et al., 2020). In contrast, changes in ecosystem composition in response to changes in fire frequency can also promote resilience, depending on the ecosystem property under consideration, e.g., increased carbon storage (Mack et al., 2021).

Additionally, the recovery to a given state depends on changes in the forcing, e.g., drought recovery might be faster if coinciding with increased water availability. An alternative way to assess recovery is to evaluate how ecosystem function is changed following a disturbance. This can be quantified by evaluating departures of the sensitivity of the target variable(s) to forcing drivers relative to a long-term sensitivity (functioning) baseline (Yu et al., 2022; Wu et al. 2022).

(iv) Memory effects and system boundaries.

As explained in point (iii), the time scale separation of an abrupt disturbance and the consequential slow recovery is an idealization. Many ecological systems exhibit memory (Ogle et al. 2015; Cranko Page et al. 2022) and specifically stress memory (Walter et al., 2013; Ruehr et al. 2019). Ecological memory might lead to altered responses to perturbations, either by conferring increased or reduced resistance (Gessler et al. 2020; Wu et al. 2022). Increased resistance might result from positive legacy effects of responses to disturbance. For example, investment in root growth during drought (McDowell et al. 2008) might attenuate impacts of a subsequent drought, or reduced competition due to mortality might increase resources available for surviving individuals (Gessler et al. 2020). Reduced resilience following disturbance might result from functional impairment (e.g., damage to the plants' hydraulic system; see e.g., McDowell et al., 2016; Ruehr et al. 2019; Wagner et al. 2023). Further, some disturbance resistance strategies might increase vulnerability to subsequent disturbances. For example, drought avoidance strategies like stomatal closure or defoliation might increase the risk of carbon starvation and long-term metabolic decline (Poyatos et al. 2013; McDowell et al., 2016), increasing tree vulnerability to biotic attacks. In addition, these effects likely depend on the magnitude of the disturbance impact (Ruehr et al. 2019).

Hence, one needs to keep in mind that concurrent impacts might not only reflect the direct disturbance effect, but also the legacy effects from similar or other types of past disturbances due to stress memory (Gessler et al. 2020; Bastos et al. 2021; Anderegg et al. 2020). It has therefore been proposed to quantify resistance by the concurrent sensitivity to a given disturbance driver ($\Delta V/\Delta F$, with state variable V and forcing F, Fu et al., (2020); Xiao et al. (2023)), which better allows to estimate "the persistence of *relationships* within a system" (Holling 1973), and allows to compare shorter periods of time (Fu et al. 2020). It should be noted, however, that these are also likely to be time-scale dependent (Linscheid et al. 2021).

Besides these challenges related to time scales, resistance and resilience can also be hard to disentangle practically because of the presence of climate-ecosystem and ecosystem-biogeochemical feedbacks. For example, the precipitation-evaporation feedback (Eltahir and Bras 1996) leads to the fact that precipitation at a certain location is not an external driver, but already affected by the response of the ecosystem to other drivers, or the same drivers at other locations. Another example is that increased carbon storage under more frequent fires reported by Mack et al. (2021) was linked to higher nutrient availability, hence, the direct effect of fires in isolation (as might be possible to be studied in models) is not well separable in observations. It is hence not always possible to observe the internal ecosystem dynamics and external stressors separately from each other.

(v) Value judgements and other subjective choices.

In highly complex (socio-)ecological systems, it also matters whether the restored ecosystems fulfill the same functions (or provide similar ecosystem services) as prior to the perturbation, i.e., if they can adapt to change in a way that is judged as beneficial (Helfgott 2015). In this sense, "resilience" has more facets than just the recovery to a previous state, or the persistence in remaining in a certain regime. It has been argued that it is a feature of resilience for systems to be self-organized (rather than just changing when being forced); and to permanently change in order to adapt to the environment by learning. For instance, species and hence ecosystems permanently change due to evolution. It is then particularly relevant to decide which function or property of the system shall be the focus ("resilience of what to what?"; Carpenter et al. 2001; Cutter 2016), and, in case of socio-ecological systems, "resilience for whom?" (Cutter 2016). If the structure and/or composition changes, but from a human perspective the ecosystem provides the same function, we might still call it resilient. In this sense, there is also a subjectivity that depends on the selected functions (or ecosystem services) and how we value them (e.g., timber production rate, species diversity, habitat for particular species). As different functions can have distinctive resilience, it matters which ones we focus on, which is necessarily a strongly human-based perspective. In this article, we however ignore these value-related issues and focus on the conceptual and practical challenges when defining and measuring resilience for a given ecosystem property or function.

3 Recovery from Individual Perturbations

As outlined above, when referring to resilience as the return rate to equilibrium after disturbances, one can hope to measure resilience by identifying large state perturbations and then estimating the recovery time empirically (van Nes and Scheffer 2007). In principle, this involves two steps: (i) detecting sufficiently large and sufficiently abrupt shifts in order to identify a clear starting time for the recovery process, and (ii) measure the time the system needs to recover. Both aspects need a conceptual choice and quantification of a reference state before the perturbation and after recovery. Closely related to measuring the recovery time, one can also on the same basis measure the recovery rate directly by fitting an exponential function to the part of the time series between the perturbation and the time when the system has fully recovered (Smith et al. 2022). As explained in Arnoldi et al. (2016), this method assumes that the recovery rate does not change during the recovery process, i.e., that the initial and asymptotic recovery are identical. This is only guaranteed in one-dimensional systems, or in case the perturbation happens in the direction of the eigenvector with the largest negative eigenvalue.

3.1 Methods for Detecting Abrupt Shifts

Sudden and large perturbations of a state variable can result from (i) disturbances from which the system can recover (Fig. 2 case 1, blue trajectory), (ii) large disturbances that push the system into an alternative state (Fig. 2 case 2, yellow trajectory) or (iii) "tipping points" originating from the internal nonlinear dynamics of the system, which can be triggered by only small and gradual changes in external conditions (Fig. 2 case 3, red trajectory), and can be practically irreversible. The first kind is useful to probe the linear stability (engineering resilience) of the system, whereas the second and third are associated with ecological resilience and result in a transition to an alternative state. In observed complex systems, it is often not clear into which of these (idealized) classes an abrupt shift falls, in particular shortly after the shift has occurred. However, all three are important events to be monitored before they can be assessed further.

Interestingly, methods to detect irreversible shifts and methods to detect large anomalies (temporary disturbances) are closely related (Flach et al. 2017; Bathiany et al. 2020). For instance, integrating a time series (i.e., computing the cumulative sum) will convert an extreme event into an abrupt persistent shift. Vice versa, differentiating (computing the difference between subsequent data points) turns an abrupt transition back into an extreme event. Consequently, a very fundamental and successful detection approach (depending on the length of the time series) works via computing the cumulative sum of a time series and then detecting extreme values (Basseville and Nikiforov 1993).

It is important to be able to detect extreme or abrupt shifts fully automatically, i.e., without the need for visual inspection to identify candidates of events. First, automatic detection allows more objective and reproducible results, and second, EO datasets are increasing in number, length, and resolution, thus increasing the amount of data that needs to be monitored and processed, which would be hardly possible by eye. A similar development is occurring for climate models of climate and parts of the Earth system.

Several detection strategies with an increasingly large number of algorithms have been developed and applied in the Earth sciences to detect abrupt shifts automatically (see, e.g., Aminikhanghahi and Cook 2017). Historically, an important application of such methods was to remove artificial offsets in observational time series, introduced by changes in the measurement process (e.g., when replacing one instrument by another). To this end, so-called change-point detection methods were introduced which detect the points in time where a sudden change in the statistical properties (usually the mean) of a time series occurs. These change-point detection frameworks involve fitting specific parametric models to time series (e.g., using ramp regression as a piecewise linear model; see Mudelsee 2000), and testing for the likelihood that different models are needed to explain the data in each segment before and after a suspected change-point (Reeves et al. 2007; Bai and Perron 2003; Zeileis et al. 2003). Such a task can also be framed as a model selection problem and information criteria can be applied (Jamali et al. 2014; Reeves et al. 2007).

Originally, several assumptions had to be made when applying these methods, which are too restrictive in the case of time series from climate or ecological systems. Consequently, change-point detection methods have been extended to be able to cope with large autocorrelation (Mudelsee 2000; Rodionov 2004, 2006; Wang 2008; Ducré-Robitaille et al. 2003), trends and non-stationary/non-iid residuals (Matyasovszky 2011; Gallagher et al. 2013), seasonality (Verbesselt et al. 2010a), non-Gaussian distributions (Lazante 1996; Chu 2002), as well as measurement and dating uncertainties (Goswami et al., 2018).

Despite these improvements, almost all methods still are restricted to univariate time series. EO data (like simulated data), in contrast, provides spatio-temporal information, allowing for detection methods that focus on a certain spatial scale of interest, and take spatial connections into account in order to improve statistical significance and to classify abrupt shifts in terms of their spatio-temporal structure (Bathiany et al. 2020). Table 1 summarizes some typical examples of applications in the climate sciences, which share the particular target of detecting real abrupt climate shifts (as opposed to artificial shifts in the measurement process).

Inspired by ecological disturbances, many techniques for detecting abrupt changes in annual time series of optical medium resolution remote sensing data have been developed. Such techniques assume that a sudden shift in the vegetation community, e.g., by fire, will result in a measurable change in the annual spectral reflectance properties. Detecting changes is mostly done by classical change-point detection algorithms, but those methods are increasingly combined with supervised classification models to filter out falsely detected breaks. Examples of such products are available for Canada (White et al. 2017) and Europe (Senf and Seidl 2021a, b).

3.2 Quantifying Recovery After Perturbations

After having identified a perturbation of the system, for instance with the methods discussed in Sect. 3.1, and under the assumption that the system is ecologically resilient (can recover to its old state), quantitative properties of the recovery can be measured in order to quantify the system's "engineering resilience". As explained in Sect. 1, in this framework, more resilient vegetation or ecosystems will return to their previous state (or a predefined state) more quickly. In practice, the disturbance can be thought of as a drought, fire, storm, or any other event that leads to an abrupt change in structure and frees resources for the development of new vegetation. In principle, it is the recovery rate which quantifies what we introduced as "engineering resilience" above, i.e., the linear stability of the system's state in terms of dynamical systems theory. However, we will discuss in this section that it matters how this property is defined precisely (in particular, if and how the time series are normalized), and how it is measured.

Table 1 Methods to detect abrupt shifts automatically in Earth system data

Method	Explanation	Original methods	Applied to ecology or climate science	Advantages	Challenges
Classical change-point detection	Fit of parametric models to univariate time series, statistical testing for break points	Basseville and Nikiforov (1993), Reeves et al. (2007), Bai and Perron (2003), Zeileis et al. (2003)	Beaulieu et al. (2012): allows for autocorrelation Verbesselt et al., (2010a): includes seasonal and trend component Bagniewski et al. (2021): Kolmogorov–Smirnov test in running window, includes varying temporal resolution (e.g., in paleo-records)	Strict statistical testing, allows extension to statistical properties beyond the mean (e.g., variance and periodicity), yields simple functions to capture time series dynamics before/after break point	Computationally expensive
Bayesian change-point detection	Based on Bayesian model calibration, models can be the same as in classical change-point detection (shift in mean, ramp regression, …)	Perreault et al. (2000), Erhardt et al. (2019)	Ducré-Robitaille et al., (2003) Riechers and Boers (2021) Schütz and Holschneider (2011)	Can deal with data uncertainties (in terms of values and dating)	A guess about the point of transition has to be made (based on investigation by eye or another method), computationally expensive
Recurrence-based	Time series are interpreted as a recurrence network, structural metrics reveal regime shifts	Marwan et al. (2009)	Donges et al. (2011) Goswami et al. (2018)	Can detect more general transitions, requires less rigid model assumptions than other methods	Results (arguably) less intuitive to understand compared to other methods
Gradient above threshold	Point in time with the largest gradient on pre-selected timescale is considered as change-point	Flach et al. (2017)	Drijfhout et al., (2015): climate model data; candidates selected by eye + post-hoc test Bathiany et al. (2020): climate model data Rousseau et al. (2017): paleodata Smith et al. (2022): EO data	Easy to understand and implement	No rigorous statistical tests

Table 1 (continued)

Method	Explanation	Original methods	Applied to ecology or climate science	Advantages	Challenges
Edge detection	Large, connected gradients in space–time are detected	Canny (1986)	Bathiany et al., (2020)	Takes spatio-temporal structure into account, computationally efficient, potentially allows classification of events based on spatio-temporal shape	Relatively complex to calibrate

The simplest way to measure recovery time is to define a reference state, and measure how long it takes for a given ecosystem to return to that reference after disturbance (Senf and Seidl 2021a, b; Wu et al. 2022; Bastos et al. 2011). This provides a time scale for recovery, which can be interpreted as a recovery rate. The main practical difficulties here are defining the reference state, and estimating the recovery time in the presence of other perturbations, which can make it difficult to estimate when the recovery process has ended. This stands in contrast to the detection methods introduced in Sect. 3.1, which typically do not require defining any baseline state a priori. Another practical difficulty in EO applications is that time series might not be long enough to robustly define a long-term reference state. Circumventing this caveat is the motivation of space-for-time approaches, discussed in Sects. 4.3 and 5.8.

To quantify the reference state the system recovers to (baseline value), one typically uses either the pre-disturbance state, or a baseline value from an undisturbed control treatment (if available). As mentioned in Sect. 2, it is a subjective choice whether the recovery is formulated in terms of absolute values, or normalized with regard to the baseline or the disturbance impact (Ingrisch and Bahn 2018). In any case, it matters whether the reference state can be well quantified using the data sources available. For instance, the recovery in NDVI or similar spectral indices will reach pre-disturbance values quite quickly after disturbances, even though this will be driven by understory plants and early vegetation (for example, see Bastos et al. 2011). Using reference states with more clearly defined ecological metrics (e.g., common definitions of forest based on tree cover and height) might overcome such problems, but require more advanced data. Here, novel developments in remote sensing might help close this gap (e.g., using LiDAR data to look specifically at structural recovery; see Stritih et al. 2023), but these are still experimental and not widely applicable to assess resilience.

In practice, a pre-defined time period after the disturbance is often selected, and the regrowth capacity is evaluated. A wide variety of measures have been applied that compare the conditions after the disturbance to a reference condition and potentially also the condition at the end of the disturbance (e.g., Frazier et al. 2018; Tao et al. 2022; Lloret et al. 2011; Van Meerbeek et al. 2021). These measures express the regrowth relative to the disturbance impact (e.g., the recovery magnitude relative to the disturbance magnitude) or to the reference state itself. Essentially, measuring regrowth in this way corresponds to measuring recovery, but with different definitions of what the reference state is.

The choice of reference pre-disturbance values varies widely across studies, some considering only the single year before disturbance (Thurm et al. 2016; Sohn et al. 2016) as reference, but most using the mean values of the 2–3 years prior to disturbance (Yi and Jackson (2021). In such short periods, it is likely that the state variable is influenced by the natural interannual variability of many ecological systems (Zhu et al. 2017; Li et al. 2022), risking to lose representativeness. Therefore, in more recent studies, longer-term reference periods have been used to quantify impacts (Flach et al. 2018; van Schaik et al. 2018; Bastos et al. 2021), yet long time series prior to disturbance might not always be available.

Another problem with using recovery time as an indicator of resilience is that it assumes that recovery will be complete after some fixed time. However, systems might take decades to centuries to reach pre-disturbance values, especially for systems composed of long-living organisms such as forests. The observation length of many data series is hence shorter than the recovery interval. This makes it especially difficult to assess changes in recovery over time, which would be of interest to detect changes in resilience caused by, e.g., climate change. Early indicators of recovery might already serve as a proxy for the long-term developmental trajectory, which yet remains to be empirically validated (Seidl and Turner 2022).

Instead of measuring the time span until a "recovery" criterion is met, it might be advantageous to directly quantify the recovery rate by fitting an exponential function to the section of a time series after a disturbance (Fig. 3) (Smith et al. 2022; Bastos et al. 2011). Based on an assumed linear dynamical systems framework, the recovery of a system after a perturbation should be exponential. Hence, the exponent of an exponential function fitted to the time series of restoration directly quantifies the recovery rate of an ecosystem.

4 Recovery from Continuous Perturbations

The problem with estimating resilience via direct estimation of the recovery rate is that one needs time series that exhibit large perturbations and that it will practically be close to impossible to measure resilience changes over time in this way. In this Section, we will discuss time series that show no large disturbances but permanent variability. Though the resilience indicators obtained from such permanent small perturbations are arguably more indirect than the recovery from single large perturbations, theory predicts a relationship between them and the advantage is that we do not need to rely on the presence and detection of large perturbations as discussed in Sect. 3.

4.1 Statistical Resilience Indicators

In the absence of large externally-induced perturbations, the internal variability of the system can also be used to gain insight in the resilience of a system to external influences. That is because the internal 'fluctuations' of a system typically contain information about

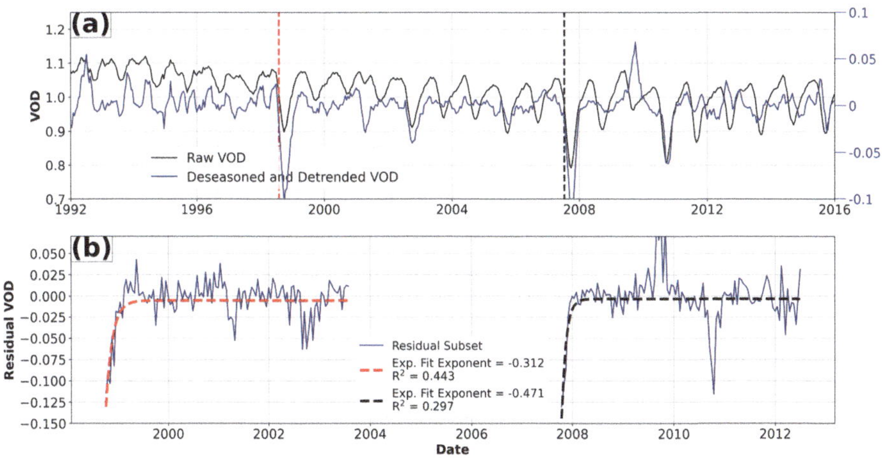

Fig. 3 Observed perturbations and recovery in vegetation optical depth (65.625°N, 155.875°W) as measured by passive microwave data (Moesinger et al. 2020). Top: full time series from 1992–2016, raw time series in black, with deseasonalized and detrended time-series residual in blue. bottom: period from 1999–2012; Large disturbances as those in 1999 and 2008 are an opportunity to estimate the recovery rate using an exponential fit

the system's response to external influences, as stated by the fluctuation–dissipation theorem (Kubo 1966) in statistical mechanics. It is worth noting that this link has also been observed in climate models (Cionni et al. 2004; Gritsun and Branstator 2007; Leith 1975). Hence, interpreting a system's variance and autocorrelation (in short, AC1, using a time lag of one resolved time step) as indicators of linear stability has become a widely adopted and straightforward approach across various fields.

Such an empirical approach is particularly relevant when monitoring systems that might be losing resilience, or even approaching a bifurcation point. When moving toward such a point, the dynamics of the system will typically exhibit critical slowing down (CSD). The initial discovery of this phenomenon can be traced back to statistical physics (Gaspard et al. 1995; Solé et al. 1996; van Kampen 1981). Subsequently, it has been investigated in the field of ecology (Wissel 1984; Oborny et al. 2005) and climate research (Kleinen et al. 2003; Held and Kleinen 2004). Critical slowing down refers to the situation where a system takes longer to return to its equilibrium state following a perturbation, indicating a decrease in its recovery rate. In this regard, indicators of CSD aim to detect a loss in "engineering resilience" (decreasing linear stability) as a proxy of closeness to a bifurcation point. It is the crossing of such a point that finally causes a catastrophic transition to an alternative state (loss of Ecological resilience), and therefore typically is seen as an impact-relevant property.

When a system approaches a bifurcation-induced tipping point, its dynamics tend to resemble a red noise process, wherein each state becomes more similar to its past state, leading to an increase in autocorrelation (Wiesenfeld 1985). Simultaneously, the slowness of the system causes a rise in variance as subsequent random fluctuations accumulate due to the slower and slower decay of these fluctuations (Wiesenfeld and McNamara 1986). Consequently, both variance and autocorrelation progressively increase until the system reaches a bifurcation point. This theory has prompted the use of rising variance and autocorrelation as "early warning signals" for impending bifurcations (Scheffer et al. 2009). Additionally, other related indicators have been proposed, such as higher statistical moments (Guttal and Jayaprakash 2008a) and spatial correlations in spatially extended systems (Dakos et al. 2010; Donangelo et al. 2010; Guttal and Jayaprakash 2008b).

There are several practical and conceptual limitations associated with these statistical resilience indicators (Bathiany et al. 2016a). First, practical limits exist because these methods require long time series with sufficiently slow and gradual change in the external forcing (i.e.,, a time-scale separation between external forcing and internal system dynamics). Second, on a conceptual level, these indicators might lack predictive power, because they might detect false positives due to, e.g., multiplicative noise (Bathiany et al. 2013), negative feedbacks in most of the measured regime, or inertial effects (Wagner and Eisenman 2015; Boulton and Lenton 2015; Bathiany et al. 2016b), the confounding effect of colored or non-stationary noise (Dakos et al. 2012), and the large spatial complexity of real-world systems (Bathiany et al. 2013; Held and Kleinen 2004). As an example, if a given climatic driver shows changes in its AC1, then the ecological metric will too (assuming sensitivity remains), which would be reflected in resilience change, while functioning was actually maintained.

There have been some recent advances however to mediate some of these problems. In particular, Boers (2021) introduced a new restoring rate parameter λ as a resilience indicator that is not prone to false alarms based on changes in the properties of the driving noise when using a more general regression method than autocorrelation (see also Boettner and Boers 2022), while Clarke et al. (2023) analyze the *ratio of spectra (ROSA)* of the system state variable relative to a forcing variable. Moreover, spatial information from EO records can be increasingly beneficial by allowing (i) to generate more data about systems than we would have in case of univariate time series (often by making a "space-for-time" assumption), and (ii) to assess spatial resilience indicators that can complement univariate indicators like autocorrelation, variance or feedback parameter λ (Dakos et al. 2010; Boulton et al. 2022).

Recent work has further linked the statistical measures of resilience to the recovery rate discussed in Sect. 3, over a globally distributed set of ecosystem recoveries. Smith et al. (2022) found that the theoretically expected relationship between AC1/Variance and recovery rate broadly held in empirical vegetation EO data, indicating that AC1 and variance can indeed be related to the recovery rate after a disturbance. This implies that AC1 and variance can be used to estimate how fast a system will recover after a shock, and provide another potential avenue to estimate recovery times and, hence, changes in resilience.

4.2 Pre-processing Time Series to Derive Resilience Indicators

Estimating the resilience of a system using Critical Slowing Down indicators (e.g., the restoring rate λ, lag1-autocorrelation (AC1) or variance) generally requires that the time series describing the system is stationary—that is, without seasonality or trends. Stationarity helps to avoid conflating normal seasonal variations with changes in short-term oscillations, which are signals of changes in system resilience, and avoids biases in the resilience indicators induced by trends in the time series. Therefore, data has to be pre-processed before statistical resilience indicators can be computed from it. It has been shown that the methodology to detect so-called anomalies is typically very important for the detection of extremes (Flach et al. 2017) as well as the estimation of resilience indicators and their trends (Smith and Boers 2023b). We note that these procedures to remove e.g., seasonality and trends are not necessary for estimating recovery times after large and rare perturbations as those discussed in Sect. 3, which focuses on the actual recovery after a perturbation (Verbesselt et al. 2010a).

A wide variety of approaches exists to remove seasonal and long-term trends from data (Smith and Boers 2023b; Verbesselt et al. 2016); the most suitable method is driven by the data set in question. For example, highly seasonal data such as NDVI or other vegetation time series can be decomposed into a trend and seasonal component with harmonic models treating seasonality as a combination of sine waves (e.g., BFAST, Verbesselt et al. 2010a; Verbesselt et al. 2010b, Masiliunas, 2021). There is no one optimal deseasoning method; the efficacy of different procedures depends strongly on the underlying signal that is to be isolated. In addition to the deseasoning method itself, additional parameters such as the length of any seasonal or annual smoother, the choice of a stable reference period with which to construct anomalies, and the number of overlapping sine functions used to represent annual seasonality can have a strong influence on the resultant deseasoned residual.

The type of long-term trend removed is also important to consider—removing a linear trend from non-linear data can leave undesirable traces in the time series that will influence estimates of changing resilience (Smith and Boers 2023b; Verbesselt et al. 2010a; Verbesselt et al. 2016). If, after detrending, the time series still contains a trend, estimating AC1 and variance (or other resilience metrics) based on a rolling window will artificially inflate local autocorrelation, as compared to a locally flat time series. Thus, a trend that fluctuates through time can induce changes in AC1 which might be confused with AC1 changes driven by high-frequency fluctuations or changes in the system dynamics. The impacts of these combined deseasoning and detrending methods can be seen in Fig. 4; not all methods

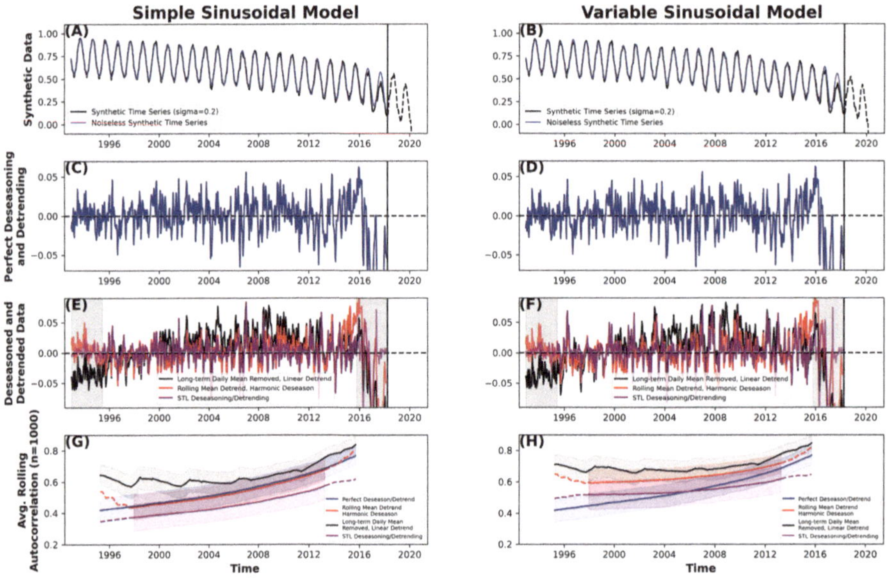

Fig. 4 Impacts of deseasoning/detrending procedures on a simple toy model (from Smith and Boers 2023b, Fig. 1) for a system moving Toward a bifurcation induced transition. Left column: seasonality modeled using a regular sine curve, right column: seasonality varies in amplitude and timing between years. **A**, **B** Time series with (black) and without (blue) noise. **C**, **D** Perfect deseasoning via removing the a-priori known seasonality curve. **E**, **F** Deseasoned time series via a harmonic fit and rolling mean (red), via long-term daily means and a linear detrender (black) and via STL (purple). Gray shading covers periods where rolling mean and STL detrending is uncertain due to incomplete time windows. **G**, **H** AC1 computed over a five-year rolling window for each method of removing seasonality and trend. While all models show an increase in AC1 as a transition approaches, not all methods preserve the monotonic increase in AC1 from (**A**, **B**). Dashed lines indicate less reliable AC1 values due to incomplete moving time windows, shaded areas cover one STD from the mean. Figure license: https://creativecommons.org/licenses/by/4.0/

perform equally well at reconstructing a stationary residual, and each leave distinct traces in the inferred changes in AC1 through time. It is important to carefully consider both the means of removing seasonality and trend, and how that procedure can influence resulting estimates of system resilience.

There also exist methods which remove both trend and seasonality in tandem, such as Seasonal Trend Decomposition by Loess (STL) (Cleveland et al. 1990). STL has the benefit of removing more complex seasonal signals (i.e.,, not sine waves) and dynamic trends (i.e.,, changing trends through time). However, overfitting, e.g., by fitting high-order splines, can remove portions of the dynamics of the system that are actually part of changing system dynamics—and hence potentially resilience, and not of the seasonality or trend (Fig. 4).

It should be emphasized that no method of deseasoning and detrending is perfect—each will have limitations that need to be considered (Smith et al. 2023; Verbesselt et al. 2016). The non-stationarity of the resilience indicators that one desires to measure can often not be separated from other non-stationary features because they occur on similar time scales. For example, if seasonality strongly increases through time, a single harmonic model fit to the whole time series will leave traces in the residual, as it is focused on removing the average signal through time. This effect could be removed by locally fitting harmonic models (e.g., every three years), with the potential downside of leaving step-changes in the residual year-on-year. In short, the potential limitations of deseasoning methods—and their knock-on effects on resilience estimates—need to be carefully considered. Therefore, careful attention is needed when fitting time series in order to explore the residuals.

4.3 Multivariate Approaches and Spatial Resilience Indicators

In the previous sections, we focused on the analysis of time series of a single state variable, so-called univariate time series. However, in practice, often more data is available, be it time series of multiple state variables or time series of spatial evolution.

Multivariate data is useful because natural systems are complex systems that have many processes acting on, and interacting across, different time scales and affecting different state variables and/or locations differently. For instance, the growth rate varies from one plant type to the other, and soil water dynamics play out on shorter time scales compared to plant dynamics. Multivariate data therefore 'samples' the full variety of processes in a system better compared to univariate data, which can lead to better resilience estimates. Mathematically, this means that multivariate data enables the estimation of multiple modes ("eigenvalues and eigenvectors") compared to univariate methods that are typically limited to one mode only.

To use multivariate data, often the univariate methods from the previous sections are extended to a multivariate setting using mode decomposition methods to derive eigenvalue–eigenvector pairs from a time series. One of the most widely-used techniques for this is a principal component analysis, for which a singular value decomposition (SVD) of the multivariate (i.e.,vector-valued) time series X forms the basis. The downside of the use of such methods is that the time ordering of the original time series is completely disregarded when constructing the eigenvectors. It is, however, possible to explicitly account for this time ordering, for example, using dynamic mode decomposition (DMD) (Kutz et al. 2016). A detailed and practically focused methodology can be found in e.g., Weinans et al. (2019).

Once such a mode decomposition has been obtained, the system's resilience can be derived from it. A naive implementation would be to look at the least stable eigenvalue–eigenvector pair (i.e., the eigenvalue with the highest real part) as this mode corresponds to the slowest recovering dynamics. However, this mode might not always correspond to the relevant dynamics. For example, when the eigenvalue has imaginary parts it could correspond to oscillatory dynamics, such as seasonality (see also Sect. 4.2). If this is not the case, it could also happen that the system has very slow dynamics that are always present but do not contain information on the system's resilience. For instance, in semi-arid areas, vegetation patterns migrate slowly over a landscape (DeBlauwe 2012; Bastiaansen et al. 2020) but that dynamics has little to do with the restorative dynamics of the vegetation. Hence, looking at all modes might be necessary to obtain the proper resilience signal.

Proper analysis of these time series using multivariate methods can lead to more accurate resilience estimates and/or reduce the required time span of the data at hand (Weinans et al. 2019; Weinans et al. 2021; see also e.g., Bastiaansen et al. 2021 for exposition of this in the context of global climate models). A number of methods have recently also been further developed and assessed that allow for the automatic detection of abrupt shifts (Bathiany et al. 2020), extreme events (Flach et al. 2017) and changing resilience indicators (Weinans et al. 2021) in multidimensional systems. There is particularly high potential to look at multivariate recovery indicators from remote sensing, with some recent evidence in the literature that multivariate indicators could indeed improve the assessment of post-disturbance recovery (Morresi et al. 2019). When monitoring forests, data on vegetation coverage, tree density, tree height, above-ground biomass, soil moisture and many more might be present (Lechner et al. 2020), and satellite data nowadays typically has high enough resolution to capture spatial patterns, such as those appearing in semi-arid regions (Bastiaansen et al. 2018; DeBlauwe et al. 2008; Veldhuis et al. 2022) or in peatlands (Eppinga et al. 2008). Also, Viana-Soto et al. (2022a) derive fractions of trees, non-treed vegetation and bare ground from remote sensing data, which allows one to study recovery trajectories in a 3D space spanned by those three land covers. A particularly promising approach in this direction is to apply machine learning methods in order to identify optimal indicators (Bury et al., 2021).

When spatial data is available, it is also possible to use the form of the eigenvector instead of the eigenvalue (or in combination with) to estimate the system's resilience. Here, the idea is that the destabilizing eigenvector of spatially extended systems has some striking characteristics that can be used as a metric (Held and Kleinen 2004; Dakos et al. 2010; Guttal and Jayaprakash 2008b). That is, when a system approaches a bifurcation-induced tipping point, it might do so in a very specific way, thus displaying specific spatial characteristics. Precisely which characteristics to look for varies from system to system. For example, in spatially homogeneous systems the spatial correlation tends to increase when a tipping point is approached (Dakos et al. 2010). In a similar vein, it is also possible to look at the spatial variation or decorrelation length. However, for spatially patterned systems, typically the spatial correlation does not precede critical shifts, and one should look for different signs (Bastiaansen et al 2020; Kéfi et al 2014). For instance, the patchiness of some dryland ecosystems, known for exhibiting spatially patterned states, seems to increase as they approach a critical shift (Kéfi et al 2014).

The strength of these spatial early warning signs is that they do not require lengthy time series to be available. However, they do require a dataset with enough spatial detail. Therefore, in particular for systems where constant monitoring is hard, or for systems that have strong seasonality, infrequent analysis of the spatial structure of a snapshot of the system might be a good alternative.

5 Challenges and Opportunities of EO-Based Resilience Monitoring

5.1 Technological Advances in Observing Terrestrial Vegetation

The challenges of resilience monitoring discussed above in principle apply to all Earth observation records. Before we highlight how these challenges manifest themselves in current datasets and applications, we here provide a very brief summary of recent developments in the retrieval of terrestrial vegetation datasets from Earth observations.

In principle, satellites measure radiative fluxes which by a retrieval algorithm are translated into products that are meant to represent a property on the ground, like vegetation coverage or biomass. The measurement can be passive or active. Passive remote sensing measures the properties of solar radiance reflected by the Earth surface, or radiance emitted by the Earth itself. Active remote sensing systems send artificially generated radiation Toward the Earth and measure its reflection properties, such as intensity and polarization. The latter category includes observation systems such as radar and lidar.

A common and largely available passive EO technology is multispectral imaging, where satellites collect data over multiple wavelength regions, such as the visible, near-infrared, and shortwave infrared part of the electromagnetic spectrum. Multispectral data form the basis of widely used vegetation indices like the Normalized Difference Vegetation Index (NDVI) or the Enhanced Vegetation Index (EVI). They are closely (though not linearly) related to photosynthesis and thus plant vigor and productivity. Yet, NDVI as well as EVI saturate in regions with high biomass (Sellers 1985; Liu et al. 2011), and the short wavelengths render them tainted by atmospheric conditions such as clouds and aerosols (Samanta et al. 2012). In addition, seasonal changes in sun-sensor geometry can lead to serious artifacts (Morton et al. 2014).

Radiometry, and its passive-microwave products such as Vegetation Optical Depth (VOD), overcome these obstacles, as biomass density drives the natural emission of radiance from the vegetation itself and the attenuation of natural emissions from the soil (Liu et al. 2011; Konings et al. 2019; Li et al. 2021; Schmidt et al. 2023). Also, improved plant surface temperature measurements from ECOSTRESS (Fisher et al., 2020) allow insights into physiological mechanisms like stomatal control and hence into the water and carbon cycle and land–atmosphere coupling. Radiometry's drawback is its low spatial resolution. In contrast, the measurement of backscatter from actively sent microwave signals, as done by synthetic aperture radar (SAR) technology allows for resolutions at meter-scale (Konings et al. 2019). SAR provides insights into terrain structure, vegetation canopy structure, and soil moisture. Besides classical multispectral and SAR-based indices, novel indices such as solar-induced chlorophyll fluorescence (SIF) can even better reveal photosynthetic activity and phenology (Zhang et al. 2022).

Finally, recent spaceborne LiDAR missions, such as GEDI and ICESat-2, provide near-global and global sampling of 3D forest information. Initial exploration of those data focused on combining GEDI samples of forest heights (relative height observations) with Landsat and Sentinel-2 images to derive global, wall-to-wall 30 m and 10 m canopy height maps, respectively (Potapov et al., 2021). These global products will enable a global analysis of canopy height time series in the future, and in turn engineering resilience studies on forest regrowth. Nevertheless, the canopy heights should be used critically as potential systematic effects in satellite LiDAR height observations may lead to an underestimation of forest regrowth (Milenkovic et al. 2022). ICESat-2, on the other hand, already provides time series of forest heights as this mission, in contrast to GEDI, samples along the same tracks and seasonally since 2018. Besides forest height, both GEDI and ICESat-2 provide information on vertical vegetation profile metrics such as leaf area index and canopy gaps (Dubayah et al. 2021), but the exploration of those variables in resilience studies is yet to come (see Stritih et al. 2023 for an example).

On the one hand, these novel developments provide us with a richer and much better resolved database on which resilience monitoring can rely. We provide a timely review of recent developments in the next section, with focus on forest monitoring. On the other hand, a number of conceptual and practical caveats need to be resolved in order to harvest the full potential of these developments. To raise awareness of the gap between current practices in the generation of EO data and the requirements of resilience monitoring, we discuss the most apparent and urgent limitations in Sect. 5.4–5.9.

5.2 Monitoring Resilience in Aquatic and Marine Ecosystems

Although examples in the rest of this review focus on terrestrial ecosystems, the notion of ecosystem resilience in general, the potential to monitor it with remote sensing, and the above conceptual and practical challenges all apply to aquatic and marine ecosystems in a similar way. For example, shallow lakes have been a prototypical example for the potential existence of alternative stable states (Scheffer and Jeppesen 2007), which differ in the turbidity of the water. In principle, earth observation by satellite-borne sensors can be useful for probing the resilience of such aquatic systems (Oezkundakci and Lehmann, 2019). A large body of research also focuses on the resilience of coastal wetlands, e.g., by observing the recovery after droughts, hurricanes and other storms, heavy precipitation and fluvial flooding. Such studies make use of a large number of remote sensing sources such as MODIS and Landsat (Tahsin et al. 2018), vegetation-based indices (Balogun et al. 2020) and sometimes use the classical statistical resilience indicators as outlined in Sect. 4.1 (Zhao and Li 2022). Moreover, similar detection methods to automatically detect abrupt shifts as outlined in Sect. 3 are being developed and applied in the marine sciences (Yang et al. 2022). Other ecosystems under investigation are giant kelp forests (Cavanaugh et al. 2019), coral reefs (Knudby et al. 2014) and seagrass ecosystems (Clemente et al., 2023), with particular concern of tipping in these systems (van de Leemput et al. 2016; van der Heide et al., 2011). While returning to terrestrial applications in the remaining sections, we point out that some of the same challenges as discussed throughout this review also apply to aquatic and marine ecosystems as well.

5.3 Developments in Earth Observations with Regard to Terrestrial Ecosystem Resilience Monitoring

In recent years, Earth observation has become a major tool in assessing resilience (Nikinmaa et al. 2020; De Keersmaecker 2022; Senf 2022), often involving the technological advances outlined in the previous section. Special focus has been put on forest ecosystems, because forests globally are changing in response to climate change and increasing land use pressures (McDowell et al. 2020). For instance, recovery from disturbance such as windthrow, bark beetle outbreaks or fire have been estimated from optical remote sensing time series in order to assess the forest's capacity to return to similar spectral conditions than before disturbance (White et al. 2019; White et al. 2022; Chirici et al. 2020). Most of those studies made use of free image archives such as Landsat (Wulder et al. 2019), which provides long enough time series to estimate recovery signals in forests. The use of Landsat also allows for the quantification of historic disturbance events (Senf and Seidl 2021a, b; Hansen et al. 2013), which in turn can be used to estimate recovery by space-for-time substitution. More specifically, combining the historic information on disturbances with recent data on forest structure, such as derived from airborne or spaceborne Light Detection and Ranging (LiDAR) data, allows to create chronosequences of structural attributes (e.g., canopy cover, height) over variable years post-disturbance (Viana-Soto et al. 2022b; Stritih et al. 2023; Senf et al. 2019). From those chronosequences, it can be estimated how quickly forests return to certain structural states or recover certain functions, which substantially increases the applicability of the result in ecosystem management (but see also discussion in Sect. 5.4). Similar results can be achieved also without the pitfalls of space-for-time substitution (Johnson and Miyanishi 2008) by recent developments in Earth observation, including machine learning regression (Senf and Seidl 2021a, b), synthetic spectral unmixing analysis (Senf et al., 2020) or artificial intelligence (Liu et al., 2023). Those novel approaches convert the spectral signals into interpretable units, such as canopy cover or biomass, allowing for example for the assessment of community change after fire disturbances (White et al. 2023; Viana-Soto et al. 2022a). Novel sensor systems, including spaceborne hyperspectral data or Radar, will likely add important information to ecosystem resilience, but their full potential hasn't been explored yet.

5.4 Signal Versus Functional Recovery

Many remote sensing based recovery studies use signal recovery as a measure of resilience. Signal recovery is derived from a time series of a single remote sensing observable, being it a spectral index, e.g., NDVI or RVI (the radar vegetation index, Saatchi 2019a, b), or a bio-geophysical variable derived from the radiance measurements, e.g., the fraction of absorbed photosynthetic radiation (fAPAR) derived from solar-reflective measurements, or vegetation optical depth (VOD) from microwave observations. Yet, by focusing on a single ecosystem property, signal recovery does not always reveal to what extent the same level of processes or structure are retained in the system. For instance, while after a disturbance the NDVI signal of a tropical forest may recover to its pre-disturbed state, the secondary forests are known to be structurally and functionally different (Caughlin et al., 2021). In addition, signal recovery does not provide information on the process behind the recovery. For instance, monocultures can be more sensitive to diseases but recover relatively fast when they are replanted. Yet, the fast signal recovery capacity of these monocultures does not necessarily reflect their resilience to disturbances. Consequently, it is important to find a proxy variable measurable by remote sensing which represents the in-situ property under investigation. If this is not directly possible, multivariate approaches based on several, independent remote sensing observables, and additional knowledge on the relationships of time scales may help constrain the recovery process quantitatively.

5.5 Spatial and Temporal Resolution

The scale of EO measurements rarely matches up with the system under study, neither in space nor in time. For example, Landsat has a theoretical revisit time of ~ 16 days (or much longer in areas with frequent cloud cover), which is too low to capture some of the dynamics of many ecosystems. To track phenological changes, or the response to extreme events like droughts or storms, daily resolution would be required. Conversely, MODIS with a daily repeat time lacks the spatial resolution to resolve small-scale features on the spatial scale of individual trees that can be picked out with Landsat data (30 m resolution). The trade-off between time and space becomes even more of an issue when very-high-resolution optical and Synthetic-aperture radar (SAR) are used, which, depending on the satellite constellation and priority of the ground target in the acquisition plan, may provide only a few acquisitions per year. It is not always clear what the optimal spatial or temporal resolution to study a given system is, and to what extent EO data captures the dynamics well enough to estimate changes in resilience through time. Multi-scale approaches, i.e., tackling a problem using various spatial and temporal resolutions, might make resilience assessments from EO data more robust. Also, it is strongly recommended to first define the scales necessary to understand a system and then choose the appropriate EO dataset, and not vice versa (Senf 2022).

5.6 Data Gaps

Gaps are common in EO data, e.g., due to cloud cover, smoke, snow and frozen ground, fog, and in the case of radar and very-high-resolution optical imagery, because of prioritization in the acquisition plan. For optical data, gaps are particularly a problem in the tropics, where cloud cover is frequent and persistent. For microwave data, large seasonal gaps may occur when the soil is frozen, covered with snow, or inundated. Gaps can cause problems both in theoretical resilience metrics (e.g., AC1, variance) and engineering metrics (recovery time). For example, gaps in a time series can bias AC1 and variance if the number of measurements over a given time period fluctuates, or if there are sporadic large jumps through time (Ben-Yami et al. 2023; Smith and Boers 2023b). Engineering resilience metrics can also be biased by missing data by either under- or over-estimating the length of a recovery due to missing data at the start or end of the perturbation or recovery period. More generally, deseasoning and detrending methods do not always take missing data into account, and can yield inconsistent results when data is sparse (Smith and Boers 2023b; Verbesselt et al. 2016). Many statistical methods can handle time series with data gaps without gap filling (e.g., by computing variance or autocorrelation from only the existing values). However, the gap properties (gap size, frequency etc.) should certainly be taken into account carefully when preprocessing time series to derive robust resilience indicators. For example, masking an artificial stationary time series with the same gap structure as in the observed data can give an estimate of gap-induced biases (Ben-Yami et al. 2023).

5.7 Time Series Length

Most resilience methods require long time series, i.e., at least long enough to cover the recovery of a single disturbance event or even a multiple of the decorrelation time scale when using statistical resilience metrics like the recovery rate λ, AC1 or variance. Especially for systems with slow internal dynamics, such as forests with long regeneration times, this might be problematic given the relatively short length of the satellite record in general, let alone individual instruments. Besides, the time series length should also allow for a joint analysis of the recovery and disturbance regime. For example, as outlined in Sect. 2.2, a system that takes 50 years to recover and is subject to a lower disturbance frequency will be able to restore most important properties, while a system with a higher disturbance frequency than the recovery period will likely have a higher chance of shifting into a state change (Senf and Seidl 2021a, b). Finally, a sufficiently long time series is required to separate cyclic changes from long-term change (see also Sect. 4.2).

However, because EO data acquisitions have become widespread and frequent only recently, many available data sets are relatively short. Very few of the early remote sensing instruments are still being used, and most missions have ended or been supplanted by newer satellites. Thus, for long-term time series analysis such as the quantification of resilience changes, data from multiple satellites need to be considered. This requires a thorough homogenization of random and systematic differences between instruments, either at the level of radiometric or spectral properties (radiances, reflectances, backscatter, or brightness temperatures; e.g., Xiong et al. 2020) or starting from biogeophysical variables derived from single sensors (e.g., Moesinger et al. 2020; Preimesberger et al. 2021).

5.8 Space-for-Time Substitution

Even though most EO products still have insufficient length to robustly assess ecosystem resilience, it is now getting easier to obtain high-resolution spatial data. Consequently, it is an appealing idea to use a space-for-time substitution. The idea to infer temporal information from spatial differences is not new; for instance, different succession stages in forest development can be identified by looking at species composition at different sites (Picket, 1989). Besides such qualitative comparisons, it is also suggested to use a similar space-for-time approach for quantitative studies (Blois et al 2013), which might also be suitable for resilience monitoring. Here, one could for instance study spatial data along an environmental gradient to get insight into the temporal evolution in accordance with the changing conditions (e.g., to assess Amazon rainforest regrowth; Milenkovic 2022). The spatial resilience indicators as outlined in Sect. 4.3, which are justified from the theory of stochastic dynamical systems, are one promising approach in this context—at least in case their validity to describe the resilience of complex ecosystems can be confirmed. Another example of a space-for-time approach is the diagnosis of multimodal distribution of tropical tree cover for a similar range of mean annual precipitation (Staver et al. 2011; Hirota et al. 2011). If the tree cover data can be understood as a representation of the same stochastic system (space-for-time replacement applies), this finding implies the existence of multiple stable states under the same annual rainfall conditions, and hence potential tipping points in tropical forests. Caveats related to the retrieval and classification of tree cover from satellites in this context have been largely resolved (Hanan et al., 2014; Staver and Hansen 2015).

However, there are also caveats and doubts regarding the validity of space-for-time replacements (Damgaard 2019). Most prominently, the fundamental assumption behind space-for-time methods is that temporal and spatial evolution are comparable and have the same drivers, and that other environmental factors, that differ spatially, can be neglected. This assumption is likely to not hold under present-day conditions where CO_2 fertilization, climate change and nutrient deposition play together, so that systems are not close to equilibrium. Also, even in equilibrium, there are usually more possible explanations for multimodality of the data distribution than the existence of multiple stable states. For example, a bimodality in tree cover states can also be imposed by bimodal environmental effects like human activity (Wuyts et al. 2017), other climate variables than annual mean precipitation (Good et al. 2016) or by nonlinear and transient growth dynamics (Yin et al. 2014). Also, the noise of a system might often have a higher spatial correlation compared to its temporal correlation. As far as we are aware, there are no comprehensive theoretical studies of this assumption in the context of resilience monitoring. Future research in this direction might reveal the merit of such approaches.

5.9 Impact of Data Artifacts on Resilience Metrics

The need to reduce gaps in satellite datasets (see Sect. 5.6) and to generate longer time series than the lifetime of a satellite, has motivated the construction of datasets from multiple instruments. In resilience research however, multi-instrument records pose serious challenges to the methodology used to detect change, e.g., how much of the change can be attributed to environmental shifts, and how much is simply due to the use of more precise instruments, different satellite retrieval schemes, changes in satellite overpass times, or other measurement changes? harmonization efforts (e.g., Moesinger et al. 2020; Preimesberger et al. 2021) usually focus on mean and variance to enable the detection of trends (e.g., changes in leaf area index through time). Unfortunately, harmonization of higher-order statistics such as autocorrelation has received less attention. This implies that satellite datasets may be biased even though they have been evaluated in terms of mean values, and extra care must be taken to ensure that changes in resilience indicators can really be attributed to environmental change (Smith et al. 2023). Further collaboration is needed between environmental scientists and the data curation and creation community to document the potential impacts of changing satellite mixes on resilience indicators, and to better document what statistical measures can be safely evaluated up to what precision.

Artificial data features that can remain after insufficient preprocessing can also be a cause of an erroneous resilience assessment. In principle, these are general issues with remote sensing products, but can be particularly relevant for resilience estimators even if the products might show unbiased mean values. For example, when aggregating overlapping images from neighboring orbits, the spectral properties of the same object may be different just because the images have different looking angles to the object (Dorigo 2012). This may cause artifacts in the time series, e.g., additional noise, and ultimately lead to a false detection of change. Thus, it is important to report what radiometric and geometric corrections were applied to the satellite data to ensure that the data is analysis ready. Sometimes, particularly in high resolution synthetic aperture radar data with strongly oblique view angles, a ground object itself causes artifacts. For example, when a patch of vegetation is completely removed, e.g., after a clear-cut event, pixels along the patch edge in the near-field of the radar satellite remain in the "shadow", i.e., are occluded by the neighboring forests. Such radar image artifacts cause an abrupt drop of the signal, which facilitates the deforestation detection (Bouvet et al. 2018), but does not recover, because the pixel remains in the radar shadow of the neighboring forest. Including such pixels in the analysis may lead to a false interpretation of forest recovery.

6 Conclusions, Recommendations, and Outlook

We have outlined conceptual and practical challenges of monitoring ecosystem resilience and resilience changes from EO data. We first discussed the long-standing diversity and subtlety in defining resilience conceptually, beyond the general classes of ecological versus "engineering" resilience. Given the large variety of resilience quantifiers, particular care should be taken in future research articles when naming and describing the particular approach taken. In this regard, "resilience" should be considered a topic or a qualitative group of system properties, and distinguished from particular quantifiers, which should be named more precisely and with more care.

We also discussed methods to detect and quantify abrupt shifts (in ecological systems: disturbances), the recovery time and statistical indicators of linear stability (recovery rate, AC1, variance, spatial correlations and others). Besides the need to acknowledge the general limitations of such methods (like the validity of the stochastic system or statistical robustness), we emphasized several additional caveats arising from data uncertainties. Many of these uncertainties arise from combining different sensors (Smith et al. 2023), from gaps in the time series (Ben-Yami et al. 2023), and from detrending and deseasoning the data (Verbesselt et al. 2016; Smith and Boers 2023b).

It has consequently been shown that for some vegetation types resilience measures are not necessarily consistent with each other even given the same data (Smith and Boers 2023b), which points to potential data uncertainties or simply unsuitable EO data for quantifying the relevant aspects of vegetation dynamics, preventing the use of these measures.

Despite the growing availability of EO data, there remain several key limitations to its use in monitoring terrestrial ecosystem resilience and its changes, particularly in near-real-time applications, maintaining a gap between the development of novel remote sensing techniques and their applications for ecosystem resilience monitoring (also see Senf 2022).

Many of these limitations are areas of active research, and can be ameliorated through improved models, data processing schemes, and longer time series.

Based on our assessment above, we make the following key recommendations:

1. There is need for more clarity in resilience studies on (i) the definitions of resilience used, (ii) what the ecological metric actually refers to (vegetation cover, biomass, composition, function, or other), and (iii) the careful choice and clear definition of reference states considered.
2. Innovations are needed in ways for dealing with inherent seasonal and phenological variations in EO time series (e.g., indices such as NDVI). Some systems have high inter-annual variability, and awareness and best practices are required to deal with it.
3. Large opportunities arise from the availability of higher spatial resolution imagery such as the Sentinels (1 and 2), and future Very High Spatial resolution satellite missions (e.g., Home—Copernicus Contributing Missions Online). Key here is to see which methods are able to measure resilience by analyzing spatial variability (heterogeneity), establish whether space-for-time assumptions can be justified, and further explore the possibility of estimating resilience via the potential of artificial intelligence, making use of the multitude of geo-spatial data, on different aspects of the ecosystem, that are currently available (Smith et al. 2023).
4. Time series are becoming longer, and are also available in near real time. They should be processed carefully regarding seasonality and the treatment of gaps and other uncertainties, and the limitations should be documented in a way that allows for quantitative uncertainty estimates of statistical indicators.
5. The suitability of remote-sensing datasets for resilience analyzes should be better verified, especially for high-biomass regions like the tropical forests. As biomass might be the most indicative parameter of rainforests, radar and radiometry in the very low frequency range (P- or L-band) should be desirable due to the deeper penetration from below the canopy. Here, upcoming satellite missions such as the European Copernicus Biomass and American NISAR mission are promising. Furthermore, future availability of in-situ measurements, for example based on terrestrial or airborne LiDAR campaigns, are recommended to evaluate the different data sets, especially in these regions.
6. Novel measurements such as land surface temperature (e.g., the future Copernicus Land Surface Temperature Monitoring satellite mission, NASA's CIRiS CubeSat, and commercial space based alternatives), vegetation structure (e.g., vegetation height via GEDI, ICESAT-2, and future BIOMASS mission) and fluorescence (e.g., via the future ESA FLEX satellite mission) provide direct biophysical variables to monitor terrestrial vegetation dynamics of the ecosystem. Before directly utilizing them in describing ecosystem change and recovery, one should first understand the uncertainties and limitations associated with those novel products as they are often optimized for other applications than resilience monitoring, or the pre-processing is simplified to foster big-data handling and processing.
7. Knowledge transfer and closer collaborations between marine, aquatic and terrestrial ecologists and remote sensing experts may create synergies in monitoring ecosystem resilience in different realms.

Overall, we conclude that it is beneficial that there are more and more applications of different resilience indicators ("early warning signals") to Earth observations (EO). These observations can provide advantages, e.g., higher spatial coverage and hence spatial indicators, potentially by leveraging space-for-time relationships. However, it is vital to be aware of the caveats associated with data generation.

We also note that the research community designing retrieval algorithms and generating EO datasets on the one hand, and the community interested in diagnosing and interpreting ecosystem resilience (often ecologists, climate scientists and applied mathematicians) on the other, are still not in sufficiently close scientific contact. This brings the risk of generating datasets without a well-defined scope of use, and the risk of (mis)using satellite datasets for resilience monitoring despite technical and statistical biases.

We hence see the need for closer collaborations that help close the gap between what EO data can offer and the requirements of EO users. Specific questions of this dialog should at least comprise: What variables are needed for what purpose? On what resolution? Which statistical properties should be optimized, and how are biases communicated? How should gaps and multi-sensor products be compiled by EO scientists? We specifically recommend better transparency and documentation in this regard. Only then can the value of long and frequent observations from space live up to their great potential, and help us get a clear view on the resilience of the Earth system, and help society navigate into a safe future.

Acknowledgements This paper is an outcome of the Workshop: 'Tipping Points and Understanding EO data needs for a Tipping Element Model Intercomparison Project' held at the International Space Science Institute (ISSI) in Bern, Switzerland (10-14 October 2022). We thank the ISSI for the great hospitality and their distinct approach to hosting. C. S. acknowledges funding from the Fachagentur Nachwachsende Rohstoffe e.V. (FNR) through the ERA-Net cofound action ForestValue (project FORECO, project nr. 2221NR088X). A.B. was funded by the European Union (ERC StG, ForExD, grant agreement No. 101039567). S. L. would like to acknowledge funding from the Earth Commission which is hosted by Future Earth and is the science component of the Global Commons Alliance. W.D. acknowledges the CONSOLIDATION funded by the Austrian Science Fund (FWF). A.B. was funded by the European Union (ERC StG, ForExD, grant agreement No. 101039567). T.S acknowledges support from the DFG STRIVE project (SM 710/2-1). S.B. and N.B. have received funding from the Volkswagen Stiftung, the European Union's Horizon 2020 research and innovation programme, and under the Marie Sklodowska-Curie grant agreement No. 956170, as well as from the Federal Ministry of Education and Research under grant No. 01LS2001A. This is ClimTip contribution #2; the ClimTip project has received funding from the European Union's Horizon Europe research and innovation programme under grant agreement No. 101137601, funded by the European Union. Views and opinions expressed are those of the authors only and do not necessarily reflect those of the European Union or the European Climate, Infrastructure and Environment Executive Agency (CINEA). Neither the European Union nor the granting authority can be held responsible for them.

Funding Open Access funding enabled and organized by Projekt DEAL.

Declarations

Conflict of interest The authors declare no conflicts of interest.

References

Aminikhanghahi S, Cook DJ (2017) A survey of methods for time series change point detection. Knowl Inf Syst 51:339–367

Anderegg WRL et al (2020) Divergent forest sensitivity to repeated extreme droughts. Nat Clim Change. https://doi.org/10.1038/s41558-020-00919-1

Angeler A (2016) Quantifying resilience. J Appl Eco! 53(3):617-624. https://besjournals.onlinelibrary.wiley.com/doi/epdf/10.1111/1365-2664.12649

Arani B et al (2021) Exit time as a measure of ecological resilience. Science. https://doi.org/10.1126/science.aay4895

Armstrong McKay D et al (2022) Exceeding 1.5 °C global warming could trigger multiple climate tipping points. Science. https://doi.org/10.1126/science.abn79

Arnoldi J-F, Loreau M, Haegmann B (2016) Resilience, reactivity and variability: a mathematical comparison of ecological stability measures. J Theor Biol 389:47–59

Bagniewski W, Ghil M, Rousseau DD (2021) Automatic detection of abrupt transitions in paleoclimate records. Chaos Interdiscip J Nonlinear Sci 31:113129. https://doi.org/10.1063/5.0062543

Bai J, Perron P (2003) Computation and analysis of multiple structural change models. J Appl Econom 18:1–22

Balogun A-L et al (2020) Spatio-temporal analysis of oil spill impact and recovery pattern of coastal vegetation and wetland using multispectral satellite landsat 8-OLI imagery and machine learning models. Remote Sens 2020(12):1225

Basseville M, Nikiforov IV (1993) Detection of abrupt changes. Theory and application/Michèle Basseville, Igor V Nikiforov, Englewood Cliffs, N.J.: Prentice Hall; London: Prentice-Hall International (Prentice Hall information and system sciences series). Available online at http://people.irisa.fr/Michele.Basseville/kniga/kniga.pdf.

Bastiaansen R, Jaïbi O, Deblauwe V, Eppinga MB, Siteur K, Siero E, Rietkerk M (2018) Multistability of model and real dryland ecosystems through spatial self-organization. Proc Natl Acad Sci 115(44):11256–11261

Bastiaansen R, Doelman A, Eppinga MB, Rietkerk M (2020) The effect of climate change on the resilience of ecosystems with adaptive spatial pattern formation. Ecol Lett 23(3):414–429. https://doi.org/10.1111/ele.13449

Bastiaansen R, Dijkstra HA, von der Heydt AS (2021) Multivariate estimations of equilibrium climate sensitivity from short transient warming simulations. Geophys Res Lett 48:e2020GL091090. https://doi.org/10.1029/2020GL091090

Bastos A, Gouveia CM, DaCamara CC, Trigo RM (2011) Modeling post-fire vegetation recovery in Portugal. Biogeosciences 8:3593–3607. https://doi.org/10.5194/bg-8-3593-2011

Bastos A, Orth R, Reichstein M, Ciais P, Viovy N, Zaehle S, Anthoni P, Arneth A, Gentine P, Joetzjer E, Lienert S (2021) Vulnerability of European ecosystems to two compound dry and hot summers in 2018 and 2019. Earth Syst Dyn 12:1015–1035. https://doi.org/10.5194/esd-12-1015-2021

Bathiany S, Claussen M, Fraedrich K (2013) Detecting hotspots of atmosphere–vegetation interaction via slowing down—Part 1: a stochastic approach. Earth Syst Dynam 4:63–78. https://doi.org/10.5194/esd-4-63-2013

Bathiany S et al (2016a) Beyond bifurcation: using complex models to understand and predict abrupt climate change. Dyn Stat Clim Syst 2016:1–31. https://doi.org/10.1093/climsys/dzw004

Bathiany S, van der Bolt B, Williamson MS et al (2016b) Statistical indicators of Arctic sea-ice stability—prospects and limitations. Cryosphere 10:1631–1645. https://doi.org/10.5194/tc-10-1631-2016

Bathiany S, Hidding J, Scheffer M (2020) Edge detection reveals abrupt and extreme climate events. J Clim 33(15):6399–6421

Beaulieu C, Chen J (2012) Sarmiento JL (2012): Change-point analysis as a tool to detect abrupt climate variations. Philos Trans Royal Soc a: Math Phys Eng Sci 370(1962):1228–1249. https://doi.org/10.1098/rsta.2011.0383

Ben-Yami M, Skiba V, Bathiany S et al (2023) Uncertainties in critical slowing down indicators of observation-based fingerprints of the Atlantic overturning circulation. Nat Commun 14:8344. https://doi.org/10.1038/s41467-023-44046-9

Blois JL, Williams JW, Fitzpatrick MC, Jackson ST, Ferrier S (2013) Space can substitute for time in predicting climate-change effects on biodiversity. Proc Natl Acad Sci 110(23):9374–9379. https://doi.org/10.1073/pnas.1220228110

Boers N (2021) Observation-based early-warning signals for a collapse of the Atlantic meridional overturning circulation. Nat Clim Chang 11:680–688

Boers N, Ghil M, Stocker T (2022) Theoretical and paleoclimatic evidence for abrupt transitions in the Earth system. Environ Res Lett 17:093006. https://doi.org/10.1088/1748-9326/ac8944
Boettner C, Boers N (2022) Critical slowing down in dynamical systems driven by nonstationary correlated noise. Phys Rev Res 4:013230. https://doi.org/10.1103/PhysRevResearch.4.013230
Boulton CA, Lenton TM (2015) Slowing down of North Pacific climate variability and its implications for abrupt ecosystem change. Proc Natl Acad Sci USA 112:11496–11501. https://doi.org/10.1073/pnas.1501781112
Boulton CA et al (2022) Pronounced loss of Amazon rainforest resilience since the early 2000s. Nat Clim Change 12:271–278. https://doi.org/10.1038/s41558-022-01287-8
Bouvet A et al (2018) Use of the SAR shadowing effect for deforestation detection with Sentinel-1 time series. Remote Sens 10(8):1250
Buma B (2012) Evaluating the utility and seasonality of NDVI values for assessing post-disturbance recovery in a subalpine forest. Environ Monit Assess 184:3849–3860. https://doi.org/10.1007/s10661-011-2228-y
Bury TM et al (2021) Deep learning for early warning signals of tipping points. PNAS 118(39):e2106140118. https://doi.org/10.1073/pnas.2106140118
Canny J (1986) A computational approach to edge detection. IEEE Trans Pattern Anal Mach Intell. https://doi.org/10.1109/TPAMI.1986.4767851
Carpenter S et al (2001) From metaphor to measurement: resilience of What to What? Ecosystems 2001(4):765–781. https://doi.org/10.1007/s10021-001-0045-9
Caughlin TT, Barber Alvarez-Buylla C, Asner GP, Glenn NF, Bohlman SA, Wilson CH (2021) Monitoring tropical forest succession at landscape scales despite uncertainty in Landsat time series. Ecol Appl 31(1):e02208. https://doi.org/10.1002/eap.2208
Cavanaugh KC, Reed DC, Bell TW, Castorani MCN, Beas-Luna R (2019) Spatial Variability in the Resistance and Resilience of Giant Kelp in Southern and Baja California to a Multiyear Heatwave. Front Mar Sci 6:413. https://doi.org/10.3389/fmars.2019.00413
Chekroun MD, Simonnet E, Ghil M (2011) Stochastic climate dynamics: random attractors and time-dependent invariant measures. Physica D 240(21):1685–1700
Chirici G, Giannetti F, Mazza E et al (2020) Monitoring clearcutting and subsequent rapid recovery in Mediterranean coppice forests with Landsat time series. Ann for Sci 77:40. https://doi.org/10.1007/s13595-020-00936-2
Chu PS (2002) Large-scale circulation features associated with decadal variations of tropical cyclone activity over the Central North Pacific. J Clim 15:2678–2689
Ciemer C et al (2019) Higher resilience to climatic disturbances in tropical vegetation exposed to more variable rainfall. Nat Geosci 12:174–179. https://doi.org/10.1038/s41561-019-0312-z
Cionni I, Visconti G, Sassi F (2004) Fluctuation dissipation theorem in a general circulation model. Geophys Res Lett 31:L09206. https://doi.org/10.1029/2004GL019739
Clarke JJ, Huntingford C, Ritchie PD, Cox PM (2023) Seeking more robust early warning signals for climate tipping points: the ratio of spectra method (ROSA). Environ Res Lett 18(3):035006. https://doi.org/10.1088/1748-9326/acbc8d
Clemente KJE, THomsen MS, Zimmermann RC (2023) Thomsen MS and Zimmermann RC (2023) The vulnerability and resilience of seagrass ecosystems to marine heatwaves in New Zealand: a remote sensing analysis of seascape metrics using PlanetScope imagery. Remote Sens Ecol Conserv 9(6):803–819
Cleveland RB et al (1990) STL: a seasonal-trend decomposition procedure based on loess. J off Stat 6(1):3–73
Cranko Page J, De Kauwe MG, Abramowitz G, Cleverly J, Hinko-Najera N, Hovenden MJ, Liu Y, Pitman AJ, Ogle K (2022) Examining the role of environmental memory in the predictability of carbon and water fluxes across Australian ecosystems. Biogeosciences 19:1913–1932. https://doi.org/10.5194/bg-19-1913-2022
Cutter SL (2016) Resilience to What? Resilience for Whom? Geogr J 182(2):110–113. https://doi.org/10.1111/geoj.12174
Dakos V, Kéfi S (2022) Ecological resilience: what to measure and how. Environ Res Lett 17:043003. https://doi.org/10.1088/1748-9326/ac5767
Dakos V, van Nes EH, Donangelo R, Fort H, Scheffer M (2010) Spatial correlation as leading indicator of catastrophic shifts. Thyroid Res 3:163–174. https://doi.org/10.1007/s12080-009-0060-6
Dakos V, van Nes EH, D'Odorico P, Scheffer M (2012) Robustness of variance and autocorrelation as indicators of critical slowing down. Ecology 93:264–271. https://doi.org/10.1890/11-0889.1
Damgaard C (2019) A critique of the space-for-time substitution practice in community ecology. Trends Ecol Evol 34(5):416–421. https://doi.org/10.1016/j.tree.2019.01.013

De Keersmaecker W et al (2022) Evaluating recovery metrics derived from optical time series over tropical forest ecosystems. Remote Sens Environ 274:112991. https://doi.org/10.1016/j.rse.2022.112991

Deblauwe V, Barbier N, Couteron P, Lejeune O, Bogaert J (2008) The global biogeography of semi-arid periodic vegetation patterns. Glob Ecol Biogeogr 17:715–723. https://doi.org/10.1111/j.1466-8238.2008.00413.x

Deblauwe V, Couteron P, Bogaert J, Barbier N (2012) Determinants and dynamics of banded vegetation pattern migration in arid climates. Ecol Monogr 82(1):3–21. https://doi.org/10.1890/11-0362.1

Donangelo R, Fort H, Dakos V, Scheffer M, van Nes EH (2010) Early warnings for catastrophic shifts in ecosystems: comparison between spatial and temporal indicators. Int J Bifurc Chaos 20:315–321. https://doi.org/10.1142/S0218127410025764

Donges J et al (2011) Nonlinear detection of paleoclimate-variability transitions possibly related to human evolution. PNAS 108(51):20422–20427. https://doi.org/10.1073/pnas.1117052108

Dorigo WA (2012) Improving the robustness of cotton status characterization by radiative transfer model inversion of multi-angular CHRIS/PROBA data. IEEE J Sel Top Appl Earth Obs Remote Sens 5(1):18–29. https://doi.org/10.1109/JSTARS.2011.2171181

Drijfhout S (2015) Catalog of abrupt shifts in intergovernmental panel on climate change climate models. Proc Natl Acad Sci USA 112:E5777–E5786. https://doi.org/10.1073/pnas.1511451112

Dubayah R, Tang H, Armston J, Luthcke S, Hofton M, Blair J: GEDI L2B canopy cover and vertical profile metrics data global footprint level V002, Publisher: NASA EOSDIS land processes DAAC, Release Date: 2021-04-16T00:00:00.000Z, https://doi.org/10.5067/GEDI/GEDI02_B.002

Ducré-Robitaille J-F, Vincent LA, Boulet G (2003) Comparison of techniques for detection of discontinuities in temperature series. Int J Climatol 23:1087–1101. https://doi.org/10.1002/joc.924

Eltahir EAB, Bras RL (1996) Precipitation recycling. Rev Geophys 34(3):367–378. https://doi.org/10.1029/96rg01927

Eppinga MB, Rietkerk M, Borren W, Lapshina ED, Bleuten W, Wassen MJ (2008) Regular surface patterning of peatlands: confronting theory with field data. Ecosystems 11:520–536. https://doi.org/10.1007/s10021-008-9138-z

Erhardt T et al (2019) Decadal-scale progression of the onset of Dansgaard-Oeschger warming events. Clim past 15:811–825

Fisher JB et al (2020) ECOSTRESS: NASA's next generation mission to measure evapotranspiration from the international space station. Water Resour Res. https://doi.org/10.1029/2019WR026058

Flach M et al (2017) Multivariate anomaly detection for Earth observations: a comparison of algorithms and feature extraction techniques. Earth Syst Dynam 8:677–696. https://doi.org/10.5194/esd-8-677-2017

Flach M, Sebastian S, Fabian G, Ana B, Alexander B, Miguel DM (2018) Contrasting biosphere responses to hydrometeorological extremes: revisiting the 2010 western Russian heatwave. Biogeosciences 15:6067–6085. https://bg.copernicus.org/articles/15/6067/2018/bg-15-6067-2018.pdf

Frazier RJ, Coops NC, Wulder MA, Hermosilla T, White JC (2018) Analyzing spatial and temporal variability in short-term rates of post-fire vegetation return from Landsat time series. Remote Sens Environ 205:32–45. https://doi.org/10.1016/j.rse.2017.11.007

Freeman ME et al (2017) Defining the fire trap: extension of the persistence equilibrium model in mesic savannas. Austral Ecol 2017(42):890–899. https://doi.org/10.1111/aec.12516

Fu Z et al (2020) Sensitivity of gross primary productivity to climatic drivers during the summer drought of 2018 in Europe. Philos Trans Royal Soc b: Biol Sci 375(1810):20190747

Gallagher C et al (2013) Changepoint detection in climate time series with long-term trends. J Clim 26(14):4994–5006

Gaspard P, Nicolas G, Provata A (1995) Spectral signature of the pitchfork bifurcation: Liousville equation approach. Phys Rev E 51:74. https://doi.org/10.1103/PhysRevE.51.74

Gessler A et al (2020) The way back: recovery of trees from drought and its implication for acclimation. New Phytol 228(6):1704–1709

Good P et al (2016) Are strong fire–vegetation feedbacks needed to explain the spatial distribution of tropical tree cover? Glob Ecol Biogr 25(1):16–25. https://doi.org/10.1111/geb.12380

Goswami B et al (2018) Abrupt transitions in time series with uncertainties. Nat Commun. https://doi.org/10.1038/s41467-017-02456-6

Grimm V, Calabrese JM (2011) What is resilience? A short introduction. In: Deffuant G, Gilbert N (eds) In: viability and resilience of complex systems understanding complex systems. Springer, Heidelberg

Gritsun A, Branstator G (2007) Climate response using a three-dimensional operator based on the fluctuation–dissipation theorem. J Atmos Sci 64:2558–2575. https://doi.org/10.1175/JAS3943.1

Guttal V, Jayaprakash C (2008a) Changing skewness: an early warning signal of regime shifts in ecosystems. Ecol Lett 11(5):450–460. https://doi.org/10.1111/j.1461-0248.2008.01160.x

Guttal V, Jayaprakash C (2008b) Spatial variance and spatial skewness: leading indicators of regime shifts in spatial ecological systems. Theor Ecol 2:3–12. https://doi.org/10.1007/s12080-008-0033-1
Hanan NP et al (2014) Analysis of stable states in global savannas: is the CART pulling the horse? Glob Ecol Biogeogr 23(3):259–263. https://doi.org/10.1111/geb.12122
Helfgott AER (2015) Operationalizing resilience: conceptual, mathematical and participatory frameworks for understanding, measuring and managing resilience, Ph.D. thesis. Adelaide, SA: The University of Adelaide
Hansen MC et al (2013) High-resolution global maps of 21st-century forest cover change. Science 342(6160):850–853. https://doi.org/10.1126/science.1244693
Hansen WD et al (2022) Global forests are influenced by the legacies of past inter-annual temperature variability. Environ Res Ecol 1:011001. https://doi.org/10.1088/2752-664X/ac6e4a
Hastings A, Wysham DB (2010) (2010) Regime shifts in ecological systems can occur with no warning. Ecol Lett 13:464–472. https://doi.org/10.1111/j.1461-0248.2010.01439.x
Held H, Kleinen T (2004a) Detection of climate system bifurcations by degenerate fingerprinting. Geophys Res Lett 31:L23207. https://doi.org/10.1029/2004GL020972
Hirota M et al (2011) Global resilience of tropical forest and savanna to critical transitions. Science 334(6053):232–235. https://doi.org/10.1126/science.1210657
Hoffmann WA (2020) Better lucky than good: How savanna trees escape the fire trap in a variable world. Ecology 101(1):e02895. https://doi.org/10.1002/ecy.2895
Holling CS (1973) Resilience and stability of ecological systems. Annu Rev Ecol Syst 4:1–23. https://doi.org/10.1146/annurev.es.04.110173.000245
Holling CS (1996) Engineering within ecological constraints. Chapter: engineering resilience versus ecological resilience. National academies of sciences, engineering, and medicine. 1996. Engineering within ecological constraints. Washington, DC: The National Academies Press. https://doi.org/10.17226/4919
Ingrisch J, Bahn M (2018) Toward a comparable quantification of resilience. Trends Ecol Evol 33(4):251–259. https://doi.org/10.1016/j.tree.2018.01.013
Jamali S, Seaquist J, Eklundh L, Ardö J (2014) Automated mapping of vegetation trends with polynomials using NDVI imagery over the Sahel. Remote Sens Environ 141:79. https://doi.org/10.1016/j.rse.2013.10.019
Jentzsch A, White P (2019) A theory of pulse dynamics and disturbance in ecology. Ecology 100(7):e02734
Johnson EA, Miyanishi K (2008) Testing the assumptions of chronosequences in succession. Ecol Lett 11(5):419–431. https://doi.org/10.1111/j.1461-0248.2008.01173.x
Kéfi S, Dakos V, Scheffer M, van Nes E, Rietkerk M (2013) Early warning signals also precede non-catastrophic transitions. Oikos 112(5):641–648
Kéfi S, Guttal V, Brock WA, Carpenter SR, Ellison AM, Livina VN, Dakos V (2014) Early warning signals of ecological transitions: methods for spatial patterns. PLoS ONE 9(3):e92097. https://doi.org/10.1371/journal.pone.0092097
Kleinen T, Held H, Petschel-Held G (2003) The potential role of spectral properties in detecting thresholds in the Earth system: application to the thermohaline circulation. Ocean Dyn 53:53–63. https://doi.org/10.1007/s10236-002-0023-6
Knudby A et al (2014) Remote sensing and modeling of coral reef resilience. In: Finkl CW, Makowski C (eds) Remote sensing and modeling: advances in coastal and marine resources, coastal research library 9. Springer, Switzerland
Konings AG, Rao K, Steele-Dunne SC (2019) Macro to micro: microwave remote sensing of plant water 1085 content for physiology and ecology. New Phytol 223:1166–1172. https://doi.org/10.1111/nph.15808
Krakovská H, Kuehn C, Longo I (2023) Resilience of dynamical systems. Eur J Appl Math. https://doi.org/10.1017/S0956792523000141
Kubo R (1966) The fluctuation-dissipation theorem. Rep Prog Phys 1966(29):255. https://doi.org/10.1088/0034-4885/29/1/306
Kuehn C (2011) A mathematical framework for critical transitions: Bifurcations, fast–slow systems and stochastic dynamics. Physica D 240(12):1020–1035. https://doi.org/10.1016/j.physd.2011.02.012
Kutz JN, Brunton SL, Brunton BW, and Proctor JL (2016) Dynamic mode decomposition: data-driven modeling of complex systems. Society for Industrial and Applied Mathematics
Kuznetsov YA (2004) Introduction to dynamical systems. In: elements of applied bifurcation theory. Applied mathematical sciences, vol 112. Springer, New York, NY. https://doi.org/10.1007/978-1-4757-3978-7_

Lazante JR (1996) Resistant, robust and non-parametric techniques for the analysis of climate data: theory and examples, including applications to historical radiosonde station data. Int J Climatol 16(11):1197–1226. https://doi.org/10.1002/(SICI)1097-0088(199611)16:11%3c1197::AID-JOC89%3e3.0.CO;2-L
Lechner AM, Foody GM, Boyd DS (2020) Applications in remote sensing to forest ecology and management. One Earth 2(5):405–412. https://doi.org/10.1016/j.oneear.2020.05.001
Leith CE (1975) Climate response and fluctuation dissipation. J Atmos Sci 32:2022–2026
Lenton TM et al (2008) Tipping elements in the Earth's climate system. PNAS 105(6):1786–1793. https://doi.org/10.1073/pnas.0705414105
Lenton TM et al (2022) A resilience sensing system for the biosphere. Phil Trans Roy Soc 377(1857):20210383. https://doi.org/10.1098/rstb.2021.0383
Lenton TM, Abrams JF, Bartsch A et al (2024) Remotely sensing potential climate change tipping points across scales. Nat Commun 15:343. https://doi.org/10.1038/s41467-023-44609-w
Li X, Wigneron JP, Frappart F, Fan L, Ciais P, Fensholt R, Moisy C (2021) Global-scale assessment and inter-comparison of recently developed/reprocessed microwave satellite vegetation optical depth products. Remote Sens Environ 253:112208. https://doi.org/10.1016/j.rse.2020.112208
Li N, Sippel S, Winkler AJ, Mahecha MD, Reichstein M, Bastos A (2022) Interannual global carbon cycle variations linked to atmospheric circulation variability. Earth Syst Dynam 13:1505–1533. https://doi.org/10.5194/esd-13-1505-2022
Linscheid N, Mahecha MD, Rammig A, Carvalhais N, Gans F, Nelson JA et al (2021) Time-scale dependent relations between Earth observation based proxies of vegetation productivity. Geophys Res Lett 48:e2021GL093285. https://doi.org/10.1029/2021GL093285
Liu YY, de Jeu RAM, McCabe MF, Evans JP, van Dijk AIJM (2011) Global long-term passive microwave satellite-based retrievals of vegetation optical depth. Geophys Res Lett. https://doi.org/10.1029/2011GL048684
Liu S et al (2023) The overlooked contribution of trees outside forests to tree cover and woody biomass across Europe. Sci Adv. https://doi.org/10.1126/sciadv.adh4097
Lloret F, Keeling EG, Sala A (2011) Components of tree resilience: effects of successive low-growth episodes in old ponderosa pine forests. Oikos 120(12):1909–1920. https://doi.org/10.1111/j.1600-0706.2011.19372.x
Mack MC et al (2021) Carbon loss from boreal forest wildfires offset by increased dominance of deciduous trees. Science 372:280–283. https://doi.org/10.1126/science.abf3903
Marwan N, Donges JF, Zou Y, Donner RV, Kurths J (2009) Complex network approach for recurrence analysis of time series. Phys Lett A 373:4246–4254. https://doi.org/10.1016/j.physleta.2009.09.042
Masiliunas D, Tsendbazar N-E, Herold M, Verbesselt J (2021) BFAST lite: a lightweight break detection method for time series analysis. Remote Sens 13:3308. https://doi.org/10.3390/rs13163308
Matyasovszky I (2011) Detecting abrupt climate changes on different time scales. Theor Appl Climatol 105:445–454. https://doi.org/10.1007/s00704-011-0401-4
McDowell NG et al (2008) Mechanisms of plant survival and mortality during drought: why do some plants survive while others succumb to drought? New Phytol 178(4):719–739
McDowell NG et al (2016) Multi-scale predictions of massive conifer mortality due to chronic temperature rise. Nat Clim Change 6:295–300
McDowell NG et al (2020) Pervasive shifts in forest dynamics in a changing world. Science 368:eaaz9463. https://doi.org/10.1126/science.aaz9463
Milenković M, Reiche J, Armston J, Neuenschwander A, De Keersmaecker W, Herold M, Verbesselt J (2022) Assessing Amazon rainforest regrowth with GEDI and ICESat-2 data. Sci Remote Sens 5:100051. https://doi.org/10.1016/j.srs.2022.100051
Mitra C, Kurths J, Donner R (2015) An integrative quantifier of multistability in complex systems based on ecological resilience. Sci Rep 5:16196. https://doi.org/10.1038/srep16196
Moesinger L, Dorigo W, de Jeu R, van der Schalie R, Scanlon T, Teubner I, Forkel M (2020) The global long-term microwave vegetation optical depth climate archive (VODCA). Earth Syst Sci Data 12:177–196. https://doi.org/10.5194/essd-12-177-2020
Morresi D, Vitali A, Urbinati C, Garbarino M (2019) Forest spectral recovery and regeneration dynamics in stand-replacing wildfires of central Apennines derived from Landsat time series. Remote Sensing 11(3):308. https://doi.org/10.3390/rs11030308
Morton DC et al (2014) Amazon forests maintain consistent canopy structure and greenness during the dry season. Nature 506:221–224. https://doi.org/10.1038/nature13006
Mudelsee M (2000) Ramp function regression: a tool for quantifying climate transitions. Comput Geosci 26:293–307

Nikinmaa L, Lindner M, Cantarello E et al (2020) Reviewing the use of resilience concepts in forest sciences. Curr for Rep 6:61–80. https://doi.org/10.1007/s40725-020-00110-x

Nimmo DG, Mac Nally R, Cunningham SC, Haslem A, Bennett AF (2015) Vive la résistance: reviving resistance for 21st century conservation. Trends Ecol Evol 30(9):516–523. https://doi.org/10.1016/j.tree.2015.07.008

Oborny B, Meszena G, Szabo G (2005) Dynamics of populations on the verge of extinction. Oikos 109:291–296

Ogle K et al (2015) Quantifying ecological memory in plant and ecosystem processes. Ecol Lett 18(3):221–235

Özkundakci D, Lehmann MK (2019) Lake resilience: concept, observation and management. NZ J Mar Freshwat Res 53(4):481–488. https://doi.org/10.1080/00288330.2019.1647855

Perreault L, Bernier J, Bobée B, Parent E (2000) Bayesian change-point analysis in hydrometeorological time series. Part 1. The normal model revisited. J Hydrol 235:221–241

Pickett ST (1989) Space-for-time substitution as an alternative to long-term studies. Long-term studies in ecology: approaches and alternatives. Springer, Newyork, pp 110–135

Pimm SL (1984) The complexity and stability of ecosystems. Nature 307:321–326. https://doi.org/10.1038/307321a0

Potapov P et al (2021) Mapping global forest canopy height through integration of GEDI and Landsat data. Remote Sens Environ 253:112165. https://doi.org/10.1016/j.rse.2020.112165

Poyatos R et al (2013) Drought-induced defoliation and long periods of near-zero gas exchange play a key role in accentuating metabolic decline of Scots pine. New Phytol 2013(200):388–401. https://doi.org/10.1111/nph.12278

Preimesberger W, Scanlon T, Su C-H, Gruber A, Dorigo W (2021) Homogenization of structural breaks in the global ESA CCI soil moisture multisatellite climate data record. IEEE Trans Geosci Remote Sens 59(4):2845–2862. https://doi.org/10.1109/TGRS.2020.3012896

Reeves J et al (2007) A review and comparison of changepoint detection techniques for climate data. J Appl Met Clim 46:900–915

Riechers K, Boers N (2021) Significance of uncertain phasing between the onsets of stadial–interstadial transitions in different Greenland ice core proxies. Clim past 17(4):1751–1775. https://doi.org/10.5194/cp-17-1751-2021

Rodionov SN (2004) A sequential algorithm for testing climate regime shifts. Geophys Res Lett. https://doi.org/10.1029/2004GL019448

Rodionov SN (2006) Use of prewhitening in climate regime shift detection. Geophys Res Lett. https://doi.org/10.1029/2006GL025904

Rousseau DD et al (2017) (MIS3 & 2) millennial oscillations in Greenland dust and Eurasian aeolian records—a paleosol perspective. Quatern Sci Rev 169:99–113. https://doi.org/10.1016/j.quascirev.2017.05.020

Ruehr NK et al (2019) Beyond the extreme: recovery of carbon and water relations in woody plants following heat and drought stress. Tree Physiol 39:1285–1299. https://doi.org/10.1093/treephys/tpz032

Saatchi S (2019) sar methods for mapping and monitoring forest biomass. Ch. 5, SAR handbook, comprehensive methodologies for forest monitoring and biomass estimation, April

Saatchi S (2019) SAR methods for mapping and monitoring forest biomass. SAR handbook: comprehensive methodologies for forest monitoring and biomass estimation. Eds. Flores A, Herndon K, Thapa R, Cherrington E. NASA. 2019. https://doi.org/10.25966/hbml-ej07

Samanta A, Ganguly S, Vermote E, Nemani RR, Myneni RB (2012) Why is remote sensing of Amazon forest greenness so challenging? Earth Interact 16:1–14. https://doi.org/10.1175/2012EI440.1

Scheffer M, Jeppesen E (2007) Regime shifts in Shallow Lakes. Ecosystems 10:1–3

Scheffer M, Bascompte J, Brock WA et al (2009) Early-warning signals for critical transitions. Nature 461:53–59. https://doi.org/10.1038/nature08227

Schmidt L et al (2023) Assessing the sensitivity of multi-frequency passive microwave vegetation optical depth to vegetation properties. Biogeosciences 20:1027–1046. https://doi.org/10.5194/bg-20-1027-2023

Schütz N, Holschneider M (2011) Detection of trend changes in time series using Bayesian inference. Phys Rev E 84:021120. https://doi.org/10.1103/PhysRevE.84.021120

Seidl R, Turner MG (2022) Post-disturbance reorganization of forest ecosystems in a changing world. PNAS 119(28):e2202190119. https://doi.org/10.1073/pnas.220219011

Sellers PJ (1985) Canopy reflectance, photosynthesis and transpiration. Int J Remote Sens 6:1335–1372. https://doi.org/10.1080/01431168508948283

Senf C (2022) Seeing the system from above: the use and potential of remote sensing for studying ecosystem dynamics. Ecosystems 25:1719–1737. https://doi.org/10.1007/s10021-022-00777-2

Senf C, Seidl R (2021a) Mapping the forest disturbance regimes of Europe. Nat Sustain 4:63–70. https://doi.org/10.1038/s41893-020-00609-y
Senf C, Seidl R (2021b) Post-disturbance canopy recovery and the resilience of Europe's forests. Glob Ecol Biogeogr. https://doi.org/10.1111/geb.13406
Senf C, Müller J, Seidl R (2019) Post-disturbance recovery of forest cover and tree height differ with management in Central Europe. Landscape Ecol 34:2837–2850. https://doi.org/10.1007/s10980-019-00921-9
Senf C et al (2020) A generalized regression-based unmixing model for mapping forest cover fractions throughout three decades of Landsat data. Remote Sens Environ 240:111691
Smith T, Boers N (2023a) Global vegetation resilience linked to water availability and variability. Nat Commun 14:498. https://doi.org/10.1038/s41467-023-36207-7
Smith T, Boers N (2023b) Reliability of vegetation resilience estimates depends on biomass density. Nat Ecol Evol. https://doi.org/10.1038/s41559-023-02194-7
Smith T, Traxl D, Boers N (2022) Empirical evidence for recent global shifts in vegetation resilience. Nat Clim Chang 12:477–484. https://doi.org/10.1038/s41558-022-01352-2
Smith T, Zotta R-M, Boulton CA, Lenton TM, Dorigo W, Boers N (2023) Reliability of resilience estimation based on multi-instrument time series. Earth Syst Dynam 14:173–183. https://doi.org/10.5194/esd-14-173-2023
Sohn JA, Saha S, Bauhus J (2016) Potential of forest thinning to mitigate drought stress: a meta-analysis. For Ecol Manage 380:261–273
Solé RV, Manrubia SC, Luque B, Delgado J, Bascompte J (1996) Phase transitions and complex systems. Complexity 1:13–26. https://doi.org/10.1002/cplx.6130010405
Staver AC, Hansen MC (2015) Analysis of stable states in global savannas: is the CART pulling the horse?—a comment. Glob Ecol Biogeogr. https://doi.org/10.1111/geb.12285
Staver AC et al (2011) The global extent and determinants of savanna and forest as alternative biome states. Science 334(6053):230–232. https://doi.org/10.1126/science.1210465
Stritih A, Senf C, Seidl R (2023) Alternative states in the structure of mountain forests across the Alps and the role of disturbance and recovery. Landscape Ecol 38:933–947. https://doi.org/10.1007/s10980-023-01597-y
Tahsin S, Medeiros SC, Singh A (2018) Assessing the resilience of coastal wetlands to extreme hydrologic events using vegetation indices: a review. Remote Sens 10:1390
Tao S, Chave J, Frison PL, Le Toan T, Ciais P, Fang J, Saatchi S (2022) Increasing and widespread vulnerability of intact tropical rainforests to repeated droughts. Proc Natl Acad Sci 119(37):e2116626119
Thurm EA, Uhl E, Pretzsch H (2016) Mixture reduces climate sensitivity of Douglas-fir stem growth. For Ecol Manage 376:205–220. https://doi.org/10.1016/j.foreco.2016.06.020
Turner MG (2010) Disturbance and landscape dynamics in a changing world. Ecology. https://doi.org/10.1890/10-0097.1
van de Leemput I et al (2016) Multiple feedbacks and the prevalence of alternate stable states on coral reefs. Coral Reefs 2016(35):857–865. https://doi.org/10.1007/s00338-016-1439-7
van der Heide T (2011) Positive feedbacks in seagrass ecosystems—evidence from large-scale empirical data. PLoS ONE 6(1):e16504
van Kampen N (1981) Stochastic Processes in Physics and Chemistry. North Holland, New York
van Nes EH, Scheffer M (2007) Slow recovery from perturbations as a generic indicator of a nearby catastrophic shift. Am Nat 169:738–747. https://doi.org/10.1086/516845
van Meerbeek K, Jucker T, Svenning J-C (2021) Unifying the concepts of stability and resilience in ecology. J Ecol 109(9):3114–3132. https://doi.org/10.1111/1365-2745.13651
van Schaik E et al (2018) Changes in surface hydrology, soil moisture and gross primary production in the Amazon during the 2015/2016 El Niño. Phil Trans Roy Soc 373(1760):20180084
Veldhuis MP, Martinez-Garcia R, Deblauwe V, Dakos V (2022) Remotely-sensed slowing down in spatially patterned dryland ecosystems. Ecography 2022:e06139. https://doi.org/10.1111/ecog.06139
Verbesselt J et al (2010a) Detecting trend and seasonal changes in satellite image time series. Remote Sens Environ 114:106–115. https://doi.org/10.1016/j.rse.2009.08.014
Verbesselt J et al (2010b) Phenological change detection while accounting for abrupt and gradual trends in satellite image time series. Remote Sens Environ 114(12):2970–2980. https://doi.org/10.1016/j.rse.2010.08.003
Verbesselt J, Umlauf N, Hirota M, Holmgren M, van Nes EH, Herold M, Scheffer M (2016) Remotely sensed resilience of tropical forests. Nat Clim Change 6(11):1028–1031. https://doi.org/10.1038/nclimate3108
Viana-Soto A, Okujeni A, Pflugmacher D, García M, Aguado I, Hostert P (2022a) Quantifying post-fire shifts in woody-vegetation cover composition in Mediterranean pine forests using Landsat time series and regression-based unmixing. Remote Sens Environ 281:113239. https://doi.org/10.1016/j.rse.2022.113239

Viana-Soto A et al (2022b) Assessing post-fire forest structure recovery by combining LiDAR data and Landsat time series in Mediterranean pine forests. Int J Appl Earth Obs Geoinf 108:102754. https://doi.org/10.1016/j.jag.2022.102754
Wagner TJW, Eisenman I (2015) False alarms: how early warning signals falsely predict abrupt sea ice loss. Geophys Res Lett 42:10333–10341. https://doi.org/10.1002/2015GL066297
Wagner Y et al (2023) Relationships between xylem embolism and tree functioning during drought, recovery, and recurring drought in Aleppo pine. Phys Plant. https://doi.org/10.1111/ppl.13995
Walker B, Holling CS, Carpenter SR, Kinzig A (2004) Resilience, adaptability and transformability in social–ecological systems. Ecol Soc 9:5
Walter J et al (2013) Ecological stress memory and cross stress tolerance in plants in the face of climate extremes. Environ Exp Bot 94:3–8
Wang XL (2008) Accounting for autocorrelation in detecting mean shifts in climate data series using the penalized maximal t or F test. J Appl Meteorol Climatol 47(9):2423–2444. https://doi.org/10.1175/2008JAMC1741.1
Weinans E, Lever JJ, Bathiany S, Quax R, Bascompte J, van Nes EH, van de Leemput IA (2019) Finding the direction of lowest resilience in multivariate complex systems. J Royal Soc Interface 16(159):20190629
Weinans E et al (2021) Evaluating the performance of multivariate indicators of resilience loss. Sci Rep. https://doi.org/10.1038/s41598-021-87839-y
White PS, Picket STA (1985) Natural disturbances and patch dynamics: an introduction. In: Pickett STA, White PS (eds) The ecology of natural disturbance and patch dynamics. Academic Press, San Diego, pp 3–13
White JC et al (2017) A nationwide annual characterization of 25 years of forest disturbance and recovery for Canada using Landsat time series. Remote Sens Environ 194:303–321. https://doi.org/10.1016/j.rse.2017.03.035
White JC et al (2019) Assessing spectral measures of post-harvest forest recovery with field plot data. Int J Appl Earth Obs Geoinf 80:102–114. https://doi.org/10.1016/j.jag.2019.04.010
White JC et al (2022) Mapping, validating, and interpreting spatio-temporal trends in post-disturbance forest recovery. Remote Sens Environ 271:11904. https://doi.org/10.1016/j.rse.2022.112904
White JC, Hermosilla T, Wulder MA (2023) Pre-fire measures of boreal forest structure and composition inform interpretation of post-fire spectral recovery rates. For Ecol Manage 537:120948
Wiesenfeld K (1985) Noisy precursors of nonlinear instabilities. J Stat Phys 38:1071–1097. https://doi.org/10.1007/BF01010430
Wiesenfeld K, McNamara B (1986) Small-signal amplification in bifurcating dynamical systems. Phys Rev A 33:629–642. https://doi.org/10.1103/PhysRevA.33.629
Wissel C (1984) A universal law of the characteristic return time near thresholds. Oecologia 65:101–107. https://doi.org/10.1007/BF00384470
Wu M et al (2022) Drought legacy in sub-seasonal vegetation state and sensitivity to climate over the northern hemisphere. Geophys Res Lett. https://doi.org/10.1029/2022GL098700
Wulder MA et al (2019) Current status of Landsat program, science, and applications. Remote Sens Environ 225:127–147. https://doi.org/10.1016/j.rse.2019.02.015
Wuyts B et al (2017) Amazonian forest-savanna bistability and human impact. Nat Commun 8:15519. https://doi.org/10.1038/ncomms15519
Xiao C, Zaehle S, Yang H, Wigneron J-P, Schmullius C, Bastos A (2023) Land cover and management effects on ecosystem resistance to drought stress. Earth Syst Dynam 14:1211–1237. https://doi.org/10.5194/esd-14-1211-2023
Xiong X et al (2020) MODIS and VIIRS calibration and characterization in support of producing long-term high-quality data products. Remote Sens 12(19):3167
Yang X et al (2022) Detection and characterization of coastal tidal wetland change in the northeastern US using Landsat time series. Remote Sens Env 276:113047
Yi C, Jackson N (2021) A review of measuring ecosystem resilience to disturbance. Environ Res Lett 16:053008
Yin Z, Dekker SC, van den Hurk BJJM, Dijkstra HA (2014) Bimodality of woody cover and biomass across the precipitation gradient in West Africa. Earth Syst Dynam 5(257–270):2014. https://doi.org/10.5194/esd-5-257-2014
Yu X et al (2022) Contrasting drought legacy effects on gross primary productivity in a mixed versus pure beech forest. Biogeosci Discus 19:4315–4329
Zeileis A et al (2003) Testing and dating of structural changes in practice. Comp Stat Data Anal 44:109–123
Zhang J et al (2022) NIRv and SIF better estimate phenology than NDVI and EVI: Effects of spring and autumn phenology on ecosystem production of planted forests. Agric Forest Meteorol 315(15):108819. https://doi.org/10.1016/j.agrformet.2022.108819
Zhao Y, Li Y (2022) Impact of fisheries footprint on an early warning indicator of resilience reduction in marine net primary productivity. ICES J Mar Sci 2022(79):2741–2751. https://doi.org/10.1093/icesjms/fsac213

Zhu Z et al (2017) The effects of teleconnections on carbon fluxes of global terrestrial ecosystems. Geophys Res Lett 44(7):3209–3218. https://doi.org/10.1002/2016GL071743

Authors and Affiliations

Sebastian Bathiany[1,2] · Robbin Bastiaansen[3,4] · Ana Bastos[5] · Lana Blaschke[1,2] · Jelle Lever[6,7] · Sina Loriani[2] · Wanda De Keersmaecker[8] · Wouter Dorigo[9] · Milutin Milenković[10] · Cornelius Senf[11] · Taylor Smith[12] · Jan Verbesselt[13,14] · Niklas Boers[1,2,15]

✉ Sebastian Bathiany
sebastian.bathiany@tum.de

1 Earth System Modeling, School of Engineering and Design, Technical University of Munich, 85521 Ottobrunn, Germany

2 Potsdam Institute for Climate Impact Research, Telegrafenberg A 31, 14473 Potsdam, Germany

3 Department of Physics, Institute for Marine and Atmospheric Research Utrecht (IMAU), Utrecht University, Heidelberglaan 8, 3584 CS Utrecht, The Netherlands

4 Department of Mathematics, Utrecht University, Heidelberglaan 8, 3584 CS Utrecht, The Netherlands

5 Max Planck Institute for Biogeochemistry, Hans-Knöll-Straße 10, 07745 Jena, Germany

6 Swiss Federal Institute for Forest, Snow and Landscape Research (WSL), Zürcherstrasse 111, 8903 Birmensdorf, Switzerland

7 Swiss Federal Institute of Aquatic Science and Technology (Eawag), Überlandstrasse 133, 8600 Dübendorf, Switzerland

8 Vlaamse Instelling Voor Technologisch Onderzoek (VITO) Research Organization, Boeretang 200, 2400 Mol, Belgium

9 Department of Geodesy and Geoinformation, TU Wien, Wiedner Hauptstraße 8-10, 1040 Vienna, Austria

10 Novel Data Ecosystems for Sustainability (NODES), International Institute for Applied Systems Analysis (IIASA), Schlossplatz 1, 2361 Laxenburg, Austria

11 Department of Life Science Systems, Chair of Ecosystem Dynamics and Forest Management, TUM School of Life Sciences, Technical University of Munich, Hans-Carl-von-Carlowitz-Platz 2, 85354 Freising, Germany

12 Institute of Geosciences, Universität Potsdam, Karl-Liebknecht-Str. 24-25, 14476 Potsdam, Germany

13 Belgian Science Policy Office (BELSPO), Simon Bolivarlaan 30 bus 7 Boulevard Simon Bolivar 30 bte 7, 1000 Brussels, Belgium

14 Laboratory of Geo-Information Science and Remote Sensing, Wageningen University and Research, Droevendaalsesteeg 4, 6708 PB Wageningen, The Netherlands

15 Department of Mathematics and Global Systems Institute, University of Exeter, North Park Road, Exeter EX4 4QE, UK

Surveys in Geophysics (2025) 46:303–326
https://doi.org/10.1007/s10712-025-09885-9

Permafrost and Freshwater Systems in the Arctic as Tipping Elements of the Climate System

Victor Brovkin[1,2] · Annett Bartsch[3] · Gustaf Hugelius[4] · Elisa Calamita[5] · J. Jelle Lever[5,6] · Eunhan Goo[7] · Hyungjun Kim[7] · Tobias Stacke[1] · Philipp de Vrese[1]

Received: 17 December 2023 / Accepted: 4 April 2025 / Published online: 2 May 2025

Abstract
The Arctic is warming several times faster than the rest of the globe. Such Arctic amplification rapidly changes hydrometeorological conditions with consequences for the structuring of cold-adapted terrestrial and aquatic ecosystems. Arctic ecosystems, which have a relatively small buffering capacity, are particularly susceptible to hydrometeorological regime shifts thus frequently undergo system-scale transitions. Abrupt ecosystem changes are often triggered by disturbances and extreme events that shift the ecosystem state beyond its buffering threshold capacity thus irreversibly changing its functioning (ecosystem tipping). The tipping depends on spatial and temporal scales. At the local scale, feedback between soil organic matter and soil physics could lead to multiple steady states and a tipping from high to low soil carbon storages. On the continental scale, local tipping is smoothed and the changes are rather gradual (no clear tipping threshold). However, due to the centennial timescale of soil carbon and vegetation dynamics, Arctic ecosystems are not in equilibrium with the changing climate, so a tipping could occur at a later time. Earth Observation (EO) is useful for monitoring ongoing changes in permafrost and freshwater systems, in particular extreme events and disturbances, as indicators of a possible tipping point. Lake change observations support gradual rather than abrupt transitions in different permafrost regions until a hydrological tipping point where lake areas start to decline leading to regional drying. Due to floodplain abundance, floodplains should be considered separately when using satellite-derived water extent records to analyse potential tipping behaviour associated with lakes. Reduction in surface water extent, increasing autocorrelation of water level of larger lakes and the impact of extreme events on ground ice can all be observed with satellite data across the Arctic. The analysis of Earth System simulations suggests significant impacts of changes in permafrost hydrology on hydroclimate in the tropics and subtropics, but there is no clear threshold in global temperature for these shifts in hydroclimate.

Keywords Permafrost · Lake · Hydrology · Earth observation · Modelling · Feedback · Tipping point

Extended author information available on the last page of the article

Article Highlights

- The presence of multiple steady states in permafrost systems suggests the possibility of local tipping of ecosystems and soil carbon storages, albeit at centennial rather than decadal scales
- Earth Observation (EO) of extreme events and disturbances is useful for monitoring ongoing changes in permafrost and freshwater systems as indicators of a possible tipping point
- The area of EO-derived open water in the Arctic both increases and decreases with a tendency of the net effect being towards a smaller area. For conclusions regarding thaw lake dynamics and tipping, floodplains should be considered separately
- An analysis of timeseries of lake water levels suggests that lake hydrology is shifting state within permafrost regions
- Earth System simulations suggest changes in permafrost hydrology can have significant effects on the hydroclimate in the tropics and subtropics, but the changes are gradual rather than abrupt

1 Introduction

Permafrost is an important component of the cryosphere. It is formally defined as ground (soil or rock) frozen for at least two consecutive years. Permafrost underlies about 14 million km^2 in the Northern Hemisphere (Obu 2021) with its largest areas in Russia, Canada, the USA, and China (Sun et al. 2018; Wu et al. 2021; Zhao et al. 2017). The deepest permafrost exists in Central Siberia, where it is more than a kilometre thick (Yershov 1990) because that region was not glaciated during the last glacial cycle (Lindgren et al. 2016) and therefore exposed to extremely cold temperatures at the surface. Cold and frozen conditions in permafrost-affected soils slow down the decomposition of soil organic matter. Organic matter accumulates in the upper layer and remains undecomposed for millennia (van Huissteden 2020). Permafrost-affected soils and sediments of the Northern Hemisphere contain 1,100 to 1500 Pg of organic carbon (Hugelius et al. 2014), about twice as much as the pre-industrial atmosphere in 1850. This soil organic matter is strongly affected by anthropogenic warming and could be released to the atmosphere as CO_2 or CH_4, well-mixed greenhouse gases (GHGs), amplifying climate change (permafrost carbon feedback, Schuur et al. 2015). To date the Arctic has warmed almost four times faster than the rest of the globe (Rantanen et al. 2022). While at present the net permafrost carbon fluxes are not very pronounced and the net permafrost region CO_2 budget is near neutral (Bruhwiler et al. 2021; Hugelius et al. 2024), this could change in the future as the current rate of warming and thaw is unprecedented in the last few thousand years (Schuur et al. 2022). Re-formation of thawed permafrost carbon is such a slow process that it makes these changes practically irreversible on human (multidecadal) timescales (de Vrese et al. 2023a). This is why a number of recent studies proposed the permafrost system as a tipping element in the Earth System (Armstrong McKay et al. 2022; Wang et al. 2023). In this paper, we focus on 1) quantitative characteristics of permafrost and freshwater systems (such as carbon or ice content, lake area) which are susceptible to hydrometeorological regime shifts thus which frequently undergo system-scale transitions; and 2) observations of these characteristics using remote sensing or modelling using process-based understanding. We follow the recent IPCC definitions of tipping element, irreversibility, and abruptness of climate changes (Box 1). These definitions are rather broad. For example, a tipping point could be interpreted as a point in time after which the system reacts much faster than before to an external forcing (abrupt change), or as a shift in the ecosystem state towards a different mode of functioning, which may be slow but irreversible (ecosystem tipping). In this paper, a tipping element is characterised by a large spatial scale that is visible on a planetary level (Armstrong McKay et al. 2022; Wang et al. 2023). Our preferred definition of tipping includes nonlinearity in response to environmental changes and/or irreversibility of changes on human (multidecadal) timescales.

- The presence of multiple steady states in permafrost systems suggests the possibility of local tipping of ecosystems and soil carbon storages, albeit at centennial rather than decadal scales
- Earth Observation (EO) of extreme events and disturbances is useful for monitoring ongoing changes in permafrost and freshwater systems as indicators of a possible tipping point
- The area of EO-derived open water in the Arctic both increases and decreases with a tendency of the net effect being towards a smaller area. For conclusions regarding thaw lake dynamics and tipping, floodplains should be considered separately
- An analysis of timeseries of lake water levels suggests that lake hydrology is shifting state within permafrost regions
- Earth System simulations suggest changes in permafrost hydrology can have significant effects on the hydroclimate in the tropics and subtropics, but the changes are gradual rather than abrupt

Box 1 Definitions of tipping terms after AR6 WGI Glossary (IPCC, 2021).

Tipping Element: A component of the Earth System that is susceptible to a Tipping Point.
Tipping Point: A critical threshold beyond which a system reorganises, often abruptly and/or irreversibly.
Abrupt climate change: A large-scale abrupt change in the climate system that takes place over a few decades or less, persists (or is anticipated to persist) for at least a few decades and causes substantial impacts in human and/or natural systems.
Irreversibility: A perturbed state of a dynamical system is defined as irreversible on a given timescale, if the recovery from this state due to natural processes takes substantially longer than the timescale of interest.

One argument that supports permafrost as a tipping element comes from high-resolution ice-core records. During the last deglaciation, the atmospheric CO_2 concentration increased abruptly 3 times with amplitudes of 10–12 ppm in each case (Marcott et al. 2014). These changes on the centennial scale occurred simultaneously to large warming events in the Earth system, presumably due to reinvigoration of the Atlantic Meridional Overturning or AMOC (Brovkin et al. 2021). The estimated amount of carbon released during each of these events is about 50 PgC (Kohler et al. 2014). The timescale of abrupt carbon increase points to the land source rather than the ocean which has a larger storage capacity but a slower response. At the peak of the Last Glacial Maximum (LGM, around 21 thousand years ago) the northern permafrost region was about 13 million km^2 larger than today, including 5 million km^2 on what are today shallow polar sea shelves (Lindgren et al. 2016). Lindgren et al. (2018) estimated that the total soil carbon stock in the LGM permafrost region was ca. 2400 PgC. Importantly, the LGM carbon stocks were dominated by permanently frozen carbon, while the present soil carbon is predominantly stored in seasonally frozen soils. The reconstructions support postglacial reductions in the permafrost carbon stock of around 1000 PgC (Lindgren et al. 2018). The abrupt release due to the simultaneous thawing of carbon frozen in the permafrost in response to abrupt warming is a plausible explanation of the geological record, indicating that it cannot be excluded in the future.

In the Arctic, organic carbon processing is controlled not only by temperature but also by hydrology with increase in GHG emissions from the inland freshwaters (lakes and rivers) (Ramage et al. 2024). The fractional area of lakes in the high latitudes of the Northern hemisphere is higher than elsewhere on the globe (Lehner and Döll 2004) though surface water extent has been decreasing across the Arctic (Webb et al. 2022). Many of these lakes are so-called thermokarst lakes (Grosse 2013), formed when the melting of massive ground ice bodies causes surface subsidence and formation of lakes in the depressions. These water bodies may increase the transport of thermal energy into the permafrost, leading to formation of year-round unfrozen ground layers (taliks) below these lakes (Burn 2002) as well as lateral erosion of lake shores (Arp et al. 2011). Significant areas of the high northern latitudes are covered by former thermokarst lake basins (Bergstedt et al. 2021; Grosse 2013; Jones et al. 2022).

The outgassing of CO_2 and CH_4 from lakes represents a significant component of the global carbon cycle (Lauerwald 2023b). Lakes in permafrost regions contribute to these emissions (Serikova et al. 2019; Wik et al. 2016). The organic carbon released from thawing permafrost is partly labile and thus it can be mineralised and emitted from these lakes to the atmosphere in the form of CO_2 and CH_4. A small fraction of the organic carbon from thawing permafrost is released into rivers and, consequently, into the ocean. About 15–20% of dissolved and particulate organic matter (DOC and POC) in Arctic freshwater systems originates from permafrost or peat (Vonk et al. 2015; Wild et al. 2019). Seasonal changes in the direction of flow may be of importance, and while the Arctic Ocean constitutes 1% of the world's ocean volume, recent estimates indicate that the Arctic Ocean receives 13% of the terrigenous DOC load delivered to the world's oceans (Liu et al. 2024; Raymond et al. 2007; Stein and Macdonald 2004), which is proportional to the pan-Arctic watershed fraction of 14% of the global land area (McGuire et al. 2009).

In addition to the land permafrost area, about 2–2.75 million km^2 of permafrost is present on the Arctic shelf seafloor; this permafrost was submerged over the last 18 thousand years during and following the deglaciation. This subaquatic permafrost stores another 2800 (1518–4982) PgC (Miesner et al. 2023) in addition to that stored in terrestrial permafrost. Subaquatic permafrost temperatures are higher than their terrestrial counterparts while the sea water temperatures are around 0 °C almost the whole year. This makes thawing at the top of subsea permafrost extremely slow, with timescales being from centuries to millennia, while the main thawing occurs at the bottom of the permafrost column due to geothermal heat. Since the next glacial period is expected to be at least 50,000 years from now (Archer and Ganopolski 2005; Berger and Loutre 2002), the subsea permafrost is expected to thaw completely within the next few millennia (Wilkenskjeld et al. 2022). Most of organic matter of terrestrial and marine origin in the subsea permafrost will be decomposed and transported as CO_2 or CH_4 through the sediment layer to the ocean water column on a millennial timescale (Puglini et al. 2020) and then to the atmosphere, amplifying anthropogenic warming.

The paper is structured as follows. Section 2 describes the changes in the Arctic permafrost and freshwater systems associated with tipping. In Sect. 2.1, we discuss abrupt changes, nonlinearity and the irreversibility of permafrost thawing processes and hypothesise that abrupt hydrological changes could lead to a local tipping point (nonlinearity) in the transient response, but will not propagate to a nonlinear instantaneous response at the pan-Arctic scale. In Sect. 2.2, we discuss observations of extreme weather events that could lead to abrupt permafrost changes. In Sect. 2.3, we continue by analysing trends in lake characteristics in permafrost regions, noting that high losses of water area are often associated with floodplains. In Sect. 2.4, we analyse early warning signals in the permafrost lake level time series and hypothesise about signals of regime shifts in lake hydrology. In the Sect. 3, we discuss a possible future shift in global hydroclimate in response to hydrological changes in the permafrost region, that could affect other Earth tipping elements such as the Amazon rainforest and the vegetation of the Sahel. We conclude in the Sect. 4.

2 Changes in Permafrost and Freshwater Systems in the Arctic

2.1 Abrupt Changes, Nonlinearity, and Irreversibility of Permafrost Thaw Processes

Permafrost thawing processes occur at several different scales, either as a landscape-wide successive increase in seasonal surface thaw (active layer deepening) or through localised abrupt surface collapse (thermokarst) caused by the melting of excess ground ice (ice that exceeds the volume of soil pore space). Thermokarst often leads to the formation of distinct landform types, including different types of thaw lakes, post-thaw peatlands and thaw slumps on hill slopes. Because permafrost that is rich in ground ice is often simultaneously rich in organic carbon, thermokarst-prone terrain holds about half of the permafrost carbon stock despite covering only 20% of the permafrost region surface area (Olefeldt et al. 2016). Thermokarst formation via abrupt thaw processes, triggered by loss of ground ice, are not yet included in global or regional scale models, but can lead to significant increases in post-thaw greenhouse gas fluxes (Turetsky et al. 2020). Under high emission scenarios, abrupt thaw carbon losses are projected to account for approximately 40% of the mean net emissions attributed to widespread gradual thaw by 2300, even if they affect only a small surface area (Turetsky et al. 2020). At policy relevant timescales, abrupt thaw processes can also be considered irreversible, as the loss of ground ice and the surface deformation it causes is (semi)permanent even if the permafrost should refreeze. It can thus be considered as a local-scale tipping element within the permafrost system.

To identify tipping points in permafrost carbon dynamics, we search for nonlinearities in the permafrost system response to changes in environmental conditions. Because the permafrost thaw and carbon decomposition combine fast and slow processes, we need to account for a range of different timescales. The carbon dynamics are different for transient responses, such as simulated by Earth System models in future scenarios with a defined end point, e.g. the year 2100 (Nitzbon et al. 2024), and for committed changes occurring in a quasi-equilibrium response arrived at after a substantial delay (Fig. 1). Areas with larger carbon storage will have larger carbon decomposition on multidecadal to centennial timescales. Therefore, our hypothesis is that the equilibrium response is more nonlinear than the transient one. Note that the existence of nonlinearities doesn't imply the existence of multiple states in the permafrost carbon system.

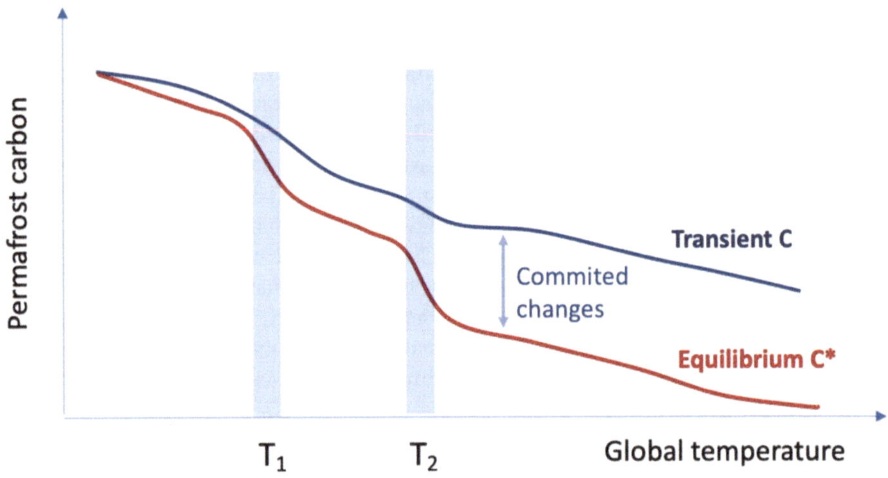

Fig. 1 Conceptual plot of permafrost carbon storage as a function of global temperature. C* is the equilibrium carbon storage for a given global temperature, C is the time-dependent storage (at the time of reaching T), C–C* is the committed carbon thaw. Due to spatially heterogeneous climate changes and permafrost carbon storages, there are hypothetical nonlinearities at the tipping points T_1 and T_2 which could be monitored by Earth Observation (EO)

Irreversibility of changes in permafrost and freshwater systems depends on timescale. Processes in permafrost-affected soils exhibit a high level of inertia in their response to changing climate conditions and many may not be reversible or even stoppable on centennial timescales (Wang et al. 2023). Rising carbon emissions from recently thawed ground are a prominent example of such committed changes (Comyn-Platt et al. 2018; Gasser et al. 2018; Natali et al. 2021). These do not only notably reduce the remaining carbon budget of the more ambitious climate goals but may even hold the potential to lead to a self-sustaining warming after all man-made greenhouse gas emissions have ceased (Randers and Goluke 2021). One reason for this inertia are the extremely long carbon turnover times which imply that it could take centuries for soil carbon uptake and emissions to reach a new steady state (Shu et al. 2020; Wu et al. 2021). The same is true for the physical state of the subsurface, where the large amounts of energy either required for or released in the phase change of water could mean that trends in soil temperatures and the soil water and ice content continue long after the climate has stabilised (de Vrese and Brovkin 2021; Eliseev et al. 2014). This latter point is also highly relevant for the surface hydrology as many of the lakes and wetlands in permafrost-affected regions depend on microtopographic features caused by ice wedges. The continued thawing of ground ice could be leading to subsidence and changes in topography (Aas et al. 2019; Ekici et al. 2019; Nitzbon et al. 2021) and thus the current trends in lake and wetland cover (Webb and Liljedahl 2023). These trends could continue even after the climate stabilises due to the hysteresis of the hydrological response to climate change (de Vrese and Brovkin 2021).

A number of these change processes could be halted if global heating and the temperature trend was partially reversed in a temperature overshoot scenario (Schwinger et al. 2022). However, the warming during the overshoot temporarily allows the decomposition of formerly frozen organic matter, thus, reducing the soil carbon content. Given that soil carbon stores and excess ice have built up over thousands of years, a reversal of the impacts of a temporary warming may still take centuries if not millennia.

In addition, simulations with land surface models including permafrost indicate that parts of the Arctic and subarctic zone may exhibit different stable steady states depending on the soil organic matter concentration at the time of climate stabilisation (de Vrese and Brovkin 2021). Therefore, even if global temperatures return to pre-industrial levels, some land areas may never return to the soil carbon stocks that existed before anthropogenic warming.

The separation between instantaneous and committed response is typical for systems with both short and long response timescales such as permafrost. In particular, the hydrological response of the land surface on local scale could be fast. With warmer and longer summers, the hydrology of the permafrost regions is changing. Ground ice is melting, new thermokarst lakes are forming, while older lakes drain and disappear (Grosse, 2013). Our working hypothesis is that abrupt hydrological changes could lead to a local tipping point (nonlinearity) in the transient response, but it will not propagate to a nonlinear instantaneous response on the pan-Arctic scale.

2.2 Extreme Weather Events as Drivers of Abrupt Permafrost Changes

With increasing warming, extreme events and disturbances such as heatwaves, floods, fires and rain-on-snow are occurring more often (Romanou et al. 2024). All extremes potentially contribute to permafrost thaw and the mobilisation of carbon. The ground warms following rain-on-snow events (Westermann et al. 2011) while ground ice content is altered, for example, after flooding (Zwieback et al. 2023) or fires (Michaelides et al. 2019). However, extremes effect on land carbon components including vegetation biomass, soil organic carbon, and sediments in freshwater systems is not well quantified. EO through use of satellite data can potentially support the identification and quantifications of the impact of extremes and eventually allow the estimation of the scale of carbon and climate consequences of extremes.

For example, heatwaves have been identified as a trigger for retrogressive thaw slumps (RTS) across the Arctic including in Canada (Jones et al. 2019; Kokelj et al. 2017; Lewkowicz and Way 2019), Alaska (Balser et al. 2014) and in Russia (Babkina et al. 2019; Bernhard et al. 2022; Kizyakov, 2013; Runge et al. 2022). RTS are a thermokarst feature that are a consequence of melting ground ice (also called cryogenic landslide (Burn and Lewkowicz 1990)) and are classified as abrupt permafrost thaw features (Turetsky et al. 2020). Melting ice lenses or ice wedges at the base of the active layer leads to ground collapse and eventually slope failure. In-situ observations for selected thaw slumps are available and sufficient data to describe the relation to unusually warm summers can be obtained through satellite data (Lewkowicz and Way 2019) primarily using Landsat data with 30 m spatial resolution (Fig. 2). RTS formation leads to vegetation removal which allows identification using the multispectral observations. For example, more than 4000 thaw slumps have been initiated since 1984 in an area of 70.000 km^2 (Lewkowicz and Way 2019), covering an area of 64 km^2 according to analyses of Landsat data. Detection capability has been recently significantly improved with the availability of Sentinel-2 with 10 m nominal resolution (Lissak et al. 2020). However, even better resolution of fine detail including topographic changes would be required for precise quantification of the amount of soil which is mobilised by RTS.

Permafrost degradation leading to loss of ground ice after extreme events such as heatwaves, flood and fires could be measured from space. Subtle changes in the order of centimetres can be captured through Synthetic Aperture Radar (SAR) data. Increased subsidence in anomalously warm summers was for example described by Bartsch et al.

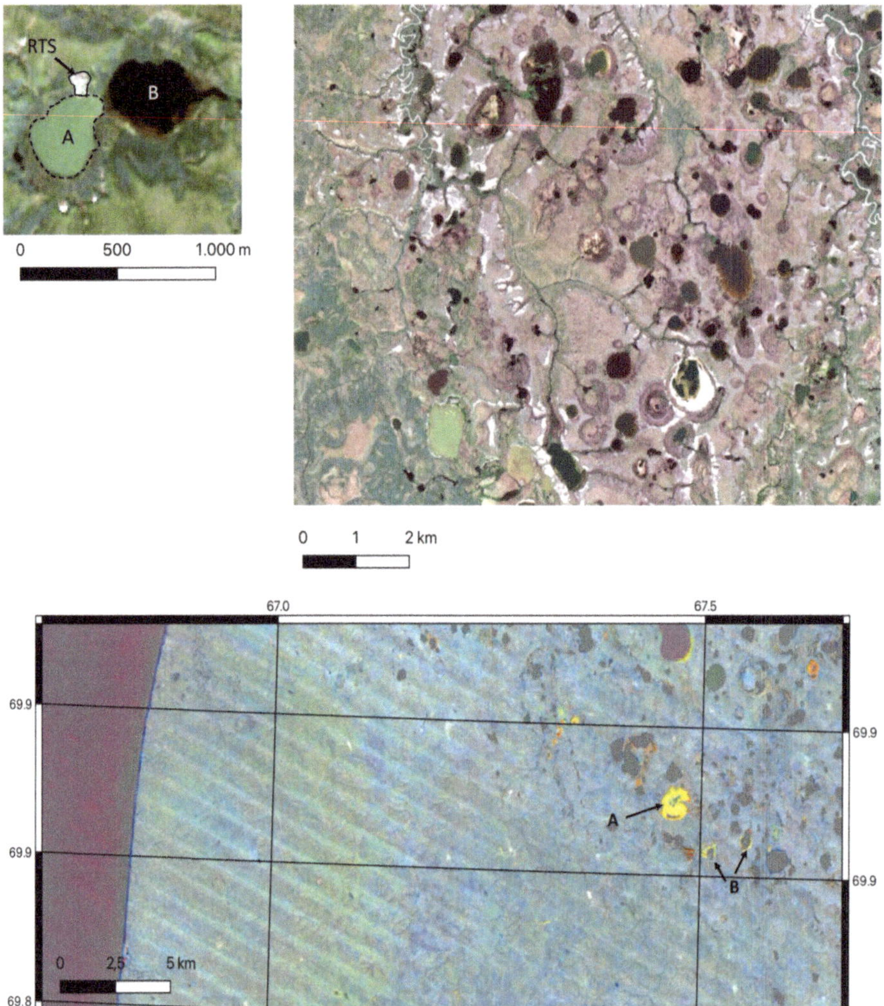

Fig. 2 Remote sensing images of abrupt thaw. **Top**: Subsets of a red–green–blue composite of an Copernicus Sentinel-2 image from central Yamal on 23.07.2017: *top left* – Example of active thaw slump (RTS, bright feature, boundary indicated through black solid line), with impact on lake colour (soil material in suspension in left lake, boundary indicated through black dashed line), compared to neighbouring lake. *Right* – Clusters of thaw lakes in different states of drainage. **Bottom**: Landsat trend visualisation for central Yamal (1999–2014, source:Nitze 2018) – the blue band north to south on the left side indicates coastal erosion, the yellow patch and circular features on the right side indicate lake drainage (A – almost complete, B – partial shrinkage with exposure at lake margins)

(2019) for central Yamal in Russia. The magnitude of the increase differed depending on soil types. Specifically the melting of tabular ground ice at the base of the active layer has been suggested to be related to warm extremes. Zwieback et al. (2021) studied this for a case on the Alaskan North Slope using actual ground ice in-situ data. Subsidence can be also triggered by flooding events (Zwieback et al. 2023). Wildfires remove vegetation and part of the surface organic layer, deepen the active layer, thaw the ground ice and increase seasonal subsidence (Michaelides et al. 2019). An increased subsidence was observed using the SAR data for more than a decade after burn activity (Michaelides et al. 2019). Overall, ground subsidence monitoring can capture the consequences of weather extremes but spatially and temporally consistent records, which are basic requirements for tipping analyses, are not available to date (Bartsch et al. 2023a). Subsidence monitoring may nevertheless support regional studies of hydrological changes and associated carbon flux alterations, but so far, most investigations in permafrost environments have focussed on technical retrieval challenges (Widhalm et al. 2025).

Gas emission craters, which are associated with warm years, have been shown to transform into lakes. Large craters are known from the Yamal Peninsula (e.g. Kizyakov, 2015). Monitoring of their formation process requires highly detailed terrain information. Pingo-like mounds form beforehand, which have been documented with repeat stereo photogrammetry (Kizyakov 2015). Lake formation and further growth as a consequence of such craters can be quantified with satellite data. Terrain and landcover changes can be however also associated with other processes and the attribution of their cause is challenging.

2.3 Changes of Lakes Properties in Permafrost Regions

In general, several physical, chemical or biological processes that may pave the way for abrupt transitions to alternative states can be identified for lakes (Calamita et al. 2024). Some of these shifts may, in turn, have strong feedbacks to the climate system. An example of such a regime shift is the abrupt change in lake water level which affects lake extent and potentially leads to lake shrinking (Zhao et al. 2017) or lake formation (Shugar et al. 2020; Stuart-Smith et al. 2021; Sun et al. 2018). Permafrost thaw, accelerating due to climate change, can create or expand thermokarst lakes with impacts on hydrology, ecology and biogeochemistry (in 't Zandt et al. 2020). In particular, carbon stored in lake sediments could be decomposed leading to the release of both CO_2 and CH_4 (in 't Zandt et al. 2020). Thus, climate change, by increasing the length of the ice-free season increases the annual water body GHG emissions at high latitudes (Wik et al. 2016). However, the quantification of GHG emissions from lakes is hampered by high uncertainties due to limited spatial and temporal coverage of observations and uncertainties in the quantification of inland water surface areas (Lauerwald et al. 2023a), especially in the case of thermokarst lakes (Olefeldt et al. 2016). While the fluxes can be estimated by data-driven estimates of the gas budget of the permafrost region (Ramage et al. 2024), the feedback of such inland waters on the climate system is not yet represented in Earth System Models (Turetsky et al. 2020).

EO has only been used to a limited extent for detecting and studying tipping points and regime shifts in lakes, especially compared to in-situ measurements, paleolimnological records or models (Calamita et al. 2024). So far, EO data has been primarily applied to detecting abrupt changes in lake water extent (Calamita et al. 2024). However, EO offers significant advantages for monitoring inland waters, serving as a strategic tool for assessing and tracking lake quality worldwide through frequent, large-scale surveys (Giardino et al. 2014). While most key variables for monitoring tipping systems are classified as Essential Climate Variables (ECVs) of the Global Climate Observing System (GCOS), some critical lake properties—such as particulate organic carbon and chlorophyll-a—remain unaccounted for (Loriani et al. 2025).

A range of remotely sensed datasets has been published recently that allow the assessment of land surface hydrology changes in permafrost regions, specifically lakes.

However, no dataset is currently available which can address changes in permafrost thaw lakes comprehensively. Current climate data records are limited to large lakes or are not specifically for lakes. Both these issues can be attributed to the lack of spatial resolution (Bartsch et al. 2023b). Parameter retrievals include lake surface temperatures (e.g. LakeCCI (v2.0.1, Carrea, 2022; Carrea et al. 2023)), lake water level (GLWS v1.1, Yao et al. 2023) and change in fractional water area (e.g. by Webb et al. (2022)). Such records allow the determination of trends and in some cases application of techniques for identification of tipping, such as the analyses of temporal autocorrelation (TAC) in time series. Current land surface temperature missions which provide time series at a global scale include for example MODIS (500 m) and Sentinel-3 SLSTR (1 km). Both use thermal sensors which leads to data gaps in case of cloud cover. An alternative are passive microwave observations, which provide data only at the tens of kilometre scale. Future thermal missions such as LSTM (Land Surface Temperature Mission, Sentinel program) will provide records with better than 100 m detail, but the frequent Arctic cloud cover remains an issue (Bartsch et al. 2025). Water level can be obtained through altimeter, however, these also provide only coarse footprint sizes of several hundred metres. Any climate data record analysis for LST and water level is therefore only feasible for larger lakes. However, such larger lakes are typically of a different origin to permafrost thaw lakes. Observed variations can, however, serve as a general indicator of increasing air temperatures and changes in the hydrological cycle.

Water surface extent can be derived independently from monitoring of specific lake objects therefore long-term analyses based on coarse spatial resolution observations is feasible (e.g. Schroeder et al. 2015). However, identification of thaw lake change and attribution to abruptness requires higher spatial and temporal resolution. Recent and future satellite missions may provide relevant data but sensors used in the past had only limited capabilities. Available circumpolar long-term change studies of surface water extent do not separate lakes from floodplains. Extensive floodplains are spread across the Arctic which regionally dominate land surface hydrology change patterns and surface water extent varies considerably seasonally (Bartsch et al. 2012; Watts et al. 2014). A multitude of lakes exists which also shows strong seasonality although not closely located to river courses (e.g. Bartsch et al. 2012; Trofaier et al. 2013). The fluvial patterns overlap with thermokarst related changes of surface water extent.

The use of trends based on the superfine water index (SWI) derived from multispectral information at 500 m (MODIS) as suggested nevertheless provides a step forward compared to the coarser global water fraction datasets based on microwave records. The SWI is positively correlated with percentage surface water cover. In the study by Webb et al. (2022), widespread water area loss across the Arctic for the last two decades (2000–2021) was found. Webb et al. (2022) further differentiated between various regions considering ground ice content, permafrost zonation and lake cover percentage. The water surface shrank on average in all of these categories. However, here too, lakes were not separated from floodplains. Due to the high spatial detail available from MODIS, an estimate of the split between these might be feasible. Smaller lakes cannot be separated within pixels but river floodplain areas can be attributed to specific pixels due to their larger extent. In order to demonstrate this we created a subset of the SWI trend dataset (Webb 2022) based on several geospatial datasets which represent non-floodplain and floodplain areas (Fig. 3). We first derived regional clusters with strong positive and negative SWI trends (areas with high change) by the following procedure: 1) extraction of pixels where SWI trends exceed one standard deviation ($> \pm 0.005$), 2) polygonalisation and application of a 2 km buffer and 3) extraction of clusters larger than 5 km^2. These clusters were semi-automatically

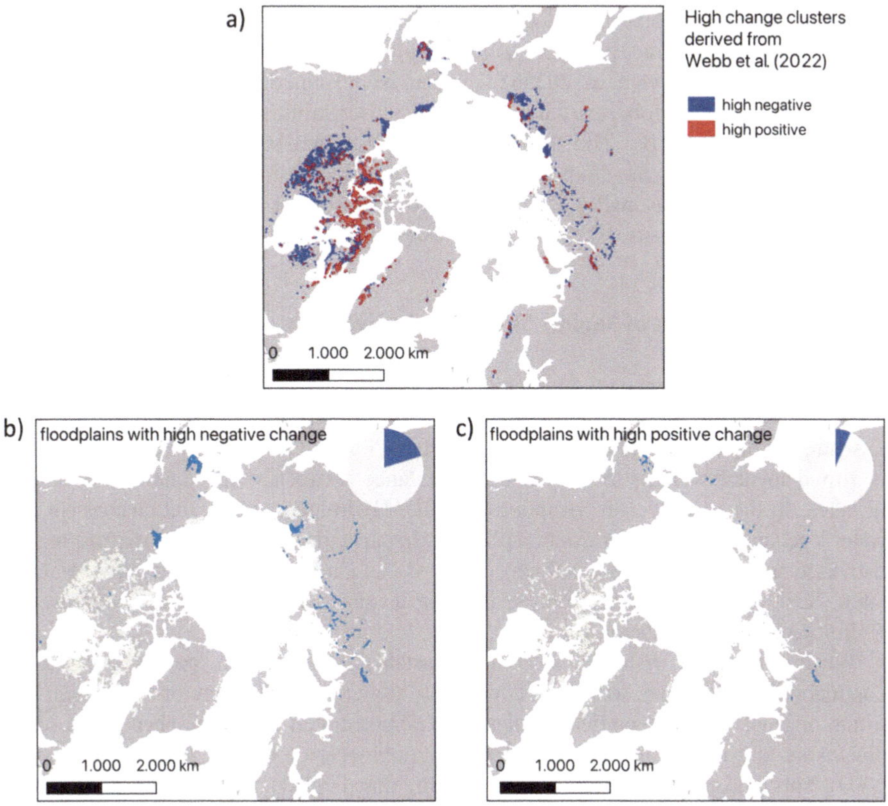

Fig. 3 Superfine water index (SWI) change (2000–2021) attribution to river floodplains. a) Clusters with SWI decline (high negative) versus increase (high positive) extracted from Webb (2022). b) Distribution (blue areas) and proportion of cluster areas with high negative trends within river floodplains. c) Distribution (blue areas) and proportion of cluster areas with high positive trends within river floodplains

assigned to floodplains or non-floodplains by considering terrain information (Copernicus Digital Elevation Model slope), the circumpolar land cover dataset of (Bartsch et al. 2024) and Google hybrid visualisations. Major coastal plains (northern part of the Alaskan North Slope, Lena Delta, Yana, Nenets reserve and Kanin coastal plain) were defined as none-floodplain. The proportion of floodplain regions which underwent high losses for 2000–2021 was 20% (and 15.5% of all areas with high change) (Fig. 3b). 7% of the areas with high water fraction gain (and 1.5% of all areas with high change) were part of floodplains. In total, at least 17% of change is floodplain-related. This needs to be considered when satellite-derived water extent records are used to analyse potential tipping behaviour linked to lakes.

Significant seasonal variations have been documented for selected Arctic deltas (Vulis et al. 2020) as well as smaller lake sizes in case of warmer climate conditions (Vulis et al. 2021). More precise accounts including actual seasonality change identification across the entire Arctic, separately for lakes in typical thermokarst landscapes and water bodies in floodplains, are needed to create datasets suitable for tipping behaviour analyses. This requires the use of as high spatial and temporal resolution as possible based on SAR data, which is unaffected by clouds. Seasonal patterns can be identified on a regional scale if sufficient data is available for selected years (example for Russia: Reschke et al. 2012). However, recent SAR missions have so far not provided regular consistent coverage of the Arctic land area (Bartsch et al. 2023b). This can be attributed to differences in acquisition strategies across the globe, partially driven by sensor failure (as for example in case of Copernicus Sentinel-1B). The recent launch of Sentinel-1C is, however, expected to advance circumpolar-scale studies. Locally, high spatial resolution optical data can also provide insight into seasonality (Cooley et al. 2019; Mullen et al. 2023). Such observations may serve as quality control for circumpolar assessments.

2.4 Emerging Signals of Regime Shifts in Lake Hydrology

Global lakes have responded dramatically to climate change in recent decades, including water level (Woolway and Merchant 2019). Lake water level (LWL) is recognised as an Essential Climate Variable (ECV) by the Global Climate Observing System (GCOS) and is a fundamental variable to understand the balance between water inputs and water loss in lakes. In the presence of disruptions in Lake Hydrology, substantial changes in Lake water level and surface water extent will occur. In particular, in ice-rich permafrost regions, increased surface temperature and ice cover loss can alter the water and energy balance in lakes. Such changes can lead to corresponding feedback mechanisms in regional to local hydrological or biogeochemical cycles.

When a complex system is becoming unstable, short-term responses to weak and continuous perturbations will slow down, thus slowing the recovery of the system back to the original state. Theoretical studies have demonstrated that these phenomena can be precursors to critical transitions as the system approaches a tipping point (Scheffer et al. 2009). This is also referred to as critical slowing down (CSD). During CSD, the increased memory in the time series, which can be quantified with increased temporal autocorrelation (TAC), has been proposed as an indicator of declining resilience (Held and Kleinen 2004; Scheffer et al. 2009). Here, resilience is defined as the capacity of the system to remain in its current state despite external perturbations.

There has been a widespread change in global lake water storage. Arctic lakes have mostly been in decline, attributed to a combination of changes in precipitation, runoff, temperature and potential evapotranspiration (PET) (Yao et al. 2023). This aligns with other observations that show the reduction of lake area across the northern permafrost zone through surface and subsurface drainage or lake infilling (Webb and Liljedahl 2023; Webb et al. 2022). On the other hand, in the Tibetan Plateau, glacier retreat and permafrost thawing has driven alpine lake expansion (Yao et al. 2023). This is consistent with the recent trends of rapidly increasing lake volumes in the Tibetan Plateau where the majority of the additional water supply is from increased net precipitation (74%), followed by glacier mass loss (13%), and ground ice melt due to permafrost degradation (12%) (Zhang et al. 2017). Along with these observations, we provide additional analysis of the signals of the hydrological system of the lakes in the permafrost zone consistently losing resilience. To explore the regime shifts in Lake Hydrology and how it has evolved over the last decade, we present here the TAC time series estimated from remotely sensed lake water levels.

Combining multiple satellite altimetry sensors, the Global database of Lake Water Storage (GLWS v1.1, Yao et al. 2023) provides information on surface areas, water levels, and storage for 1972 of Earth's large water bodies spanning from 1992 to 2020 at a monthly frequency. From the globally distributed lakes, we selected 97 lakes located in the permafrost regions (permafrost probability >0.5) of the Northern Hemisphere.

This includes both the continuous and discontinuous permafrost regions where recent findings (e.g. Webb and Liljedahl 2023; Webb et al. 2022) have demonstrated surface water decline, especially in discontinuous permafrost regions (permafrost probability 0.5–0.9). Of these, 69 lakes were selected from the Northern Arctic, and 28 from the Tibetan Plateau. A lake showing spurious drop in water level, attributed to dam failure, was removed (Liu et al. 2016). To obtain Lake water level anomalies, we first removed the long-term linear trend by performing a simple linear regression on the monthly time series and subtracting the fitted trend from the original data. To capture interannual variability, we computed annual mean z-score anomalies based on monthly time series after removing seasonality, which ensures the analysis is robust, as the observations are only available during the boreal summer.

The 15-year (2006–2020) mean annual TAC time series of lake water level anomalies shows an increase, particularly since 2010 (Fig. 3a, c). To account for differing geographic regions and permafrost extents, we grouped the lakes into four categories based on combinations of region (i.e. Arctic or Tibetan Plateau) and permafrost probability (i.e. continuous or discontinuous) (Fig. 3b). In the Arctic lakes, both the continuous and discontinuous permafrost regions show an increasing TAC time series from 2010 (continuous: $\tau=0.64$, $P=0.006$; discontinuous: $\tau=0.53$, $P=0.03$). However, since there was also a decline in TAC values until 2010, we cannot clearly disentangle if the increasing and decreasing signals are due to internal natural variability or if they are fingerprints of climate change. Multidecadal oscillations affecting Arctic air temperature (Chylek et al. 2009; Fang et al. 2022) and alterations in thermal or mixing regimes in lakes may affect these temporal changes in TAC. In the Tibetan Plateau, trends of TAC time series for both the continuous and discontinuous permafrost regions are insignificant ($P=0.087$). Unlike Arctic lakes, many lakes on the Tibetan Plateau seem to have already entered a transient phase, where lake water levels exhibit strongly increasing trends, and thus their TAC time series might not show significant trends in recent periods. Of the 95 lakes, 32 lakes (33.7%) exhibit positive trends, 10 lakes (10.5%) exhibit negative trends and 55 lakes (57.8%) show insignificant trends (Fig. 3d).

Our findings suggest that lake hydrology is shifting state within permafrost regions, which agrees with various regional studies (Andresen and Lougheed 2015; Liu et al., 2024a; Saros et al. 2025; Su et al. 2023). For example, sub-lake talik formation or expansion can accelerate permafrost thaw by thermokarst processes, and increased runoff can induce river-to-ocean heat transport, which will create a positive feedback loop in sea-ice loss (Nitzbon et al. 2024). Also, changes in snow cover on the Tibetan Plateau can dynamically influence climate variabilities in other regions (Li et al. 2018; Zhang et al. 2023a) and establish teleconnections to other tipping elements (Liu et al. 2023). As freshwater systems become more sensitive and approach tipping points, they can possibly trigger nonlinear responses, leading to widespread impacts on hydrology, ecology, and biogeochemistry (Anthony et al. 2018; de Vrese and Brovkin 2021; Hessen et al. 2024; Wunderling et al. 2024).

It should be noted that most of the lakes considered here are located outside the identified clusters with high surface water index change as derived from Webb et al. (2022) (Fig. 3). They are mostly glacially formed, larger lakes, outside of typical thermokarst regions. Altimeter data which can provide the lake levels for the relatively small lake sizes typical for thaw lakes, would be required to extend the above analyses to such small thaw lakes. The capabilities of state-of-the-art SAR-based altimeter data (e.g. Corpenicus Sentinel-3 with along-track footprint of 330 m) are still limited although an improvement.

3 Remote Effects of the Permafrost Thaw on the Earth System

The feedback between permafrost and climate is usually considered by studying permafrost carbon emissions to the atmosphere in the form of well-mixed GHGs (Schuur et al., 2022). There is another way how permafrost changes could affect the climate, and this is through changes in land surface physics and hydrology (Bonan 2016). Increased permafrost thaw in the future is expected to lead to a shift in land surface hydrology towards greater subsurface flow and increased Arctic river discharge (Rawlins and Karmalkar 2024; Zhang et al. 2023b), affecting the oceanic freshwater budget and ocean circulation. The Arctic is a cloudy region (Chernokulsky and Esau 2019). The Arctic clouds have a strong impact on the surface energy budget, generally warming the surface in winter and cooling it in summer (Kay and L'Ecuyer 2013), altering the climate feedbacks (Goosse et al. 2018). During plant growth periods, evapotranspiration from plants leads to higher moisture fluxes to the atmosphere, more clouds, and less radiation reaching the surface. The presence of permafrost limits the amount of water that ca' be transpired (van Huissteden 2020), and less transpiration means a less moist atmosphere and less clouds, warming the surface (Bonan 2016). Changes in Arctic vegetation affect climate through both transpiration and albedo changes (Swann et al. 2010). Such biogeophysical feedbacks from permafrost to climate have rarely been considered in analyses of the Earth System.

Recent analyses suggested that changes in energy and water cycling in the Arctic will have an effect far beyond the Arctic region, providing biogeophysical teleconnections to the other tipping elements such as Atlantic meridional overturning circulation, Amazon, or Sahel (de Vrese et al. 2023a) and through change in wetland distribution, to methane emissions from tropical regions (de Vrese et al. 2023b). In these studies, the hydrological modelling parameters in high latitudes were changed between the limits of a plausible range of parameters configurations, namely between the Dry and Wet setups (Fig. 4). In the Dry setup, enhanced drainage and runoff lead to a higher water flux to the ocean than to the atmosphere, resulting in less moisture recycling between land and atmosphere. In boreal summer, when the high latitudes receive almost the same cumulative radiations as the tropics, this leads to less cloud cover and a relatively warmer and drier land surface. In the Wet setup, more moisture was recycled between land and atmosphere, leading to a higher cloud fraction, and wetter and cooler conditions at the land surface in summer (de Vrese et al. 2023a). For Wet and Dry simulations using a coupled land–atmosphere-ocean model, the MPI Earth System model (MPI-ESM), run until 2100 under the RCP8.5 scenario of greenhouse and other forcings, it is observed that in both simulations the Arctic region got warmer, but with different regional patterns. For a valid comparison, the differences between the climate change patterns have to be taken at the same global temperature level. In Fig. 5, the regional patterns of differences in annual mean surface temperature are presented for 2, 3, and 4 °C global warming levels. Because the differences are taken at the same temperature level for both, the Dry and Wet simulations, the average change in the temperature for each plot is zero.

For both setups, the high northern latitudes show a strong warming especially due to lower cloud cover in summer, compensated by cooling elsewhere. The relative cooling over land in South America and North Africa is very pronounced and statistically significant (Fig. 5). This relative cooling for these two regions is especially strong at the 3 °C of global warming level. While there is some change in patterns between 2 and 3 °C warming levels, generally the difference between the Wet and Dry simulations remains of the same magnitude. Although there are some changes in spatial patterns with increasing

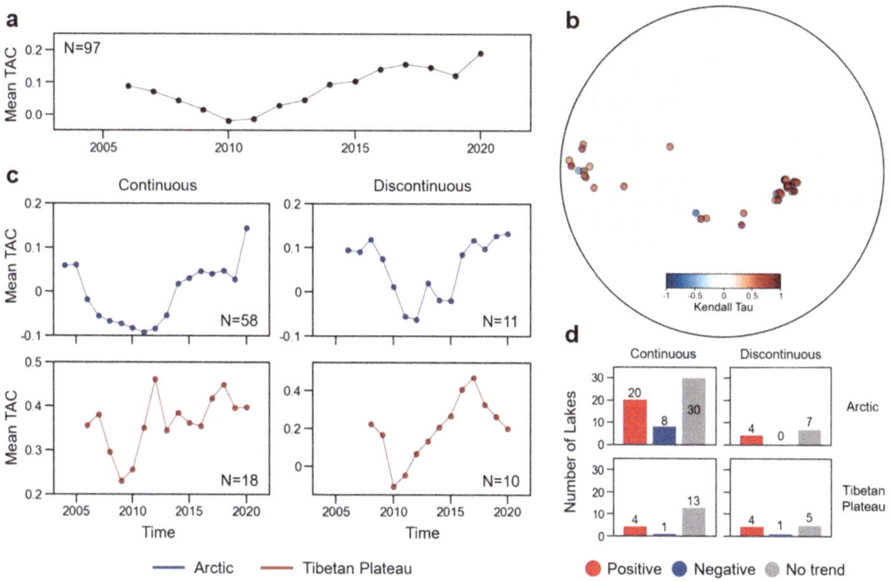

Fig. 4 **a.** Mean TAC time series for lake water level anomalies calculated from every selected lake (N = 97). **b.** A circumpolar map of the Kendall tau (τ) values of each lake's TAC time series from 2010 ($P < 0.05$). Insignificant lakes are not shown. **c.** Same as (a), but the lakes are divided into four categories: based on permafrost probability (continuous/discontinuous), and geographic regions (Arctic/Tibetan Plateau). Only the years having TAC values for more than 90% of the samples are shown for each case (**a.**, **c.**) **d.** Trends of TAC time series for selected lakes based on τ values from 2010 divided into four categories. Positive trends mean increasing TAC time series, which implies a decline in resilience, and vice versa for negative trends. No trend implies insignificant τ values

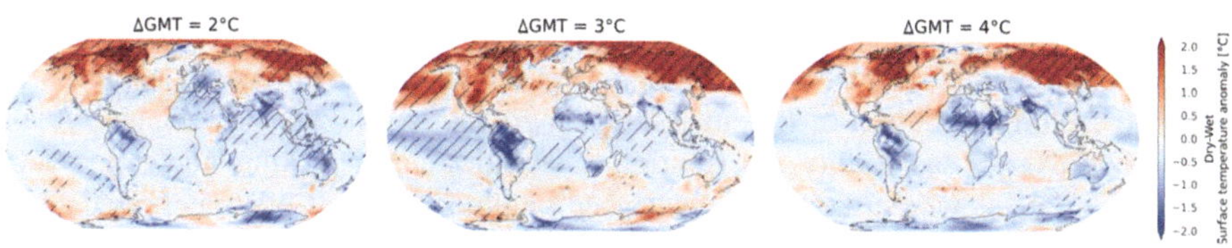

Fig. 5 Remote effects of changes in permafrost hydrology on changes in global mean temperature, GMT (hatched areas indicate significant differences). Temperature differences between the Dry and Wet permafrost setups are shown for 2, 3, and 4 °C of global warming (±0.25 °C). Data from (de Vrese et al. 2023b)

forcing, they are gradual rather than abrupt. We can interpret this as a gradual tipping with increasing global temperature. Although the changes are gradual, they indicate the future hydrological regime in the permafrost region will have a strong impact on other parts of the climate system, including the Sahel and Amazon, which have themselves already been identified as sensitive tipping elements (McKay et al. 2022). The analysis of permafrost's role as a tipping element in the climate system is going to be continued in the framework of the Tipping Point Modelling Intercomparison Project (TIPMIP) (Winkelmann et al., in preparation). Similarly, the role of the cloud physics in tipping elements of the climate system is studied in the Cloud Feedback Model Intercomparison Project (CFMIP)

(Webb et al. 2017) with the aim of improving understanding of cloud–climate feedback mechanisms, in particular in the permafrost region (de Vrese et al. 2024). Together, model intercomparison projects help to understand linkages between different components of the Earth System and potential nonlinearities in the climate system dynamics.

4 Summary and Conclusions

Permafrost in the Arctic was formed during the geological past and served as a storage of organic carbon during relatively cold phases. It could have released carbon during abrupt thaw events during the last deglaciation, contributing to the reconstructed increase in atmospheric CO_2 and CH_4 concentrations. Current understanding is that substantial amounts of permafrost carbon could be released to the atmosphere in the future in response to anthropogenic warming, amplifying climate change. This carbon release, as simulated by existing Earth System and land surface models, is rather gradual and proportional to ongoing warming. This does not qualify permafrost carbon to be a tipping element. However, many local feedbacks are neglected in ESMs. Existence of alternative steady states supports the possibility of ecosystem tipping, although on centennial rather than decadal scale. Study of the equilibrium response of permafrost carbon to climate change on centennial to millennial timescale has not yet been done, and the difference between the instantaneous and equilibrium response could be pronounced.

Hydrological processes in lakes have internal feedbacks, which can lead to nonlinear responses to ongoing anthropogenic warming. The area of lakes in the Arctic both increases and decreases with a tendency of the net effect towards a smaller area which means effective drying of the land surface. At the same time, there is an increase in autocorrelation in the amount of water in lakes, with a caveat that only data for large lakes is available. As an early warning signal of changes in the water level, increased autocorrelation could indicate potential shift towards another state, but a possibility remains that this change is an artefact of the method. Higher resolution data and driver analysis are needed for further clarification. Earth observation can potentially also offer the means to capture the impact of extreme events on ground ice and to study changes in land surface hydrology's seasonality. Observation capabilities currently limit studies to a local to regional scale, however, the recently launched satellites such as Copernicus Sentinel-1C will help to enable circumpolar-scale studies.

To conclude, the crucial question about changes in the Arctic is whether the Arctic will be drier or wetter in the future. Using Earth System model simulations, we demonstrate that this difference in wetness is important for hydrometeorological changes far beyond Arctic. We have not identified nonlinearities in the climate response. However, we cannot rule out that even linear changes in climate will be translated into nonlinear changes in CO_2 and CH_4 emissions because of nonlinearities in the ecosystem response. In addition to the potential of permafrost and freshwater ecosystem systems to regime changes on local and regional scale, a possibility also exists for a nonlinear response on a pan-Arctic scale.

Acknowledgements This paper is an outcome of the Workshop on "Tipping Points and Understanding EO data needs for a Tipping element Model Intercomparison Project" held at the International Space Science Institute (ISSI) in Bern, Switzerland (10-14 October 2022). EC was supported by the European Space Agency (ESA) Climate Change Initiative fellowship (ESA ESRIN/Contract No. 4000133621/20/I/NB). This work was funded by the German Research Foundation as part of the CLICCS Clusters of Excellence (DFG EXC 2037), by the European Research Council under the European Union's 691 Horizon 2020 Research and Innovation programme as part of the Q-Arctic project (grant agreement number 951288), by the European

Union's Horizon 2020 Research and Innovation programme as part of the ESM2025 project (grant number 101003536), European Space Agency projects CCI+ Permafrost (4000123681/18/I-NB) and AMPAC-Net (4000137912/22/I-DT), Brain Pool Plus program funded by the Ministry of Science and ICT through the National Research Foundation of Korea (2021H1D3A2A03097768), and the National Research Foundation of Korea (NRF) grant funded by the Korean government (MSIT) (NRF-2022R1I1A4071660). We thank Thomas Kleinen and Thomas Riddick for providing detailed and helpful comments on the manuscript. We also thank five anonymous reviewers for their thoughtful and constructive comments.

Funding Open Access funding enabled and organized by Projekt DEAL.

Declarations

Conflict of interests The authors declare that they have no competing financial or non-financial interests.

References

Aas KS, Martin L, Nitzbon J, Langer M, Boike J, Lee H, Berntsen TK, Westermann S (2019) Thaw processes in ice-rich permafrost landscapes represented with laterally coupled tiles in a land surface model. Cryosphere 13:591–609. https://doi.org/10.5194/tc-13-591-2019

Andresen CG, Lougheed VL (2015) Disappearing Arctic tundra ponds: Fine-scale analysis of surface hydrology in drained thaw lake basins over a 65year period (1948–2013). J Geophys Res-Biogeosci 120:466–479. https://doi.org/10.1002/2014jg002778

Anthony KW, von Deimling TS, Nitze I, Frolking S, Emond A, Daanen R, Anthony P, Lindgren P, Jones B, Grosse G (2018) 21st-century modeled permafrost carbon emissions accelerated by abrupt thaw beneath lakes. Nature Commun 9:57389. https://doi.org/10.1038/s41467-018-05738-9

Archer D, Ganopolski A (2005) A movable trigger: fossil fuel CO_2 and the onset of the next glaciation. Geochem Geophys Geosyst 6:7. https://doi.org/10.1029/2004gc000891

Armstrong McKay DIA, Staal A, Abrams JF, Winkelmann R, Sakschewski B, Loriani S, Fetzer I, Cornell SE, Rockstrom J, Lenton TM (2022) Exceeding 15°C global warming could trigger multiple climate tipping points. Science 377:1171. https://doi.org/10.1126/science.abn7950

Arp CD, Jones BM, Urban FE, Grosse G (2011) Hydrogeomorphic processes of thermokarst lakes with grounded-ice and floating-ice regimes on the Arctic coastal plain. Alaska, Hydrolog Process 25:2422–2438. https://doi.org/10.1002/hyp.8019

Babkina EA, Leibman MO, Dvornikov YA, Fakashchuk NY, Khairullin RR, Khomutov AV (2019) Activation of cryogenic processes in central yamal as a result of regional and local change in climate and thermal state of permafrost. Russ Meteorol Hydrol 44:283–290. https://doi.org/10.3103/s1068373919040083

Balser AW, Jones JB, Gens R (2014) Timing of retrogressive thaw slump initiation in the Noatak Basin, northwest Alaska, USA. J Geophys Res Earth Surf 119:1106–1120. https://doi.org/10.1002/2013jf002889

Bartsch A, Trofaier AM, Hayman G, Sabel D, Schlaffer S, Clark DB, Blyth E (2012) Detection of open water dynamics with ENVISAT ASAR in support of land surface modelling at high latitudes. Biogeosciences 9:703–714. https://doi.org/10.5194/bg-9-703-2012

Bartsch A, Leibman M, Strozzi T, Khomutov A, Widhalm B, Babkina E, Mullanurov D, Ermokhina K, Kroisleitner C, Bergstedt H (2019) Seasonal progression of ground displacement identified with satellite radar interferometry and the impact of unusually warm conditions on permafrost at the yamal peninsula in 2016. Remote Sens 11:25. https://doi.org/10.3390/rs11161865

Bartsch A, Bergstedt H, Pointner G, Muri X, Rautiainen K, Leppänen L, Joly K, Sokolov A, Orekhov P, Ehrich D, Soininen EM (2023a) Towards long-term records of rain-on-snow events across the Arctic from satellite data. Cryosphere 17:889–915. https://doi.org/10.5194/tc-17-889-2023
Bartsch A, Strozzi T, Nitze I (2023b) Permafrost monitoring from space. Surv Geophys 44:1579–1613. https://doi.org/10.1007/s10712-023-09770-3
Bartsch A, Efimova A, Widhalm B, Muri X, von Baeckmann C, Bergstedt H, Ermokhina K, Hugelius G, Heim B, Leibman M (2024) Circumarctic land cover diversity considering wetness gradients. Hydrol Earth Syst Sci 28:2421–2481. https://doi.org/10.5194/hess-28-2421-2024
Bartsch A, Muri X, Hetzenecker M, Rautiainen K, Bergstedt H, Wuite J, Nagler T, Nicolsky D (2025) Benchmarking passive-microwave-satellite-derived freeze–thaw datasets. Cryosphere 19:459–483. https://doi.org/10.5194/tc-19-459-2025
Berger A, Loutre MF (2002) An exceptionally long interglacial ahead? Science 297:1287–1288. https://doi.org/10.1126/science.1076120
Bergstedt H, Jones BM, Hinkel K, Farquharson L, Gaglioti BV, Parsekian AD, Kanevskiy M, Ohara N, Breen AL, Rangel RC, Grosse G, Nitze I (2021) Remote sensing-based statistical approach for defining drained lake basins in a continuous permafrost region, north slope of alaska. Remote Sens 13:26. https://doi.org/10.3390/rs13132539
Bernhard P, Zwieback S, Bergner N, Hajnsek I (2022) Assessing volumetric change distributions and scaling relations of retrogressive thaw slumps across the Arctic. Cryosphere 16:1–15. https://doi.org/10.5194/tc-16-1-2022
Bonan G (2016) Ecological climatology: concepts and applications. 3rd Edn. Ecological climatology: concepts and applications, 3rd Edition, pp 1–692
Brovkin V, Brook E, Williams JW, Bathiany S, Lenton TM, Barton M, DeConto RM, Donges JF, Ganopolski A, McManus J, Praetorius S, de Vernal A, Abe-Ouchi A, Cheng H, Claussen M, Crucifix M, Gallopin G, Iglesias V, Kaufman DS, Kleinen T, Lambert F, van der Leeuw S, Liddy H, Loutre MF, McGee D, Rehfeld K, Rhodes R, Seddon AWR, Trauth MH, Vanderveken L, Yu ZC (2021) Past abrupt changes, tipping points and cascading impacts in the Earth system. Nat Geosci 14:550–558. https://doi.org/10.1038/s41561-021-00790-5
Bruhwiler L, Parmentier FJW, Crill P, Leonard M, Palmer PI (2021) The arctic carbon cycle and its response to changing climate. Curr Clim Chang Rep 7:14–34. https://doi.org/10.1007/s40641-020-00169-5
Burn CR (2002) Tundra lakes and permafrost, Richards Island, western Arctic coast, Canada. Can J Earth Sci 39:1281–1298. https://doi.org/10.1139/e02-035
Burn CR, Lewkowicz AG (1990) Canadian landform examples 17 Retrogressive thaw slumps. Can Geogr Geogr Can 34:273–276. https://doi.org/10.1111/j.1541-0064.1990.tb01092.x
Calamita E, Lever JJ, Albergel C, Woolway RL, Odermatt D (2024) Detecting climate-related shifts in lakes: a review of the use of satellite Earth Observation. Limnol Oceanogr 69:723–741
Carrea L, Crétaux JF, Liu XH, Wu YH, Calmettes B, Duguay CR, Merchant CJ, Selmes N, Simis SGH, Warren M, Yesou H, Müller D, Jiang DL, Embury O, Bergé-Nguyen M, Albergel C (2023) Satellite-derived multivariate world-wide lake physical variable timeseries for climate studies. Sci Data 10:28. https://doi.org/10.1038/s41597-022-01889-z
Carrea L, JF Crétaux, X Liu et al (2022) ESA Lakes Climate Change Initiative (Lakes_cci): Lake products, Version 2.0.1., ESA (Ed..
Chernokulsky A, Esau I (2019) Cloud cover and cloud types in the Eurasian Arctic in 1936–2012. Int J Climatol 39:5771–5790. https://doi.org/10.1002/joc.6187
Chylek P, Folland CK, Lesins G, Dubey MK, Wang MY (2009) Arctic air temperature change amplification and the atlantic multidecadal oscillation. Geophys Res Lett. https://doi.org/10.1029/2009gl038777
Comyn-Platt E, Hayman G, Huntingford C, Chadburn SE, Burke EJ, Harper AB, Collins WJ, Webber CP, Powell T, Cox PM, Gedney N, Sitch S (2018) Carbon budgets for 15 and 2 °C targets lowered by natural wetland and permafrost feedbacks. Nature Geosci 11:568. https://doi.org/10.1038/s41561-018-0174-9
Cooley SW, Smith LC, Ryan JC, Pitcher LH, Pavelsky TM (2019) Arctic-boreal lake dynamics revealed using cubesat imagery. Geophys Res Lett 46:2111–2120. https://doi.org/10.1029/2018gl081584
de Vrese P, Brovkin V (2021) Timescales of the permafrost carbon cycle and legacy effects of temperature overshoot scenarios. Nature Commun. https://doi.org/10.1038/s41467-021-23010-5
de Vrese P, Georgievski G, Rouco JFG, Notz D, Stacke T, Steinert NJ, Wilkenskjeld S, Brovkin V (2023a) Representation of soil hydrology in permafrost regions may explain large part of intermodel spread in simulated Arctic and subarctic climate. Cryosphere 17:2095–2118. https://doi.org/10.5194/tc-17-2095-2023

de Vrese P, Beckebanze L, Galera LD, Holl D, Kleinen T, Kutzbach L, Rehder Z, Brovkin V (2023) Sensitivity of Arctic CH4 emissions to landscape wetness diminished by atmospheric feedbacks. Nature Climate Change 13:832. https://doi.org/10.1038/s41558-023-01715-3

de Vrese P, Stacke T, Gayler V, Brovkin V (2024) Permafrost cloud feedback may amplify climate change. Geophys Res Lett. https://doi.org/10.1029/2024gl109034

Ekici A, Lee H, Lawrence DM, Swenson SC, Prigent C (2019) Ground subsidence effects on simulating dynamic high-latitude surface inundation under permafrost thaw using CLM5. Geosci Model Develop 12:5291–5300. https://doi.org/10.5194/gmd-12-5291-2019

Eliseev AV, Demchenko PF, Arzhanov MM, Mokhov II (2014) Transient hysteresis of near-surface permafrost response to external forcing. Clim Dyn 42:1203–1215. https://doi.org/10.1007/s00382-013-1672-5

Fang M, Li X, Chen HW, Chen DL (2022) Arctic amplification modulated by Atlantic Multidecadal Oscillation and greenhouse forcing on multidecadal to century scales. Nature Commun. https://doi.org/10.1038/s41467-022-29523-x

Gasser T, Kechiar M, Ciais P, Burke EJ, Kleinen T, Zhu D, Huang Y, Ekici A, Obersteiner M (2018) Path-dependent reductions in CO2 emission budgets caused by permafrost carbon release. Nature Geosci 11:830. https://doi.org/10.1038/s41561-018-0227-0

Giardino C, Bresciani M, Stroppiana D, Oggioni A, Morabito G (2014) Optical remote sensing of lakes: an overview on Lake Maggiore. J Limnol 73:201–214. https://doi.org/10.4081/jlimnol.2014.817

Goosse H, Kay JE, Armour KC, Bodas-Salcedo A, Chepfer H, Docquier D, Jonko A, Kushner PJ, Lecomte O, Massonnet F, Park HS, Pithan F, Svensson G, Vancoppenolle M (2018) Quantifying climate feedbacks in polar regions. Nat Commun 9:13. https://doi.org/10.1038/s41467-018-04173-0

Grosse GB, Jones C (2013) Arp thermokarst lakes, drainage, and drained basins, in: treatise on geomorphology, edited by: Shroder, JF Academic Press, 325–353

Held H, Kleinen T (2004) Detection of climate system bifurcations by degenerate fingerprinting. Geophys Res Lett. https://doi.org/10.1029/2004gl020972

Hessen DO, Andersen T, McKay DA, Kosten S, Meerhoff M, Pickard A, Spears BM (2024) Lake ecosystem tipping points and climate feedbacks. Earth Syst Dynam 15:653–669. https://doi.org/10.5194/esd-15-653-2024

Hugelius G, Strauss J, Zubrzycki S, Harden JW, Schuur EAG, Ping CL, Schirrmeister L, Grosse G, Michaelson GJ, Koven CD, O'Donnell JA, Elberling B, Mishra U, Camill P, Yu Z, Palmtag J, Kuhry P (2014) Estimated stocks of circumpolar permafrost carbon with quantified uncertainty ranges and identified data gaps. Biogeosciences 11:6573–6593. https://doi.org/10.5194/bg-11-6573-2014

Hugelius G, Ramage J, Burke E, Chatterjee A, Smallman TL, Aalto T, Bastos A, Biasi C, Canadell JG, Chandra N, Chevallier F, Ciais P, Chang J, Feng L, Jones MW, Kleinen T, Kuhn M, Lauerwald R, Liu J, López-Blanco E, Luijkx IT, Marushchak ME, Natali SM, Niwa Y, Olefeldt D, Palmer PI, Patra PK, Peters W, Potter S, Poulter B, Rogers BM, Riley WJ, Saunois M, Schuur EAG, Thompson RL, Treat C, Tsuruta A, Turetsky MR, Virkkala AM, Voigt C, Watts J, Zhu Q, Zheng B (2024) Permafrost region greenhouse gas budgets suggest a weak CO2 sink and CH4 and N2O sources, but magnitudes differ between top-down and bottom-up methods. Glob Biogeochem Cycle. https://doi.org/10.1029/2023gb007969

van Huissteden J (2020) Thawing permafrost permafrost carbon in a warming arctic introduction, THAWING PERMAFROST: permafrost carbon in a Warming Arctic, pp 1–50

IPCC (2021) Annex VII: Glossary, in: Climate Change 2021: the physical science basis. Contribution of working group i to the sixth assessment report of the intergovernmental panel on climate change [Masson-Delmotte V, P Zhai A Pirani, SL Connors, C Péan, S Berger, N Caud, Y Chen, L. Goldfarb, M.I. Gomis, M. Huang, K. Leitzell, E. Lonnoy, J.B.R. Matthews, T.K. Maycock, T. Waterfield, O. Yelekçi, R. Yu, and B. Zhou (eds.)]. edited by: Matthews, J. B. R., V. Möller, R. van Diemen, J.S. Fuglestvedt, V. Masson-Delmotte,, and C. Méndez, S. S., A. Reisinger Cambridge University Press,, Cambridge, United Kingdom and New York, NY, USA, pp 2215–2256

Jones MKW, Pollard WH, Jones BM (2019) Rapid initialization of retrogressive thaw slumps in the Canadian high Arctic and their response to climate and terrain factors. Environ Res Lett 14:14. https://doi.org/10.1088/1748-9326/ab12fd

Jones BM, Grosse G, Farquharson LM, Roy-Léveillée P, Veremeeva A, Kanevskiy MZ, Gaglioti BV, Breen AL, Parsekian AD, Ulrich M, Hinkel KM (2022) Lake and drained lake basin systems in lowland permafrost regions. Nature Rev Earth Environ 3:85–98. https://doi.org/10.1038/s43017-021-00238-9

Kay JE, L'Ecuyer T (2013) Observational constraints on Arctic Ocean clouds and radiative fluxes during the early 21st century. J Geophys Res-Atmos 118:7219–7236. https://doi.org/10.1002/jgrd.50489

Kizyakov A, Zimin M, Leibman M, Pravikova N (2013) Monitoring of the rate of thermal denudation and thermal abrasion on the western coast of Kolguev Island, using high resolution satellite images. Kriosfera Zemli 17:36–47
Kizyakov A, Sonyushkin AV, Leibman MO, Zimin MV, Khomutov AV (2015) Geomorphological conditions of the gas-emission crater and its dynamics in Central Yamal. Earth's Cryosphere XIX, 15–25
Kohler P, Knorr G, Bard E (2014) Permafrost thawing as a possible source of abrupt carbon release at the onset of the Bolling/Allerod. Nature Commun 5:6520. https://doi.org/10.1038/ncomms6520
Kokelj SV, Lantz TC, Tunnicliffe J, Segal R, Lacelle D (2017) Climate-driven thaw of permafrost preserved glacial landscapes, northwestern Canada. Geology 45:371–374. https://doi.org/10.1130/g38626.1
Lauerwald R, Allen GH, Deemer BR, Liu SD, Maavara T, Raymond P, Alcott L, Bastviken D, Hastie A, Holgerson MA, Johnson MS, Lehner B, Lin PR, Marzadri A, Ran LS, Tian HQ, Yang X, Yao YZ, Regnier P (2023) Inland water greenhouse gas budgets for RECCAP2: 1 state-of-the-art of global scale assessments. Glob Biogeochem Cycle 37:32. https://doi.org/10.1029/2022gb007657
Lauerwald R, Allen GH, Deemer BR (2023) Inland water greenhouse gas budgets for RECCAP2: 2. Regionalization and Homogenization of Estimates. Global Biogeochem Cycles 37:1–16. https://doi.org/10.1029/2022GB007658
Lehner B, Döll P (2004) Development and validation of a global database of lakes, reservoirs and wetlands. J Hydrol 296:1–22. https://doi.org/10.1016/j.jhydrol.2004.03.028
Lewkowicz AG, Way RG (2019) Extremes of summer climate trigger thousands of thermokarst landslides in a High Arctic environment. Nat Commun 10:11. https://doi.org/10.1038/s41467-019-09314-7
Li WK, Guo WD, Qiu B, Xue YK, Hsu PC, Wei JF (2018) Influence of Tibetan Plateau snow cover on East Asian atmospheric circulation at medium-range time scales. Nature Commun 9:6762. https://doi.org/10.1038/s41467-018-06762-5
Lindgren A, Hugelius G, Kuhry P, Christensen TR, Vandenberghe J (2016) GIS-based maps and area estimates of northern hemisphere permafrost extent during the last glacial maximum. Permaf Periglac Process 27:6–16. https://doi.org/10.1002/ppp.1851
Lindgren A, Hugelius G, Kuhry P (2018) Extensive loss of past permafrost carbon but a net accumulation into present-day soils. Nature 560:219. https://doi.org/10.1038/s41586-018-0371-0
Lissak C, Bartsch A, De Michele M, Gomez C, Maquaire O, Raucoules D, Roulland T (2020) Remote sensing for assessing landslides and associated hazards. Surv Geophys 41:1391–1435. https://doi.org/10.1007/s10712-020-09609-1
Liu BK, Du YE, Li L, Feng QS, Xie HJ, Liang TG, Hou FJ, Ren JZ (2016) Outburst flooding of the moraine-dammed Zhuonai Lake on Tibetan plateau: causes and impacts. IEEE Geosci Remote Sens Lett 13:570–574. https://doi.org/10.1109/lgrs.2016.2525778
Liu T, Chen D, Yang L, Meng J, Wang ZCL, Ludescher J, Fan JF, Yang SN, Chen DL, Kurths J, Chen XS, Havlin S, Schellnhuber HJ (2023) Teleconnections among tipping elements in the Earth system. Nature Climate Change 13:67. https://doi.org/10.1038/s41558-022-01558-4
Liu M, Raymond PA, Lauerwald R, Zhang Q, Trapp-Müller G, Davis KL, Moosdorf N, Xiao C, Middelburg JJ, Bouwman AF, Beusen AHW, Peng C, Lacroix F, Tian H, Wang J, Li M, Zhu Q, Cohen S, van Hoek WJ, Li Y, Li Y, Yao Y, Regnier P (2024) Global riverine land-to-ocean carbon export constrained by observations and multi-model assessment. Nat Geosci 17:896–904. https://doi.org/10.1038/s41561-024-01524-z
Liu AB, Cheng X, Wang CX, Chen YT (2024) Evidence of ecosystem tipping point on st. Lawrence Island: widespread lake drainage events after 2018. Geophys Res Lett. https://doi.org/10.1029/2024gl110161
Loriani S, Bartsch A, Calamita E, Donges JF, Hebden S, Hirota M, Landolfi A, Nagler T, Sakschewski B, Staal A, Verbesselt J, Winkelmann R, Wood R, Wunderling N (2025) Monitoring the multiple stages of climate tipping systems from space: do the GCOS essential climate variables meet the needs? Surv Geophys. https://doi.org/10.1007/s10712-024-09866-4
Marcott SA, Bauska TK, Buizert C, Steig EJ, Rosen JL, Cuffey KM, Fudge TJ, Severinghaus JP, Ahn J, Kalk ML, McConnell JR, Sowers T, Taylor KC, White JWC, Brook EJ (2014) Centennial-scale changes in the global carbon cycle during the last deglaciation. Nature 514:616. https://doi.org/10.1038/nature13799
McGuire AD, Anderson LG, Christensen TR, Dallimore S, Guo LD, Hayes DJ, Heimann M, Lorenson TD, Macdonald RW, Roulet N (2009) Sensitivity of the carbon cycle in the Arctic to climate change. Ecol Monogr 79:523–555. https://doi.org/10.1890/08-2025.1
McKay DIA, Staal A, Abrams JF, Winkelmann R, Sakschewski B, Loriani S, Fetzer I, Cornell SE, Rockstrom J, Lenton TM (2022) Exceeding 15°C global warming could trigger multiple climate tipping points. Science 377:1171. https://doi.org/10.1126/science.abn7950

Michaelides RJ, Schaefer K, Zebker HA, Parsekian A, Liu L, Chen JY, Natali S, Ludwig S, Schaefer SR (2019) Inference of the impact of wildfire on permafrost and active layer thickness in a discontinuous permafrost region using the remotely sensed active layer thickness (ReSALT) algorithm. Environ Res Lett 14:12. https://doi.org/10.1088/1748-9326/aaf932
Miesner F, Overduin PP, Grosse G, Strauss J, Langer M, Westermann S, von Deimling TS, Brovkin V, Arndt S (2023) Subsea permafrost organic carbon stocks are large and of dominantly low reactivity. Sci Rep 13:36471. https://doi.org/10.1038/s41598-023-36471-z
Mullen AL, Watts JD, Rogers BM, Carroll ML, Elder CD, Noomah J, Williams Z, Caraballo-Vega JA, Bredder A, Rickenbaugh E, Levenson E, Cooley SW, Hung JKY, Fiske G, Potter S, Yang YL, Miller CE, Natali SM, Douglas TA, Kyzivat ED (2023) Using high-resolution satellite imagery and deep learning to track dynamic seasonality in small water bodies. Geophys Res Lett 50:1102327. https://doi.org/10.1029/2022gl102327
Natali SM, Holdren JP, Rogers BM, Treharne R, Duffy PB, Pomerance R, MacDonald E (2021) Permafrost carbon feedbacks threaten global climate goals. Proc Natl Acad Sci USA 118:3. https://doi.org/10.1073/pnas.2100163118
Nitzbon J, Langer M, Martin LCP, Westermann S, von Deimling TS, Boike J (2021) Effects of multi-scale heterogeneity on the simulated evolution of ice-rich permafrost lowlands under a warming climate. Cryosphere 15:1399–1422. https://doi.org/10.5194/tc-15-1399-2021
Nitzbon J, von Deimling TS, Aliyeva M, Chadburn SE, Grosse G, Laboor S, Lee HN, Lohmann G, Steinert NJ, Stuenzi SM, Werner M, Westermann S, Langer M (2024) No respite from permafrost-thaw impacts in the absence of a global tipping point. Nat Clim Chang 14:573–585. https://doi.org/10.1038/s41558-024-02011-4
Nitze, I (2018) Trends of land surface change from Landsat time-series 1999-2014, Transect T1,Western Siberia. . Alfred Wegener Institute, H. C. f. P. a. M. R. (Ed.), PANGAEA, Bremerhaven
Obu J (2021) How Much of the Earth's Surface is Underlain by Permafrost? J Geophys Res Earth Surf 126:5. https://doi.org/10.1029/2021jf006123
Olefeldt D, Goswami S, Grosse G, Hayes D, Hugelius G, Kuhry P, McGuire AD, Romanovsky VE, Sannel ABK, Schuur EAG, Turetsky MR (2016) Circumpolar distribution and carbon storage of thermokarst landscapes. Nat Commun 7:11. https://doi.org/10.1038/ncomms13043
Puglini M, Brovkin V, Regnier P, Arndt S (2020) Assessing the potential for non-turbulent methane escape from the East Siberian Arctic Shelf. Biogeosciences 17:3247–3275. https://doi.org/10.5194/bg-17-3247-2020
Ramage J, Kuhn M, Virkkala AM, Voigt C, Marushchak ME, Bastos A, Biasi C, Canadell JG, Ciais P, López-Blanco E, Natali SM, Olefeldt D, Potter S, Poulter B, Rogers BM, Schuur EAG, Treat C, Turetsky MR, Watts J, Hugelius G (2024) The Net GHG balance and budget of the permafrost region (2000–2020) from ecosystem flux upscaling. Glob Biogeochem Cycle 38:7953. https://doi.org/10.1029/2023gb007953
Randers J, Goluke U (2021) An earth system model shows self-sustained thawing of permafrost even if all man-made GHG emissions stop in 2020 (vol 10, 18456, 2020). Sci Rep 11:1. https://doi.org/10.1038/s41598-021-81695-6
Rantanen M, Karpechko AY, Lipponen A, Nordling K, Hyvärinen O, Ruosteenoja K, Vihma T, Laaksonen A (2022) The Arctic has warmed nearly four times faster than the globe since 1979. Commun Earth Environ 3:10. https://doi.org/10.1038/s43247-022-00498-3
Rawlins MA, Karmalkar AV (2024) Regime shifts in Arctic terrestrial hydrology manifested from impacts of climate warming. Cryosphere 18:1033–1052. https://doi.org/10.5194/tc-18-1033-2024
Raymond PA, McClelland JW, Holmes RM, Zhulidov AV, Mull K, Peterson BJ, Striegl RG, Aiken GR, Gurtovaya TY (2007) Flux and age of dissolved organic carbon exported to the Arctic Ocean: a carbon isotopic study of the five largest arctic rivers. Glob Biogeochem Cycle 21:9. https://doi.org/10.1029/2007gb002934
Reschke J, Bartsch A, Schlaffer S, Schepaschenko D (2012) Capability of C-Band SAR for operational wetland monitoring at high latitudes. Remote Sens 4:2923–2943. https://doi.org/10.3390/rs4102923
Romanou A, Hegerl GC, Seneviratne SI, Abis B, Bastos A, Conversi A, Landolfi A, Kim H, Lerner PE, Mekus J, Otto-Bliesner BL, Pausata FSR, Pinto I, Suarez-Guiterrez L (2024) Extreme events contributing to tipping elements and tipping points. Surv Geophys. https://doi.org/10.1007/s10712-024-09863-7
Runge A, Nitze I, Grosse G (2022) Remote sensing annual dynamics of rapid permafrost thaw disturbances with LandTrendr. Remote Sens Environ 268:18. https://doi.org/10.1016/j.rse.2021.112752
Saros JE, Hazuková V, Northington RM, Huston GP, Lamb A, Birkel S, Pereira R, Bourdin G, Jiang BB, McGowan S (2025) Abrupt transformation of West Greenland lakes following compound climate

extremes associated with atmospheric rivers. Proc Natl Acad Sci USA 122:122. https://doi.org/10.1073/pnas.2413855122
Scheffer M, Bascompte J, Brock WA, Brovkin V, Carpenter SR, Dakos V, Held H, van Nes EH, Rietkerk M, Sugihara G (2009) Early-warning signals for critical transitions. Nature 461:53–59. https://doi.org/10.1038/nature08227
Schroeder R, McDonald KC, Chapman BD, Jensen K, Podest E, Tessler ZD, Bohn TJ, Zimmermann R (2015) Development and evaluation of a multi-year fractional surface water data set derived from active/passive microwave remote sensing data. Remote Sens 7:16688–16732. https://doi.org/10.3390/rs71215843
Schuur EAG, McGuire AD, Schädel C, Grosse G, Harden JW, Hayes DJ, Hugelius G, Koven CD, Kuhry P, Lawrence DM, Natali SM, Olefeldt D, Romanovsky VE, Schaefer K, Turetsky MR, Treat CC, Vonk JE (2015) Climate change and the permafrost carbon feedback. Nature 520:171–179. https://doi.org/10.1038/nature14338
Schuur EAG, Abbott BW, Commane R, Ernakovich J, Euskirchen E, Hugelius G, Grosse G, Jones M, Koven C, Leshyk V, Lawrence D, Loranty MM, Mauritz M, Olefeldt D, Natali S, Rodenhizer H, Salmon V, Schädel C, Strauss J, Treat C, Turetsky M (2022) Permafrost and climate change: carbon cycle feedbacks from the warming arctic. Annu Rev Environ Resour 47:343–371. https://doi.org/10.1146/annurev-environ-012220-011847
Schwinger J, Asaadi A, Goris N, Lee H (2022) Possibility for strong northern hemisphere high-latitude cooling under negative emissions. Nat Commun 13:9. https://doi.org/10.1038/s41467-022-28573-5
Serikova S, Pokrovsky OS, Laudon H, Krickov IV, Lim AG, Manasypov RM, Karlsson J (2019) High carbon emissions from thermokarst lakes of Western Siberia. Nat Commun 10:7. https://doi.org/10.1038/s41467-019-09592-1
Shu SJ, Jain AK, Koven CD, Mishra U (2020) Estimation of permafrost SOC stock and turnover time using a land surface model with vertical heterogeneity of permafrost soils. Glob Biogeochem Cycle 34:23. https://doi.org/10.1029/2020gb006585
Shugar DH, Burr A, Haritashya UK, Kargel JS, Watson CS, Kennedy MC, Bevington AR, Betts RA, Harrison S, Strattman K (2020) Rapid worldwide growth of glacial lakes since 1990. Nature Climate Change 10:939. https://doi.org/10.1038/s41558-020-0855-4
Stein R and Macdonald RW (2004) Organic carbon budget: arctic ocean vs. global ocean in: the organic carbon cycle in the arctic ocean edited by: Stein R, MacDonald RW, Springer, Berlin, Heidelberg
Stuart-Smith RF, Roe GH, Li S, Allen MR (2021) Increased outburst flood hazard from Lake Palcacocha due to human-induced glacier retreat. Nat Geosci 14:9. https://doi.org/10.1038/s41561-021-00686-4
Su Y, Ran YH, Zhang GQ, Li X (2023) Remotely sensed lake area changes in permafrost regions of the Arctic and the Tibetan Plateau between 1987 and 2017. Sci Total Environ 880:163355. https://doi.org/10.1016/j.scitotenv.2023.163355
Sun J, Zhou TC, Liu M, Chen YC, Shang H, Zhu LP, Shedayi AA, Yu H, Cheng GW, Liu GH, Xu M, Deng W, Fan JH, Lu XY, Sha YK (2018) Linkages of the dynamics of glaciers and lakes with the climate elements over the Tibetan Plateau. Earth-Sci Rev 185:308–324. https://doi.org/10.1016/j.earscirev.2018.06.012
Swann AL, Fung IY, Levis S, Bonan GB, Doney SC (2010) Changes in Arctic vegetation amplify high-latitude warming through the greenhouse effect. Proc Natl Acad Sci 107:1295–1300. https://doi.org/10.1073/pnas.0913846107
Trofaier AM, Bartsch A, Rees WG, Leibman MO (2013) Assessment of spring floods and surface water extent over the Yamalo-Nenets Autonomous District. Environ Res Lett 8:9. https://doi.org/10.1088/1748-9326/8/4/045026
Turetsky MR, Abbott BW, Jones MC, Anthony KW, Olefeldt D, Schuur EAG, Grosse G, Kuhry P, Hugelius G, Koven C, Lawrence DM, Gibson C, Sannel ABK, McGuire AD (2020) Carbon release through abrupt permafrost thaw. Nature Geosci 13:138. https://doi.org/10.1038/s41561-019-0526-0
Vonk JE, Tank SE, Bowden WB, Laurion I, Vincent WF, Alekseychik P, Amyot M, Billet MF, Canário J, Cory RM, Deshpande BN, Helbig M, Jammet M, Karlsson J, Larouche J, MacMillan G, Rautio M, Anthony KMW, Wickland KP (2015) Reviews and syntheses: effects of permafrost thaw on Arctic aquatic ecosystems. Biogeosciences 12:7129–7167. https://doi.org/10.5194/bg-12-7129-2015
Vulis L, Tejedor A, Schwenk J, Piliouras A, Rowland J, Foufoula-Georgiou E (2020) Channel network control on seasonal lake area dynamics in arctic deltas. Geophys Res Lett 47:710. https://doi.org/10.1029/2019gl086710
Vulis L, Tejedor A, Zaliapin I, Rowland JC, Foufoula-Georgiou E (2021) Climate signatures on lake and wetland size distributions in arctic deltas. Geophys Res Lett. https://doi.org/10.1029/2021gl094437
Wang SV, Foster A, Lenz EA, Kessler JD, Stroeve JC, Anderson LO, Turetsky M, Betts R, Zou SJ, Liu W, Boos WR, Hausfather Z (2023) Mechanisms and impacts of earth system tipping elements. Rev Geophys 61:81. https://doi.org/10.1029/2021rg000757

Watts JD, Kimball JS, Bartsch A, McDonald KC (2014) Surface water inundation in the boreal-Arctic: potential impacts on regional methane emissions. Environ Res Lett 9:13. https://doi.org/10.1088/1748-9326/9/7/075001

Webb EE, Liljedahl AK (2023) Diminishing lake area across the northern permafrost zone. Nat Geosci 16:202–209

Webb MJ, Andrews T, Bodas-Salcedo A, Bony S, Bretherton CS, Chadwick R, Chepfer H, Douville H, Good P, Kay JE, Klein SA, Marchand R, Medeiros B, Siebesma AP, Skinner CB, Stevens B, Tselioudis G, Tsushima Y, Watanabe M (2017) The cloud feedback model intercomparison project (CFMIP) contribution to CMIP6. Geosci Model Dev 10:359–384. https://doi.org/10.5194/gmd-10-359-2017

Webb EE, Liljedahl AK, Cordeiro JA, Loranty MM, Witharana C, Lichstein JW (2022) Permafrost thaw drives surface water decline across lake-rich regions of the Arctic. Nature Climate Change 12:841. https://doi.org/10.1038/s41558-022-01455-w

Webb E (2022) Pan-Arctic surface water (yearly and trend over time) 2000–2021. Arctic Data Center

Westermann S, Boike J, Langer M, Schuler TV, Etzelmüller B (2011) Modeling the impact of wintertime rain events on the thermal regime of permafrost. Cryosphere 5:945–959. https://doi.org/10.5194/tc-5-945-2011

Widhalm B, Bartsch A, Strozzi T, Jones N, Khomutov A, Babkina E, Leibman M, Khairullin R, Göckede M, Bergstedt H, von Baeckmann C, Muri X (2025) InSAR-derived seasonal subsidence reflects spatial soil moisture patterns in Arctic lowland permafrost regions. Cryosphere 19:1103–1133. https://doi.org/10.5194/tc-19-1103-2025

Wik M, Varner RK, Anthony KW, MacIntyre S, Bastviken D (2016) Climate-sensitive northern lakes and ponds are critical components of methane release. Nature Geosci 9:99. https://doi.org/10.1038/ngeo2578

Wild B, Andersson A, Bröder L, Vonk J, Hugelius G, McClelland JW, Song WJ, Raymond PA, Gustafsson Ö (2019) Rivers across the Siberian Arctic unearth the patterns of carbon release from thawing permafrost. Proc Natl Acad Sci USA 116:10280–10285. https://doi.org/10.1073/pnas.1811797116

Wilkenskjeld S, Miesner F, Overduin PP, Puglini M, Brovkin V (2022) Strong increase in thawing of subsea permafrost in the 22nd century caused by anthropogenic climate change. Cryosphere 16:1057–1069. https://doi.org/10.5194/tc-16-1057-2022

Woolway RI, Merchant CJ (2019) Worldwide alteration of lake mixing regimes in response to climate change. Nature Geosci 12:271. https://doi.org/10.1038/s41561-019-0322-x

Wu DH, Liu D, Wang T, Ding JZ, He YJ, Ciais P, Zhang GX, Piao SL (2021) Carbon turnover times shape topsoil carbon difference between Tibetan Plateau and Arctic tundra. Sci Bull 66:1698–1704. https://doi.org/10.1016/j.scib.2021.04.019

Wunderling N, von der Heydt AS, Aksenov Y, Barker S, Bastiaansen R, Brovkin V, Brunetti M, Couplet V, Kleinen T, Lear CH, Lohmann J, Roman-Cuesta RM, Sinet S, Swingedouw D, Winkelmann R, Anand P, Barichivich J, Bathiany S, Baudena M, Bruun JT, Chiessi CM, Coxall HK, Docquier D, Donges JF, Falkena SKJ, Klose AK, Obura D, Rocha J, Rynders S, Steinert NJ, Willeit M (2024) Climate tipping point interactions and cascades: a review. Earth Syst Dynam 15:41–74. https://doi.org/10.5194/esd-15-41-2024

Yao FF, Livneh B, Rajagopalan B, Wang JD, Crétaux JF, Wada Y, Berge-Nguyen M (2023) Satellites reveal widespread decline in global lake water storage. Science 380:743–749. https://doi.org/10.1126/science.abo2812

Yershov ED (1998) General geocryology (Studies in Polar Research). First published in Russian as obshchaya geokriologiya by Nedra, 1990, Cambridge University Press Cambridge, New York, Melbourne, p 604

Zandt in't MH, Liebner S, Welte CU (2020) Roles of thermokarst lakes in a warming world. Trends Microbiol 28:769–779. https://doi.org/10.1016/j.tim.2020.04.002

Zhang GQ, Yao TD, Shum CK, Yi S, Yang K, Xie HJ, Feng W, Bolch T, Wang L, Behrangi A, Zhang HB, Wang WC, Xiang Y, Yu JY (2017) Lake volume and groundwater storage variations in Tibetan Plateau's endorheic basin. Geophys Res Lett 44:5550–5560. https://doi.org/10.1002/2017gl073773

Zhang C, Duan AM, Jia XJ, Wang ZB, Pan ZL (2023) A dynamic link between spring Arctic sea ice and the Tibetan Plateau snow increment indicator. Npj Clim Atmosph Sci 6:505. https://doi.org/10.1038/s41612-023-00505-0

Zhang S, Gan TY, Bush ABG, Zhang G (2023b) Evaluation of the impact of climate change on the streamflow of major pan-Arctic river basins through machine learning models. J Hydrol 619:129295. https://doi.org/10.1016/j.jhydrol.2023.129295

Zhao ZL, Liu FG, Zhang YL, Liu LS, Qi W (2017) The dynamic response of lakes in the tuohepingco basin of the tibetan plateau to climate change. Environ Earth Sci 76:15. https://doi.org/10.1007/s12665-017-6446-7

Zwieback S, Meyer FJ (2021) Top-of-permafrost ground ice indicated by remotely sensed late-season subsidence. Cryosphere 15:2041–2055. https://doi.org/10.5194/tc-15-2041-2021

Zwieback S, McClernan M, Kanevskiy M, Jorgenson MT, Walker DA, Chang QY, Bergstedt H, Toniolo H, Romanovsky VE, Meyer FJ (2023) Disparate permafrost terrain changes after a large flood observed from space. Permaf Periglac Process 16:451. https://doi.org/10.1002/ppp.2208

Publisher's Note Springer Nature remains neutral with regard to jurisdictional claims in published maps and institutional affiliations.

Authors and Affiliations

Victor Brovkin[1,2] · Annett Bartsch[3] · Gustaf Hugelius[4] · Elisa Calamita[5] · J. Jelle Lever[5,6] · Eunhan Goo[7] · Hyungjun Kim[7] · Tobias Stacke[1] · Philipp de Vrese[1]

✉ Victor Brovkin
victor.brovkin@mpimet.mpg.de

1 Max Planck Institute for Meteorology, 20146 Hamburg, Germany

2 CEN, Universität Hamburg, 20146 Hamburg, Germany

3 b. Geos GmbH, Industriestr., 1A, 2100 Korneuburg, Austria

4 Department of Physical Geography and Bolin Centre for Climate Research, Stockholm University, 106 91 Stockholm, Sweden

5 Swiss Federal Institute of Aquatic Science and Technology (Eawag), Überlandstrasse 133, 8600 Dübendorf, Switzerland

6 Swiss Federal Institute for Forest, Snow and Landscape Research (WSL), Zürcherstrasse 111, 8903 Birmensdorf, Switzerland

7 Korea Advanced Institute of Science and Technology, 291 Daehak-ro, Yuseong-gu, Daejeon 34141, South Korea

Surveys in Geophysics (2025) 46:443–502
https://doi.org/10.1007/s10712-024-09859-3

Opportunities for Earth Observation to Inform Risk Management for Ocean Tipping Points

Richard A. Wood[1] · Jonathan A. Baker[1] · Grégory Beaugrand[2] · Jacqueline Boutin[3] · Alessandra Conversi[4] · Reik V. Donner[5,6] · Ivy Frenger[7] · Eric Goberville[8] · Hakase Hayashida[9] · Wolfgang Koeve[7] · Karin Kvale[10] · Angela Landolfi[11] · Wieslaw Maslowski[12] · Andreas Oschlies[7] · Anastasia Romanou[13] · Christopher J. Somes[7] · Thomas F. Stocker[14] · Didier Swingedouw[15]

Received: 10 December 2023 / Accepted: 22 August 2024 / Published online: 6 November 2024

Abstract
As climate change continues, the likelihood of passing critical thresholds or tipping points increases. Hence, there is a need to advance the science for detecting such thresholds. In this paper, we assess the needs and opportunities for Earth Observation (EO, here understood to refer to satellite observations) to inform society in responding to the risks associated with ten potential large-scale ocean tipping elements: Atlantic Meridional Overturning Circulation; Atlantic Subpolar Gyre; Beaufort Gyre; Arctic halocline; Kuroshio Large Meander; deoxygenation; phytoplankton; zooplankton; higher level ecosystems (including fisheries); and marine biodiversity. We review current scientific understanding and identify specific EO and related modelling needs for each of these tipping elements. We draw out some generic points that apply across several of the elements. These common points include the importance of maintaining long-term, consistent time series; the need to combine EO data consistently with in situ data types (including subsurface), for example through data assimilation; and the need to reduce or work with current mismatches in resolution (in both directions) between climate models and EO datasets. Our analysis shows that developing EO, modelling and prediction systems together, with understanding of the strengths and limitations of each, provides many promising paths towards monitoring and early warning systems for tipping, and towards the development of the next generation of climate models.

Keywords Earth observation · Ocean circulation · Ocean deoxygenation · Ocean ecosystems · Tipping point

Abbreviations

ACSs	Abrupt community shifts
AI	Artificial intelligence
AIS	Automatic identification system
AMOC	Atlantic meridional overturning circulation

Extended author information available on the last page of the article

ASI	Infrared atmospheric sounding interferometer
bbp	Particulate back scattering
BG	Beaufort gyre
BH	Beaufort high
CH	Cold halocline
CIMR	Copernicus imaging microwave radiometer
CI	Climate change initiative
CMIP	Climate model intercomparison project
DMS	Dimethylsulfide
DMSP	Dimethylsulfoniopropionate
ENSO	El Niño Southern Oscillation
EO	Earth observation
ESM	Earth system models
EWIs	Early warning indicators
GCMs	General circulation models
GHGs	Greenhouse gases
HILL	High impact low likelihood
IL	Icelandic low
ISSI	International space science institute
ITCZ	Intertropical convergence zone
LGM	Last glacial maximum
METAL	MacroEcological theory on the arrangement of life
ML	Machine learning
NAHosMIP	North Atlantic Hosing model intercomparison project
NAO	North Atlantic Oscillation
nNLS	Near-shore non-large meander
OAEs	Oceanic anoxic events
ODZs	Oxygen deficient zones
oNLS	Off-shore non-large meander
POC	Particulate organic carbon
RFI	Radio frequency interference
RMSE	Root mean square error
Rrs	Remote sensing reflectances
SCICEX	Scientific ice expeditions
SPG	Atlantic sub-polar gyre
SMAP	Moisture active passive
SMOS	Soil moisture and ocean salinity
SROCC	Special report on the ocean and cryosphere in a changing climate
SSH	Sea surface height
SSS	Sea Surface salinity
SST	Sea surface temperature
TE	Tipping element
TipMIP	Tipping point modelling intercomparison project
tLS	Typical large meander
TP	Tipping point
TPD	Transpolar drift

Article Highlights

- We review scientific understanding of ten potential tipping points in ocean circulation, biogeochemistry, and ecosystems
- We identify opportunities for Earth Observation to support improved models, predictions and early warning of these tipping points
- We identify how EO data can be combined with in situ data and models to maximise the value of each

1 Introduction

The possibility of tipping behaviour in the ocean has a long history, heading back at least to the pioneering work of Stommel (1961), who identified the potential for bistability and threshold behaviour in an idealised two-box ocean circulation model that is controlled by density differences arising from temperature and salinity variations. At the time, Stommel did not have a specific application in mind, although he noted the important role of salinity in the circulations of the Mediterranean and Red Seas. Only much later did oceanographers realise the potential applicability of these ideas to global-scale flows such as the Atlantic Meridional Overturning Circulation (AMOC), and hypothesise their role in major palaeoclimatic variations (Rooth 1982; Broecker et al. 1985; Broecker 1987, see also Stocker et al. 2024). In more recent times, the potential of tipping behaviour or regime shift has been recognised in a much wider range of elements of the ocean-climate system, including more regional circulation elements, elements of ocean biogeochemical cycles, and ocean ecosystems (e.g. Swingedouw 2020).

While a number of definitions of 'tipping elements' (TEs) and 'tipping points' (TPs) have been used in the literature, we consciously adopt a rather loose definition of a TE here as an element of the Earth System whose response to a smoothly varying change in forcing includes an abrupt, discontinuous or irreversible change when the forcing passes a certain threshold (the TP). This definition brings together a number of responses that are similar at the coarse level, but it will remain important when discussing each individual phenomenon in depth to characterise the tipping behaviour more precisely. Some key questions for each TE include: what is the relevant control parameter/forcing?; is there a critical threshold in the forcing (bifurcation) or simply a nonlinear response to forcing?; if a threshold is crossed, is there an element of irreversibility/hysteresis if the forcing is returned below the critical value?; what is the timescale for adjustment to the new state if a threshold is crossed?; if the forcing is reduced back to a subcritical value quickly enough, could tipping be avoided (resilience time)?; could internal 'noise' or particularly rapid forcing changes make transition to a new state more likely?

Many of the TPs discussed here can be considered examples of 'high impact, low likelihood' (HILL) climate hazards. In many cases uncertainty in current scientific understanding is so deep that it is not possible to provide robust estimates of how likely the events are to occur. Such hazards pose particular risks to society because they may imply changes that are greater than society's planned capacity to adapt to climate change, in some cases being even in the opposite direction to the expected changes. Adaptation measures may be extremely costly or have undesirable side effects, and investing in such measures when they probably will not be needed is not attractive. Despite these uncertainties, it may be possible to develop tools that allow management of such HILL risks, such as storylines of the impacts that would occur if the TP were crossed, and observable early warning indicators (EWIs) that give advance notice that the TP is being approached (Wood et al. 2023). The availability of reliable early warning of such events would enable greater flexibility in responding to the risks, while avoiding the costs of building resilience to an event that might never occur (see e.g. Marchau et al. 2019). Earth Observation (EO, taken here to refer to satellite observations) has the potential to provide important elements of these tools, including direct monitoring of the TE itself, monitoring of EWIs, and improved understanding of the processes and mechanisms at play in the TE, that can be used to constrain and refine models.

To derive the full value from EO, it is frequently important to combine the EO data with in situ observations and/or models. This can be done through: multivariate analysis of observations (which typically requires some form of simple underlying process-based model); assimilation of multiple observable types into more complex models such as General Circulation Models (GCMs), to provide either a self-consistent estimate of past variations (reanalysis), or an initial state to initialise forecasts/hindcasts of the ocean state; or confrontation of free-running GCMs (no data assimilated) with the observations, in order to evaluate the models themselves and point to paths for improvement. Assimilation of multiple data types is challenging as the error and covariance characteristics of the different variables need to be consistently specified. In this paper we aim to focus on EO variables, while addressing the developments needed in modelling, data assimilation and in situ datasets to deliver the full scientific potential.

This paper arises from discussion among a subset of the authors at the International Space Science Institute (ISSI) Workshop on “Tipping Points and Understanding EO data needs for a Tipping Element Model Intercomparison Project” in October 2022. We present a perspective of ocean TPs, focussing on opportunities to exploit existing EO data or potential new missions, to: monitor ocean tipping elements; provide useful early warning (EW) of tipping; and evaluate and improve models. The paper builds on the previous study of Swingedouw et al., (2020), by considering more (ocean) TEs, and more potential EO applications (not just early warning). We consider ten potential TEs (see below). The selection of TEs reflects to some extent the expertise of the author team, but there is a focus on TEs that are either intrinsically large-scale, or where more localised TEs are expected to be relevant over a wide geographical range. We are nevertheless conscious that some other potential ocean TEs or abrupt changes have been noted in the literature, e.g. acidification and carbon cycle (Heinze et al. 2021; Heinze et al. 2023), Mediterranean convection sites (Schneider et al. 2014), North Sea circulation (Holt et al. 2018), occurrences of marine heat waves (Benedetti-Cecchi 2021), tropical reefs (Lenton et al. 2019). We also do not discuss here impacts thresholds in coastal systems such as mangroves, sea grasses and coral reefs.

In Sect. 2, we consider six potential tipping elements of large (at least basin) scale ocean circulation and biogeochemistry: Atlantic Meridional Overturning Circulation (Sub-Sect. 2.1); Atlantic Subpolar Gyre convection and circulation (2.2); Beaufort Gyre fresh water storage (2.3); Arctic halocline (2.4); Kuroshio large meanders (2.5); and large-scale deoxygenation (2.6). Section 3 discusses four additional potential tipping elements in pelagic ecosystems that are themselves more local/regional, but where processes may be applicable to multiple geographical regions: phytoplankton (3.1); zooplankton (3.2); higher trophic levels and fisheries (3.3) and overall biodiversity (3.4). For each of the ten tipping elements considered, we provide a perspective on current knowledge of the nature of the tipping element (processes, timescales, drivers and impacts of tipping), and needs and opportunities for EO to advance knowledge of and resilience to the risks of tipping. In

Sect. 4 we synthesise the results of the previous sections, and suggest possible ways forward for the use of EO data in this field.

2 Tipping Points of Large-Scale Ocean Circulation and Biogeochemistry

2.1 Atlantic Meridional Overturning Circulation (AMOC)

2.1.1 Nature of Tipping Point

The Atlantic Meridional Overturning Circulation (AMOC) is a global-scale system of currents which transports warm water northwards in the Atlantic Basin. Much of the heat is lost to the atmosphere in the North Atlantic high latitudes, and consequently the AMOC has a fundamental role in shaping Northern Hemisphere climate (e.g. Bonnet et al. 2021). The AMOC is often characterised as a global-scale 'conveyor belt' circulation (Broecker 1991), driven by heat and fresh water exchanges with the atmosphere, and the consequent density changes. While this view is undoubtedly a gross simplification, model studies suggest that on decadal and longer timescales, coherent basin- or global-scale variations in the circulation are possible. In this section, we focus on large-scale variations whereas North Atlantic subpolar gyre variability is discussed in Sect. 2.2.

The AMOC has been long recognised as one of the major tipping elements of the climate system (Stommel 1961; Rooth 1982; Rahmstorf 1996, Lenton et al 2008 Swingedouw 2020; Armstrong McKay et al. 2022; Boers et al 2022). AMOC transitions (tipping) and multiple stable states have been found in a hierarchy of models, ranging from simple box models through to comprehensive General Circulation Models (GCMs). However not all climate models exhibit bistability (e.g. Schiller et al. 1997), highlighting considerable variation among climate models which persists to the current (CMIP6) generation (Jackson et al 2023). See Weijer et al. (2019) for a review.

Abrupt or gradual shifts in AMOC strength have been also identified in the paleoclimate record. For example, AMOC bi-stability has been proposed as a mechanism for rapid climate variability between glacial and interglacial periods in the Pleistocene (Dansgaard et al. 1993; Blunier et al 1998; Renssen et al. 2001; de Abreu et al. 2003; Lynch-Stieglitz 2017; Moffa-Sanchez et al. 2019), and during the last interglacial (Galaasen et al. 2014). The Younger Dryas cold event has been associated with AMOC collapse following the discharge of glacial lakes (Broecker et al 1985, 1987; Rooth 1982; Lehman and Keigwin 1992; O' Hare et al. 2011; Rind et al 2018). AMOC variations are also thought to drive the "bipolar seesaw" leading to changes of opposite phase in the Arctic and Antarctic climates (Broecker 1998; Stocker 1998; Rahmstorf 2002; Stocker and Johnsen 2003, Skinner and Elderfield 2007; Pedro et al. 2011).

The wide range of model behaviour, and plausibility of abrupt AMOC changes, leads to a pressing need for observations to constrain the models, and to provide independent warning if such an event is approaching.

2.1.1.1 Drivers of AMOC Tipping The main mechanism that drives abrupt AMOC changes in models is the salt advection feedback: if the overturning circulation carries relatively salty waters into the North Atlantic, an AMOC slowdown would tend to make the Atlantic basin fresher, amplifying the slowdown itself. Early comprehensive climate models were judged to have a 'too-stable' AMOC due to biases in the fresh water transport by the AMOC (e.g. Liu et al. 2017). These biases are still present in more recent climate models, although advances in modelling and resolution have reduced them to some extent (Deshayes et al. 2013; Mecking et al. 2016; Hirschi et al 2020).

However overall AMOC stability depends on the combined effect of several positive and negative feedbacks (e.g. Vellinga and Wood 2002; Yin et al. 2006; Swingedouw et al. 2007; Hofmann and Rahmstorf 2009; Hirschi et al. 2020). The AMOC only tips when the destabilising effect of the salt advection feedback overcomes the stabilising feedbacks (Jackson et al. 2017).

Key additional processes include the freshwater transport by the wind-driven gyre circulation in the South Atlantic (Thomas and Fedorov 2019) and North Atlantic (Wood et al. 2019), discharges of excess freshwater through Fram or Davis straits, triggered by changing winds (Haine et al 2015), cooling over the Atlantic Sub-polar Gyre (SPG) due to storm track changes (Li et al 2022a), North Atlantic Oscillation (NAO) variability (Klus et al 2018), sea-ice interactions (Swingedouw et al. 2007), the strength of the 'warming hole' (Keil et al 2020), the rate of freshwater input (Wood et al 2019; Kim et al 2021) and other atmosphere–ocean feedbacks which may be subdued, underestimated or biased in many models (Gent 2018). Some such processes (e.g. fresh water input from the Greenland Ice Sheet) are still only incompletely represented in current climate models (Swingedouw et al. 2007; Bakker et al. 2016).

The location and likelihood of tipping in climate models depends on model parameters that are poorly known and need to be constrained using parametric uncertainty analysis. Simpler models can help here, provided they can be shown to capture the key AMOC tipping processes and feedbacks. Parameters can be estimated either by formulating the simpler models in terms of observable quantities (Wood et al. 2019), or by Bayesian ensemble inversion methods (e.g. Lux et al (2022), which showed that suitable paleo-proxy time-series may contain enough information to constrain the likely position of AMOC TPs). By identifying the crucial parameters to constrain the position of TPs, it may be possible to focus observational efforts to provide early warning. But these problems are still very much at the research stage.

It is not clear whether there is a critical level of global warming that would trigger AMOC collapse. Based on a literature survey, Armstrong McKay et al (2022) estimate a very broad range of 1.4 to 8 °C for such a threshold, if it exists. Other factors not strongly correlated with warming level (e.g. response of the hydrological cycle to warming) may play a key role, partly explaining the wide range of warming levels suggested by Armstrong McKay et al. The results of Romanou et al (2023) suggest that even moderate climate forcing (such as SSP2-4.5) may be sufficient to drive the AMOC over a threshold.

Noise-induced transitions, in which internal climate variability causes a temporary AMOC collapse, have recently been shown to be possible in future climate simulations (Castellana and Dijkstra 2020). Romanou et al. (2023) in a fully coupled climate model and without applying external freshwater forcing, found evidence of such noise-induced transitions in a state-of-the-art climate model (NASA-GISS-E2-1-G), spontaneously arising from stochastic variability of high latitude freshwater fluxes into the ocean and with an AMOC off state lasting for about 800 years. Observational requirements to initialise prediction of such behaviour are not currently known.

Rate-induced tipping (Stocker and Schmittner 1997; Alkhayuon et al. 2019; Ritchie et al. 2022) might also occur: too-fast forcing variations (e.g. greenhouse gases or fresh water input) may cause the AMOC to tip, even before the forcing reaches a critical level. This implies that safe climate mitigation pathways need to consider rates of change as well as the final destination. Further, given that future mitigation strategies may eventually lead to drastic reductions to GHGs, more understanding is needed of potential abrupt changes under 'overshoot' or negative emissions scenarios (e.g. Schwinger et al 2022).

2.1.1.2 Timescales of Tipping Based on a literature survey, Armstrong McKay et al. (2022) suggest a timescale for AMOC collapse, once a threshold is passed, of 15–300 years, consistent with timescales estimated from a range of model studies (Bryan 1986; Jackson and Wood 2017; Weijer et al. 2019; Sinet et al. 2022).

As discussed earlier, models might be too stable (Liu et al. 2017), and the processes that spread the freshwater discharge from GIS might be too slow (Swingedouw et al. 2021), hence the timescales from CMIP6 (Coupled Model Intercomparison Project, Phase 6) models are possibly too long. Shorter timescales were inferred from the paleoclimate record; for example the termination of the Younger Dryas cold event took only 10 years (Alley and Agustdottir 2005). The Intergovernmental Panel on Climate Change's (IPCC's) 6th Assessment Report (IPCC 2021, referred to as AR6) concludes that the AMOC will not collapse within this century, but only assigns 'medium confidence' to that conclusion, highlighting the poor capabilities of present-day models to represent a number of crucial processes.

2.1.1.3 Impacts of Tipping A shut-down of the AMOC would have global impacts on climate. As well as the well-known impacts over the Euro-Atlantic sector (e.g. Jackson et al. 2015; Bellomo et al. 2023), it would lead to more extensive Arctic ice and warming of the Southern Hemisphere. The Inter-tropical Convergence Zone (ITCZ) would be expected to migrate southward and the global water cycle perturbed with weaker Indian and Asian summer monsoons (Buckley and Marshall 2016; Wassenburg et al 2021). El Nino-Southern Oscillation (ENSO) variability may be reduced (Orihuela-Pinto 2022) and shifted eastwards (Williamson et al 2018). Sea levels would rise on the western side of the North Atlantic and Atlantic storm tracks shift northward and become more intense (IPCC 2019a, b).

These effects would be expected to lead to a decrease in marine productivity in the North Atlantic, a reduction in Sahelian, South Asian and part of Amazonian rainfall, and a decrease in the number of tropical cyclones in the Atlantic (IPCC 2019a, b). While detailed sectoral impact studies are few (e.g. Ritchie et al. 2020), the societal impacts would be expected to be far-reaching (e.g. DeFrance et al. 2017).

2.1.1.4 AMOC Early Warning Signals and Proxies As discussed in Swingedouw et al. (2020), to monitor AMOC shifts and identify possible TPs we need adequately long records, e.g. of temperature and salinity in the North Atlantic and transports across various latitudes (as in Frajka-Williams et al. 2019). However that may still not be sufficient to obtain early warning signals (EWS). Since direct AMOC observations have only become available fairly recently, it is likely that practical early warning signals will need to use various proxies or fingerprints of the AMOC to obtain long enough samples.

Low order dynamical system theory suggests that near a bifurcation point systems become more sluggish in response to small perturbations (Held and Kleinen 2004; Carpenter and Brock 2006; Boulton et al 2014) i.e. increasing autocorrelation and variance. A few studies have highlighted that recent changes in various AMOC proxies/fingerprints, might be exhibiting such a change in autocorrelation properties (Boers 2021; Michel et al. 2022), highlighting the potential approach of the AMOC to a tipping point. Because a long enough historical time series of the AMOC is not available, these studies unavoidably depend on proxies or 'fingerprints', i.e. spatio-temporal observations of other variables that are correlated with AMOC changes, e.g. SST (Dima and Lohmann (2010), Rahmstorf et al. (2015), and Caesar et al. (2018)), subsurface hydrography (Chen and Tung 2018; Fraser and Cunningham 2021), EO gravity data (Landerer et al. 2015), and surface altimetry in combination with various in situ observations (Willis 2010; Frajka-Williams 2015; Mercier et al. 2015). Both the consistency of the proxies (e.g. Little et al. 2020; Kilbourne et al. 2022) and their relevance for AMOC tipping as opposed to recent variability (Jackson and Wood 2020) remain the subject of debate. Meanwhile more process-based early warning indicators (e.g. Alkhayuon et al. 2019) are at an early stage of research.

2.1.2 EO needs and opportunities

EO can play a valuable role in monitoring the AMOC and providing insights into key processes. However, at least on the 5–10 year time horizon, it will be essential to use EO alongside key in situ datasets, (trans-basin arrays, Argo, Arctic Ocean exchanges), which observe sub-surface signatures of the AMOC that are inaccessible to satellite instruments.

Long-term consistency and continuity of existing datasets is particularly critical because the (at least) decadal timescale of AMOC tipping means that early warning indicators could be contaminated by data inhomogeneities.

2.1.2.1 Proxies Proxy indicators ('fingerprints') of the AMOC give the opportunity to monitor in future at locations where trans-basin sections are not available or practical, and potentially allow longer past timeseries to be constructed, opening up greater options for early warning. SST, SSH and gravity observations have been identified as particular opportunities for EO. However there is substantial disagreement between different proxy timeseries, while modelling (Jackson and Wood 2020) suggests that no single proxy can capture all aspects of AMOC variability (natural variations, response to global warming, tipping).

- Work is therefore needed to develop *robust, multi-variate* AMOC proxies, probably combining EO and in situ data, for monitoring and attribution of future AMOC change.
- Machine learning, trained on a range of model data, potentially provides novel opportunities to develop such multivariate proxies (e.g. DelSole and Nedza 2022; Solodoch et al. 2023; Michel et al. 2023).

2.1.2.2 Processes Improved estimates of heat and fresh water exchanges with the subpolar basin will allow improved monitoring of the drivers of AMOC change and hence potential early warning of major changes. This includes surface heat and freshwater fluxes, water inputs from rivers and the Greenland ice sheet, and Arctic exchanges.

- Key regions for monitoring include freshwater and ice transports across Fram/Denmark and Davis Straits as well as storage of freshwater in the different subpolar basins (Haine et al 2015) and ice edge/melting rates, in order to estimate changes in freshwater budgets of the deep water formation sites (Lobelle et al 2022). The potential for EO to inform ice transports in particular is discussed in Sect. 2.3.2.
- Polar regions, western boundary currents, storm tracks, and the subtropical-subpolar transition zone in the North Atlantic (Buckley and Marshall 2016 and references therein) are key regions for surface heat and freshwater exchanges. Advances may

best be delivered through intensive, localised multi-platform campaigns, combining EO with in situ observations and high-resolution models to evaluate climate models and develop improved parametrisations.
- Improved observational resolution of estuaries and coastal regions is needed, to better define their role in basin scale heat and regional freshwater budgets

2.1.2.3 Modelling A range of model experimentation is needed to inform future monitoring systems. Much of this could potentially be done as part of the emerging TipMIP programme (Tipping Point Model Intercomparison Project, https://tipmip.pik-potsdam.de/), building on recent experiments in the North Atlantic Hosing Model Intercomparison Project NAHosMIP (Jackson et al 2023). Modelling developments include:

- Generating plausible scenarios (storylines) of AMOC tipping, moving from traditional 'hosing' experiments to a more direct link with greenhouse gas forcing. This should include scenarios of bifurcation-induced, noise-induced and rate-induced tipping. These scenarios can be used to identify mechanisms of tipping and to develop physically-based early warning indicators.
- Understanding why different climate models have different AMOC tipping behaviour, and identifying circumstances in which the AMOC is likely to tip (either properties of the model climate system, or particular emissions pathways likely to result in tipping). This can inform missions to target key areas of process uncertainty.

2.1.2.4 Data assimilation, reanalysis and prediction Assimilation of multiple data types into advanced models (reanalysis) provides a synthesis of observations, which are themselves sparse and incomplete, with physical process understanding. Current reanalysis products capture many aspects of AMOC variability seen at the three trans-basin in situ AMOC sections (Fig. 1), and allow the observations from the sections to be extended back in time, providing a longer historical context (Fig. 1d). However, current reanalyses by no means capture all features seen in the sections. Comparison with the SAMBA (South Atlantic MOC Basin-wide Array) section (Fig. 1c) is quite poor; however, Baker et al. (2023) show that the reanalysis is well correlated with an alternative estimate of monthly-mean AMOC variability on this section, derived from combined altimeter SSH and in-situ hydrographic data (Dong et al. 2021).

Future development may consider the following:

- Assimilation of new data types, e.g. surface salinity measurements provided by SMAP (Soil Moisture Active Passive) and SMOS (Soil Moisture and Ocean Salinity) (Vinogradova et al 2019; Boutin et al. 2023), perhaps also with higher spatial resolution, and improved coverage at high latitudes (Estella-Perez et al. 2020)
- Improved data assimilation methods (e.g. Polkova et al. 2019; Barthélémy et al. 2022), simulators and Observing System Simulation Experiments are needed to make the most of the information in existing EO datasets and guide future missions, e.g. most current data assimilation systems give low or zero weight to altimeter data outside the subtropics, due to the challenge of consistently combining multiple data types, especially in the subpolar region where stratification is weak but dynamically important.

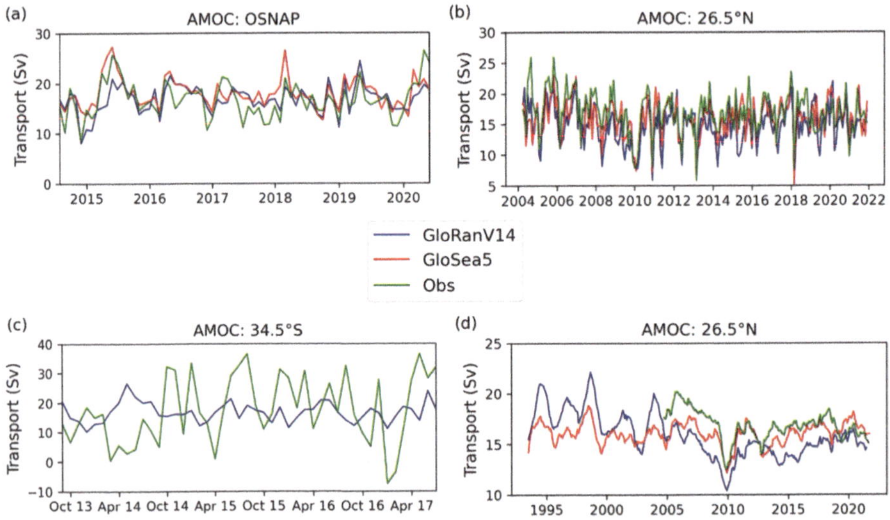

Fig. 1 Reconstructions of AMOC from two global ocean reanalysis products: GloSea5 (Jackson et al. 2016, 2019) and GloRanV14 (Baker et al. 2022, 2023), compared with direct estimates on three in-situ sections: **a** OSNAP in the subpolar gyre (Fu et al. 2023a, b), **b** RAPID-MOCHA around 26.5°N (Moat et al. 2023) and **c** SAMBA around 34.5°S (Kersalé et al 2020, 2021). In each case the period shown is that for which direct observations are available. In **c** only the GloRanV14 product is shown, as issues with the altimeter assimilation in the earlier GloSea5 product caused spurious MOC features at this latitude. In **d** is shown the 12-month running mean AMOC at RAPID-MOCHA, back to the start of the reanalysis (1993, coinciding with the start of precision altimetry), illustrating the potential of reanalysis to provide a longer historical perspective on limited in situ timeseries

- While AMOC tipping is likely to take longer than decadal timescales, decadal prediction systems, initialised with a wide range of observations, may give early indications of whether observed changes show the beginning of a longer term trend or simply represent random variability.

2.2 Atlantic Subpolar Gyre

2.2.1 Nature of Tipping Point

Convection in the Labrador and Irminger seas south of Greenland and the associated Subpolar Gyre (SPG) circulation form a regional ocean circulation element, that is related to the larger scale AMOC, but is not the same thing, as a substantial component of the SPG circulations recirculate regionally without a direct impact on the global circulation. The SPG circulation has been proposed as a climate tipping element with medium confidence by Armstrong McKay et al. (2022). That study was based on analyses of CMIP5 (Sgubin et al. 2017) and CMIP6 (Swingedouw et al. 2021) databases and the IPCC Special Report on Oceans, Cryosphere and Climate Change (SROCCC, Collins et al. 2019a, b). Armstrong McKay et al. (2022) estimated with high confidence that the SPG might have a global warming threshold of ~ 1.8 °C (1.1–3.8 °C) and a tipping timescale of ~ 10 years (5–50 years). The impact of such a collapse of the SPG might be substantial for the climate in the bordering regions (Collins et al. 2019b, a; Oltmanns et al. 2020; Sgubin et al. 2019) but also with wider large-scale teleconnections in the tropical areas, as well as for marine biogeochemistry (e.g. Gary et al. 2020). The Little Ice Age has been linked to weakening of the SPG (Michel et al. 2022) possibly due to large exports of sea ice that followed atmospheric blocking relaxation (Lapointe and Bradley 2021).

Nevertheless, we argue here that there is large uncertainty concerning the risk of this system to tip, given the high dependency of the collapse of the SPG on the model considered. Generally speaking, there are fewer models that do show a rapid collapse of the SPG than ones showing none or very gradual ones. Also, the Arctic and North Atlantic basins exhibit a high degree of variability and complexity, implying interactions between the ocean, the atmosphere and the cryosphere as well as fine-scale processes (Yashayev et al. 2015; Yashayev and Loder 2016; Romanou et al 2023), which even further decrease the confidence we might have in the current generation of climate models. The models showing a tipping are among the best in terms of ocean stratification both in CMIP5 and CMIP6. Stratification is indeed a key variable of ocean dynamics, in particular for what concerns convection, a central element of the SPG. It is at the heart of the positive salt advection feedback arising from this system: if convection ceases in the Labrador and Irminger Seas, the North Atlantic drift is no longer steered into these basins, and the supply of tropical salty waters in these seas is reduced, leading to a strong halocline that further suppresses convection.

This positive feedback can lead to a very stable new steady state with the risk of irreversibility and hysteresis behaviour (Born et al. 2013), which has nonetheless been poorly evaluated in the literature up to now. The threshold of convection is related to surface density, mainly driven by Sea Surface Salinity SSS (Swingedouw et al. 2020; Romanou et al 2023). For a given critical SSS threshold, the temperature cooling in winter is not sufficient to destabilise the water column: the freezing point is reached before convection can occur (as e.g. in the Arctic), so that sea ice is forming in place of convection. As shown in Swingedouw et al. (2020), in CMIP5 models, the threshold might be relatively close to the present-day observed state.

A strong freshwater anomaly is building in the Beaufort gyre from the Arctic, so that there is now considerable risk that this freshwater might flush in the SPG (Zhang et al. 2021, Lin et al. 2023, see Sect. 2.3). Given that the SPG system already recently experienced its largest freshening for the last 120 years in its eastern side (Holliday et al. 2020), this is raising serious concerns.

2.2.2 EO Needs and Opportunities

2.2.2.1 Early Warning The potential proximity to a threshold in SSS in the SPG highlights the need to use models initialised towards observations to include the recent freshening of the North Atlantic—Arctic sector. Such models starting from observed initial conditions constitute the decadal climate prediction systems. More than 10 such systems are developed around the world (WCRP decadal prediction) and provide predictions every year for the next decade (Polkova et al. 2023). Those models are assimilating the most recent EO (typically SST and altimetry) alongside in situ data such as Argo, and are therefore making a good integration of those data into a potentially useful early warning system, but should also integrate new SSS products in their assimilation scheme. The increase in error of SSS products with increasing latitudes is a limitation for data assimilation, which we hope might be improved in the near future. Particular opportunities include:

- Refinement of some components of the radiative transfer modelling used in the salinity retrieval chain, such as the dielectric constant or sea surface roughness modelling (e.g. Boutin et al. 2023)
- Application of new Radio Frequency Interference (RFI) mitigation techniques (e.g. Bonjean et al. 2024), planned to be applied in the future version of CCI v5 SSS.
- Develop corrections for contamination by the proximity of sea ice (Jiménez et al 2021)

In the longer term, the Copernicus Imaging Microwave Radiometer (CIMR), scheduled for launch in 2028, should achieve better radiometric resolution and stability, powerful filtering of RFI-contaminated measurements, and provide multi-frequency measurements that should enable improved corrections of temperature- and wind-related effects.

The use of lower frequencies than L-Band that are more sensitive to SSS in cold waters is another potential way for reducing SSS errors in cold waters (Johnson et al. 2021). The CRYORAD mission idea, recently selected by the European Space Agency ESA, is proposing such a concept. Nevertheless, the decadal prediction systems usually exhibit very large drift when starting from an observed state (Bilbao et al. 2021). This is due to the model biases and the fact that, starting from an observed state, the model usually drifts towards its own mean state. In the SPG, this can be crucial, since CMIP6 climate models usually exhibit a strong cool bias, so that the drift is usually a cooling drift, which might prevent the possibility of predicting a strong cooling that might really occur. The value of bias correcting the forecasts to remove drift, in a case where a tipping point is about to be crossed, is therefore limited. Improvements concerning the initial shock of a decadal prediction system is therefore strongly needed in order to provide skillful prediction of the SPG fate (Polkova et al. 2023). Such improvements might be achieved through reduction of model bias, through assimilation of new data types such as SSS, or through improved methods of assimilating multiple data types into biased models, especially at depth (e.g. Polkova et al. 2019; see also altimetry at high latitudes, see Sect. 2.1).

The direct use of EO observations as independent early warning systems might be very relevant for the SPG. In this respect, existing systems that monitor SSS (e.g. SMAP and SMOS, Vinagradova et al. 2019; Boutin et al. 2023) may provide important ground-truth information. The recent SSS data from EO observations (CCI) are shown in Fig. 2 as compared to in situ ones from EN4 and ISAS. The different SSS products are mainly consistent, they do not entirely agree with each other for the recent trends in terms of their best guess. While EN4 and ISAS show a relatively stable SSS evolution in the 2010s, the CCI product exhibits a large increase, mainly due to an upward shift in the years around 2013. This upward shift is reduced with the most recent CCI version which was better filtered for radio frequency interferences even though inconsistencies remain. Understanding and resolving this type of inconsistencies between the observation products might be required to better assess the crossing of a SSS tipping point in the SPG system, which remains also to be better estimated in climate models.

Another avenue of research might be to better analyse models showing abrupt changes in their SPG and the way their patterns in some observable variables evolve just before the collapse. This approach, which can be seen as a space-for-time substitution (Dakos et al. 2010) has been used for the Amazon rainforest (Verbesselt et al. 2016) and might be very promising for the ocean realm. In this respect the SSH, SST and SSS observations might be very relevant and useful, notably given their already relatively high resolution in space (~25 km) as compared to CMIP6 models (gridpoints usually larger than 50 km). Observations of the Arctic, in terms of SSS and sea ice cover and transport might be also informative, given the risk of a massive flush of freshwater from the Arctic towards the SPG. In this respect, reduction in the

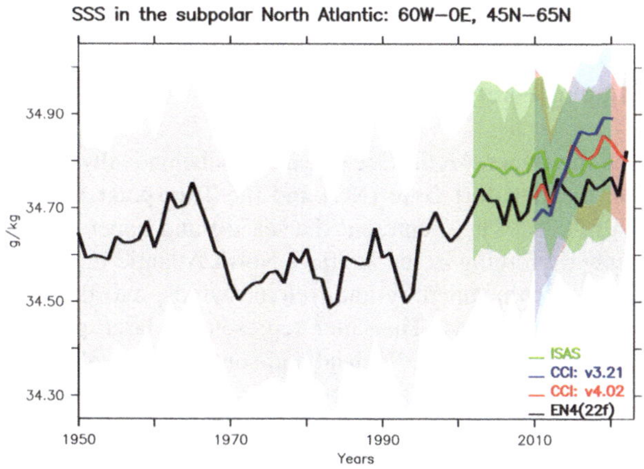

Fig. 2 SSS time series since 1950 averaged the subpolar gyre defined as the box 60–0°W, 45–65°N for different monitoring systems. In black is represented the EN4 product and in green the ISAS products, both of which are mainly based on in situ observation records, notably from ARGO from 2004. In blue and red are recent products from CCI in two different versions (Boutin et al.2021a, b). The overlap represents an upper estimate of the error for each product. It is estimated as the annual and spatial mean of the uncertainty provided at the gridpoint level. The mask from CCI v4 products using only gridpoints without sea ice has been applied to all products to allow a correct comparison of the data products

errors of SSS at such very high latitudes (Estella-Perez et al. 2020) should be considered as a key requirement for future EO products.

2.2.2.2 Model Development and Evaluation Present-day climate models are missing key processes of the SPG dynamics (Le Corre et al. 2020). In particular, CMIP6 models poorly represent the role of eddies, notably for the spread of freshwater anomalies from Greenland Ice Sheet (GIS) melting (Martin and Biastoch 2023, Swingedouw et al. 2021), but more generally concerning water transformation, a key process for stratification properties of the SPG (Desbruyeres et al. 2019; Hirschi et al. 2020). In this respect, next generation models should try to better represent fine scale processes, either through improved parametrization or grid-zooming in some key locations (Hewitt et al. 2020; Zanna and Bolton 2020). The challenge in modelling is compounded by the fact that seasonal sea ice cover in the SPG area plays a crucial role in modifying atmosphere–ocean buoyancy fluxes. Improved resolution of EO products will be decisive in this respect to better assess the dynamics of those processes, and can be used also for spatial early warning, or to evaluate freshening on and off the Greenland coast (e.g. Castelao et al. 2023), relative to EO-based estimates of meltwater from the ice sheet (e.g. Bamber, et al. 2018). Finally, consistent assimilation of multiple EO and in situ data types into models is a major challenge, especially in regions such as the SPG where stratification is weak. Improvements in this area have the potential to extract greater value from the available datasets for modelling and prediction systems.

2.3 Beaufort Gyre Freshwater Accumulation and Release

2.3.1 Nature of Tipping Point

The circulation of the upper Arctic Ocean can be schematically characterised by two main features, i.e. the Beaufort Gyre (BG) and the Transpolar Drift (TPD) (Timmermans and Marshall 2020). They represent the sea ice and upper ocean response to the large-scale atmospheric forcing in the Arctic – North Atlantic region dominated by the Beaufort high (BH), accompanied by anticyclonic winds, and the Icelandic low (IL), accompanied by cyclonic winds. The latter represents a large persistent atmospheric low-pressure centre located between Iceland and southern Greenland and it is the main atmospheric driver of the North Atlantic Subpolar Gyre (SPG) discussed in Sect. 2.2. Notably, the North Atlantic storm track originating from the IL has a strong effect on winter weather over Europe and North America (Serreze and Barry 2005) and its northerly shift represents the positive phase of the North Atlantic Oscillation (NAO), which can affect the BG (Moore et al. 2018), hence the BG.

The BG anticyclonic circulation tends to trap both sea ice and relatively fresh water within, which sets the vertical stratification in the western Arctic (Timmermans and Toole 2023). The sea ice present in the BG for most of the year is not only a source of freshwater in summer but also modulates the wind momentum transfer, which drives the gyre. Its thinning and decline over the past decades (Ding et al. 2017) increase the solar radiation and wind momentum (Rampal et al. 2009) transfer to the upper ocean, hence the intensification of the BG, an increased accumulation of freshwater within (Fig. 3), and a stronger cold halocline (CH, see Sect. 2.4 for more discussion). Estimates of freshwater content in the BG show significant interannual variability superimposed over the overall increasing trend (Proshutinsky et al. 2019). The BG freshwater content over the seasonal cycle varies by approximately $4–5\times10^3$ km^3 between maximum in December and minimum in April (Proshutinsky et al. 2009), while it has increased by approximately 6.4×10^3 km^3 from 2003 to 2018, which represents a 40% increase relative to the climatology of the 1970s (Proshutinsky et al. 2019).

Any significant release of freshwater from this reservoir into the North Atlantic could impact the wintertime deep-water formation and stability of AMOC and SPG, as discussed in Sects. 2.1, 2.2. Export of such "Great Salinity Anomalies" (GSAs) out of the Arctic and its propagation in the subpolar North Atlantic has been observed in the past (Dickson et al. 1988; Belkin 2004), including more recently (Bilo et al. 2022). While the consequent positive buoyancy flux at the surface of the subpolar Atlantic has been shown to freshen subpolar waters produced by deep convection, or to suppress deep convection all together (Dickson et al. 1988; Bilo et al. 2022), its impact on the circulation of the SPG has been less clear. Comparison of the advection rate of the observed GSA events in the 1970s, 1980s, and 1990s, reveals a substantial intensification of the circulation in the SPG (Belkin 2004).

Recent analysis by Lin et al. (2023) of an extensive hydrographic dataset from 2003 to 2019 and satellite-derived dynamic ocean topography data shows that the BG has transitioned to a quasi-stable state over the past decade, as represented by the sea surface height and freshwater content. They also find that the centre of BG has shifted southeastward (also see Fig. 3) due to regional wind forcing, and the CH layer within the BG has thinned significantly. The southeastward movement of the BG moves its freshwater closer to the Arctic gateways to the subpolar Atlantic (Canadian Arctic Archipelago and

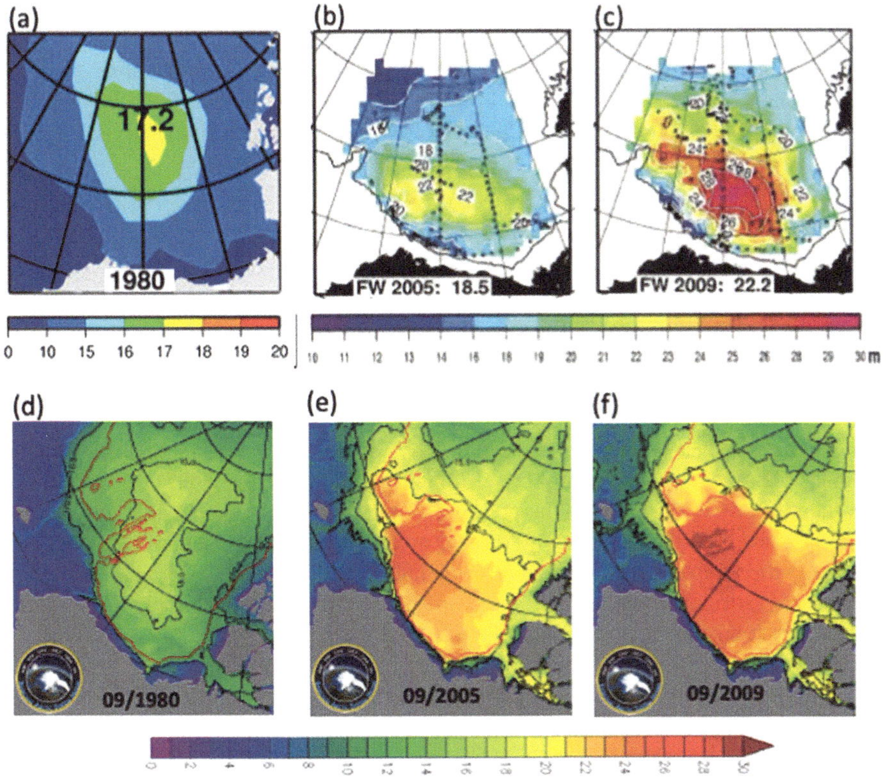

Fig. 3 Estimates of the BG freshwater content (in meters) from observations (top; https://www2.whoi.edu/site/beaufortgyre/) and the high-resolution Regional Arctic System Model (RASM; https://nps.edu/web/rasm) (bottom), in 1980 (**a**, **d**), 2005 (**b**, **e**), and 2009 (**c**, **f**)

Fram Strait). If continued, such changes could change the current stable state of the BG and allow a release of freshwater into the Atlantic.

2.3.2 EO Needs and Opportunities

The primary hydrographic data on the evolution of the BG and its freshwater content have been provided by *in-situ* measurements as part of the Beaufort Gyre Observing System (BGOS; https://www2.whoi.edu/site/beaufortgyre/). The BGOS, a collaborative project between the Woods Hole Oceanographic Institution (WHOI) in the U.S. and the Institute of Ocean Sciences (IOS) in Canada, was established in 2003 to measure fresh water and heat content / fluxes in the BG using moorings, drifting buoys, and remote sensing, with support from the U.S. National Science Foundation (NSF). In addition to the altimeter and laser measurements of sea surface height, snow, and sea ice thickness discussed in Sect. 2.4 below, measurements of sea ice drift and sea surface salinity derived from scatterometers and radiometers have been helpful to investigate the complex dynamics of the BG. However, their relatively coarse resolution has limited their application to study many important ocean and sea ice processes and interactions occurring at mesoscale (e.g. Manucharyan et al. 2022). Increasingly, km-scale and sub-daily measurements are in demand, in part due to the advancement of regional (Maslowski et al. 2012, 2014; Clement Kinney et al. 2022) and global (Wang et al. 2020; Jeong et al. 2023) high-resolution and process-resolving Earth system models (see Figs. 2, 3, 1) accelerated by the exascale computing capability (https://www.exascaleproject.org/). Such high-resolution observations will allow the complex air-sea-ice interactions in the region to be better quantified, and transports through the narrow gateways between the Arctic and Atlantic basins to be monitored.

2.4 Arctic Halocline Stability

2.4.1 Nature of Tipping Point

The vertical stratification in the Arctic Ocean is commonly characterised by a cold halocline (CH) separating a relatively fresh and cold water layer above from the warm and salty Atlantic Water (AW) below (Aagaard et al. 1981). The CH represents a heat sink for the underlaying AW and its strength determined by salinity driven vertical density gradient regulates the rate of upward heat release from AW to the surface mixed layer and sea ice. The stability of CH is critical to the presence of the overlaying sea ice, as it inhibits deep convection and it partly explains why vertical mixing rates in the upper Arctic Ocean away from steep topography are lower than in midlatitudes (Rippeth et al. 2015). The low mixing rates are also attributed to low tidal energy and sea ice limiting the wind momentum transfer from the atmosphere (Dosser and Rainville 2016).

However, changes in the distribution and strength of the CH layer in the Arctic Ocean have been observed in the past. Comparing the submarine hydrographic data from the Scientific Ice Expeditions (SCICEX) in the 1990s against the long-term climatology, Steele and Boyd (1998) revealed the retreat of CH layer from the Amundsen Basin back into the Makarov Basin. This change was attributed to a weakening of the Beaufort High, which in turn resulted in a cyclonic shift of the sea ice Transpolar Drift and freshwater export from the Siberian shelves into the deep basins of the Arctic Ocean (Maslowski et al. 2001). A more recent analysis of the 2002–2015 mooring data from the eastern Eurasian Basin by Polyakov et al. (2020) has shown weakening and shoaling of CH, which was linked to winter convection due to brine rejection during sea ice formation. The decrease of the CH depth in the Eurasian Basin at the rate of -2.5 m per decade has been recently confirmed by Muilwijk et al. (2023). However, their analysis of observed mean CH depth in the Amerasian basin (or western Arctic) between 1970 and 2018 indicates an opposite, deepening CH depth trend of 7.3 m per decade (Fig. 4). The latter trend could in part be related to the increasing freshwater content in the Beaufort Gyre, as discussed in Sect. 2.3 above, and it emphasises the need to understand critical processes and feedbacks operating in different parts of the Arctic Ocean.

Given the observed summer sea ice retreat over the satellite record (Druckenmiller et al. 2022), a more efficient and direct air–sea coupling is expected to increase vertical

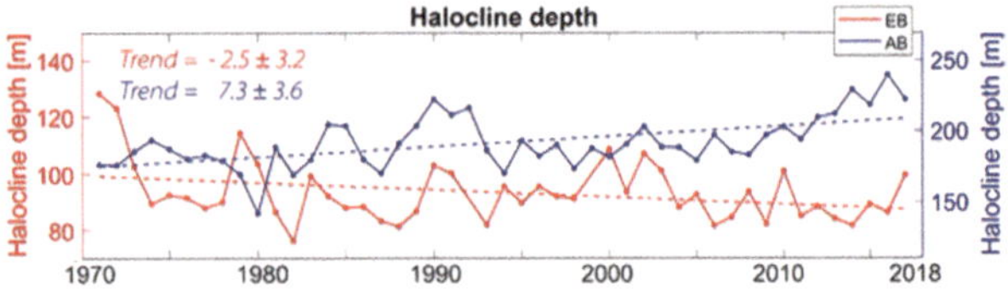

Fig. 4 Observed annual mean depth of halocline base in the Eurasian basin (EB; red) and Amerasian basin (AB; blue) regions (adapted from Fig. 1b of Muilwijk et al. 2023)

mixing, via generation of surface and internal waves (Liu et al. 2016), and the upward heat flux, thus constituting positive wind-ice-ocean feedback (Fine and Cole 2022). Another positive feedback linked to ice-ocean dynamics relates the thinning (Kwok 2018) and reduced strength (Hibler 1979; Rothrock 1975) of the Arctic sea ice to increased ice drift (Gascard et al. 2008; Rampal et al. 2009), ice-ocean stress, and the upper ocean mixing (McPhee 2008), With increased mixing, the heat accumulated in the upper ocean due to summer insolation (Maykut and McPhee 1995) can be entrained into the surface mixed layer (Maslowski et al. 2012) to reduce ice growth in winter and melt more ice in summer. In both cases, the ice-ocean coupling augments the summer-dominated ice-albedo feedback.

Impacts of such dynamically controlled feedbacks have been recently investigated by Beer et al. (2023) using an idealised climate model with varying specified mixing rates for ice-covered and ice-free surface conditions. They find that the wind-ice-ocean feedback can result in a novel bistability in the climate system, which is associated with a hysteresis loop within forced warming and cooling for a limited range of parameters. Furthermore, this study shows that an abrupt tipping point can occur with an irreversible transition to ice-free conditions when crossing the bifurcation point during forced warming. While not conclusive, this study points to the need for improved models of the Arctic and continued and new data to better constrain them and to further investigate potential tipping points associated with the stability of Arctic cold halocline (CH).

2.4.2 EO Needs and Opportunities

One example of EO measurements in support of the CH stability is altimeter data from polar orbiting missions, i.e. *ERS-1/2, Envisat, and CryoSat-2*, to measure sea surface height, snow, and sea ice freeboard/thickness. Their application to investigation of wave and wind climate in the Arctic Ocean was reported by Liu et al. (2016). However, only CryoSat-2 is still active, while well beyond its design lifetime and likely approaching the end of its mission. Note, that CryoSat-2 measurements of ice freeboard overlapping the NASA IceSat-2 mission, have also allowed improved estimates of snow and sea ice distribution and thickness in the Arctic Ocean (Kacimi and Kwok 2022). A major challenge for EO missions is providing a year-long data on the ocean surface (salinity, temperature, height), sea ice (thickness, deformation, drift) and snow at km and daily-to-weekly timescales. Hence, future EO high spatial and temporal resolution measurements of the Arctic Ocean surface, snow, and sea ice are critically needed.

In parallel, models of the Arctic Ocean and climate need significant improvements to reduce persistent biases, especially in representing the coupling across the air-ice-ocean boundary layer (Maslowski et al. 2012), sea ice thickness distribution (Watts et al. 2021), and the upper ocean vertical stratification (Shu et al. 2023). Some progress is being made in these areas in part due to increased spatio-temporal resolution applied to regional Earth System models (Lee et al. 2023a, 2023b) or global models using unstructured grid with regional refinements (Veneziani et al. 2022; Wang 2024). However, expanded measurements, better understanding, and model representation of critical physical processes, for example involved in shelf-basin exchanges or vertical mixing, are needed and require a combination of in-situ and remote observations.

2.5 The Kuroshio Large Meander—an Anomalous State of a Multi-Stable Western Boundary Current

2.5.1 Nature of Tipping Point

Boundary currents are an essential part of the global ocean circulation. Among them, especially western boundary currents like the Gulf Stream in the North Atlantic and the Kuroshio in the North Pacific can exhibit a rich variety of spatio-temporal flow patterns due to their interactions with atmospheric forcing factors in the presence of complex coastlines and bathymetry. At the same time, their pathways impact downstream currents and associated Sea Surface Temperature (SST) patterns (Qiu et al. 2020) with profound impacts on the marine biosphere (Chang et al. 2019; Lizarbe Baretto et al. 2021), ocean–atmosphere heat fluxes and resulting atmospheric processes potentially alternating air temperature, wind and precipitation patterns over large regions.

While for many well-studied parts of such boundary currents (e.g. the Gulf Stream and North Atlantic Current), there seems to exist a continuum of possible pathways, the Kuroshio south of Japan presents a rather unique case with a clear clustering of observed current pathways indicating multistability of the regional ocean circulation system. Specifically, extending previous works already starting in the 1960s, Kawabe (1995) suggested the existence of three distinct major pathways of the Kuroshio termed near-shore and off-shore non-large meander and typical large meander (nNLS, oNLS and tLS, respectively). Especially the large meander state with its recirculation gyre and enclosed cold-surface water anomaly has attracted great interest during the last decades due to its complex dynamics and relevance for the climate and marine biosphere around the Japanese islands. The existence of discrete states opens the possibility of a discontinuous response to external forcing such as climate change; however the research reported here represents attempts to understand these inherently discrete states of the system, which is a pre-requisite for anticipating climate change responses.

Between 1950 and present-day, eight Kuroshio large meander events took place at apparently random occurrence times (Qiu and Chen 2021) and with a broad distribution of lifetimes of the large meander state, the latest being the longest on record and continuing since August 2017. By combining sea-surface height (SSH) observations from satellite altimetry with reanalysis based surface wind stress patterns and comparing the recent patterns with the conditions during the previous (2004–2005) large meander event, Qiu et al. (2023) attributed the unusual stability of the ongoing event to an exceptionally stable dynamic state of the Kuroshio Extension, which minimizes the westward propagation of eddies, potentially destabilising the upstream Kuroshio path while in parallel fostering the maintenance of the recirculation gyre.

2.5.1.1 Impacts The presence of a Kuroshio large meander has a profound impact on the climate conditions in the Northwest Pacific region around Japan. Sugimoto et al. (2021) reported marked coastal warming during summer off the Kanto-Tokai district of Japan during the present large meander. Using a regional atmospheric model, they could show that a warming of the near-coastal surface waters in the wake of the recirculation gyre leads to an increase of water vapour in the low-level atmosphere due to enhanced evaporation near the coast. Along with altered near-surface winds this results in a net increase of downward longwave radiation over the Kanto region and, hence, increasingly hot and humid summers. By contrast, for winter conditions during the large meander event of 2004–05, Xu et al.

(2010) already reported reduced winds and precipitation resulting from the cool water pool between the Kuroshio and the Japanese coast. In general, the cool water pool along with coastal warming off the Tokai province of Japan appears as a common fingerprint of large meander events (Sugimoto et al. 2019). Other related effects of the wintertime presence of a Kuroshio large meander include a southward shift of primary extratropical cyclone tracks along with reduced latent heat fluxes south of Japan (Nakamura et al. 2012; Hayasaki et al. 2013). Based on regional climate modelling, Murazaki et al. (2015) showed that the cold SST anomaly of the large meander consistently leads to reduced upward surface turbulent heat flux, lower precipitation frequency, and less frequent steep horizontal gradients in equivalent potential temperature over the ocean during both summer and winter.

In addition to its pivotal effects on regional atmospheric processes, the presence of a Kuroshio large meander also exerts direct influence on ocean dynamics, resulting in distinct patterns of eddy kinetic energy, potential vorticity, relative vorticity, and eddy-mean flow interaction south of Japan (Ma 2019). Based on a new multi-scale statistical analysis tool, Yang and Liang (2019) identified a spatially coherent inverse cascade of kinetic energy from synoptic eddies to the slowly varying mean flow over the whole large meander region, along with a particular relevance of an initial influx of mesoscale eddy energy from upstream regions responsible for the emergence of the inverse cascade. In another study, Qiu and Chen (2021) identified intense anticyclonic eddies emanating from the Subtropical Counter-Current interacting with Kuroshio path perturbations southeast of Kyushu, generating positive relative vorticity along the coast finally leading to the formation of the large meander. Not surprisingly, the reported effects on ocean circulation at various spatial and temporal scales can also have a profound impact on biological and biogeochemical processes in the Northwest Pacific (Hayashida et al. 2023).

2.5.2 EO Needs and Opportunities

From the aforementioned results, it may be noted that especially the turbulent flow patterns in the adjacent Pacific Ocean south of Japan are crucial for understanding the processes leading to the formation and maintenance of the Kuroshio large meander. While atmospheric impacts can be well traced with state of the art in situ and remote sensing based products, the importance of upscale kinetic energy transfer from synoptic scales (Yang and Liang 2019) points to a particular area high-frequency EO data (SST, sea surface height SSH, and potentially SSS, on timescales shorter than 15 days, and especially in near-coastal regions) is important to fully quantify the dynamics preceding the emergence of a large meander. Based on such improved real-time observational datasets along with better process understanding (generated by exploiting physics-based models and/or modern artificial intelligence methods), forecasts of future state transitions of the Kuroshio path and its regional climatic impacts could potentially be developed at societally relevant lead times.

2.6 Ocean Deoxygenation

2.6.1 Nature of Tipping Point

With progressing global warming there is an increased risk that climate-relevant biogeochemical ocean elements might cross critical thresholds and tip into novel states, with severe consequences for the climate system, marine ecosystems and human societies

(Heinze et al. 2021; Rockström et al. 2009; Lenton et al. 2008). Among the biogeochemical elements that may undergo large and possibly irreversible (on human time scales) changes in response to the ongoing anthropogenic perturbations is the ocean oxygen (O_2) content, with potentially large effects on marine biogeochemistry and ecosystems (Watson 2016; Rockström et al. 2009; Lenton et al. 2008; Heinze et al. 2021). The oceanic O_2 content has undergone large variations throughout the earth's history, driving the evolution of novel metabolic pathways that shaped ocean biogeochemistry as we see it today (Canfield et al., 2010). Dissolved O_2 is necessary for sustaining aerobic life in the ocean and regulates redox-sensitive biogeochemical processes. The cycling of O_2 is strongly linked with the nitrogen (N), phosphorus (P), iron (Fe) and the carbon (C) cycles. Changes in oxygen will impact the cycling of any of the other elements and can potentially lead to an amplification/reduction of the initial perturbation, with stabilising or destabilising feedbacks to the climate system.

Currently, ocean warming is affecting the oceanic dissolved oxygen content. Oxygen concentrations have decreased by about 2% over the last 50 years (Ito, et al. 2017; Schmidtko et al. 2017), a process termed ocean deoxygenation. Tropical Oxygen Deficient Zones (ODZs), where oxygen concentrations fall below values critical for the onset of the anaerobic microbial processes and/or have sublethal effects (hypoxia) on animals, are expanding (Stramma et al., 2008). These oxygen changes are due to a combination of chemical, physical and biological drivers (Bopp et al. 2017; Oschlies et al. 2018) and show a significant regional and depth variability (Ito et al. 2017). The reduction in temperature-dependent O_2 solubility is well quantified and estimated to account only for about 15% of the observed oceanic global O_2 loss (Helm et al. 2011). The contribution from changes in circulation and associated reduced ventilation is less certain, potentially explaining the majority of the deep ocean O_2 changes (Breitburg et al. 2018). O_2 consumption due to biological respiration is reduced in the low-latitude nutrient-limited regions that experience reduced productivity, but enhanced in high-latitude regions where biological production has intensified under global warming (Laufkötter et al., 2015, 2016; Steinacher et al. 2010).

Next to uncertainty of driving processes, detectability of climate-driven O_2 trends is challenged by natural short-term variability of oxygen (Frölicher et al. 2009; Long et al. 2016; Bopp et al. 2017) and poor spatiotemporal coverage of O_2 observations (Grégoire et al. 2021). The time required for anthropogenically forced signals to emerge above background noise (time of emergence) is found to be regionally variable. Longer timescales are required in regions where processes cause compensating effects (compensation of the thermal and non-thermal (biological productivity and ventilation) changes), with a resulting net weak trend (Bopp et al. 2017) such as in the tropical regions (Cabre et al. 2015). Model projections suggest that while stopping CO_2 emissions will stop ocean surface warming and near surface ocean deoxygenation, the deep ocean layers are expected to continue losing oxygen for centuries (Oschlies et al. 2021; Frölicher et al. 2020). This committed deoxygenation results from the reduced solubility of warm deep waters that have been last in contact with the atmosphere during the high CO_2 emission era, and the longer ventilation ages in the ocean interior due to the weakened overturning circulation inertia (Oschlies et al. 2021). However, high uncertainties exist if current steady ocean deoxygenation can evolve into abrupt widespread O_2 loss over human timescales scales (centennial).

Paleoclimate records indicate that widespread ocean deoxygenation events, known as Oceanic anoxic events (OAEs), associated with profound geochemical changes, occurred multiple times during the Earth's history (Jenkyns 2010). The occurrence of these events has generally been associated with an abrupt temperature rise induced by rapid atmospheric CO_2 increase (e.g. by volcanic activity) and associated reduced O_2 solubility, ocean circulation changes and, on time scales longer than 100 kyears, enhanced continental nutrient weathering (Kiehl and Shields, 2005). It is hypothesised that amplification by positive biogeochemical feedbacks such as phosphorus (Jenkyns 2010; Monterio et al., 2012; Watson et al. 2017) or iron (Fe) sediment regeneration (Wallmann et al. 2019), from low-oxygen sediments, is required to further boost the biological demand of oxygen, and meet the conditions for wide-spread OAEs. Characteristics of localised ODZs (nowadays in the tropical low latitude ocean basins) have changed over time. ODZs expanded and contracted not only over millennial but also on shorter, centennial and decadal timescales in response to ocean ventilation changes, due to climate fluctuations (Jaccard and Galbraith 2012; Galbraith et al., 2013) and natural variability (Stramma et al. 2020; Segschneider et al. 2018), and ecological shifts that control the depth of organic particle remineralization (Lu et al. 2018, Hülse et al. 2021; Crichton et al. 2023). Continental configuration is also thought to influence the sensitivity of the ocean system to biogeochemical feedbacks (Monteiro et al. 2012, Meyer et al. 2008).

Currently the drivers, the sign of change and time scale of O_2 tipping are not well known (Armstrong McKay et al. 2022). The projected O_2 decline under global warming scenarios is found to rebound within several thousand years in multi-millennial model studies due to circulation reinvigoration (Yamamoto et al., 2015; Frölicher et al. 2020) or to lagged responses in biogeochemical feedbacks (Oschlies et al. 2019; Kvale et al., 2019). O_2 tipping behaviour has emerged in modelling studies associated with the weakening of the overturning circulation, under past atmospheric and/or orbital forcings over millennial timescales (Pohl et al. 2021; Segschneider et al. 2018; Brown and Galbraith 2016; Schmittner et al. 2007), changes to continental configuration (Pohl et al. 2022), and from complex biogeochemical feedbacks, both in box-models (Watson et al. 2017, Wallmann et al., 2003, 2019, 2022; Handoh and Lenton 2003) and in intermediate-complexity Earth system model (ESM) experiments (Kemena et al. 2019; Niemeyer et al. 2017, Lu et al. 2018, Meyer et al. 2008, Hülse et al. 2021, Crichton et al. 2021, Frölicher et al. 2020.). However, a comprehensive assessment of the potential physical and biogeochemical drivers of tipping behaviour in ocean oxygen is lacking. This is mostly due to the limited knowledge of key biochemical complexity and elevated computational costs of long-times scale (centennial to multi-millennial) modelling experiments. As such, our understanding and ability to identify the potentially tipping behaviour of O_2 is currently limited.

2.6.1.1 Drivers and Timescales More than 99% of all of the molecular oxygen on the planet is in the atmosphere with less than 1% being dissolved in the ocean (Oschlies 2019). The oceanic O_2 content is supplied by uptake from the atmosphere at the surface ocean, mostly driven by O_2 temperature-dependent solubility and ocean circulation that transports cold surface O_2-rich waters into the ocean interior during deep water formation at high latitudes. Dissolved O_2 is lost in the ocean by biological consumption via biological respiration. Factors that affect O_2 solubility, ocean circulation and biological oxygen consumption can drive ocean deoxygenation (Fig. 5) and trigger O_2 tipping behaviour.

2.6.1.2 Physical–Chemical Drivers In model simulations a general pattern emerges on millennial timescales that is consistent with paleoproxy data. Under cold climates, such as the Last Glacial Maximum, the reduction in deep water formation leads to an increasingly separated well-oxygenated upper ocean and a poorly ventilated oxygen-depleted deep ocean (Jaccard and Galbraith 2012; Somes et al. 2017; Pohl et al. 2021). Under warming climates, ocean surface warming leads to a reduction of O_2 solubility and to the intensification

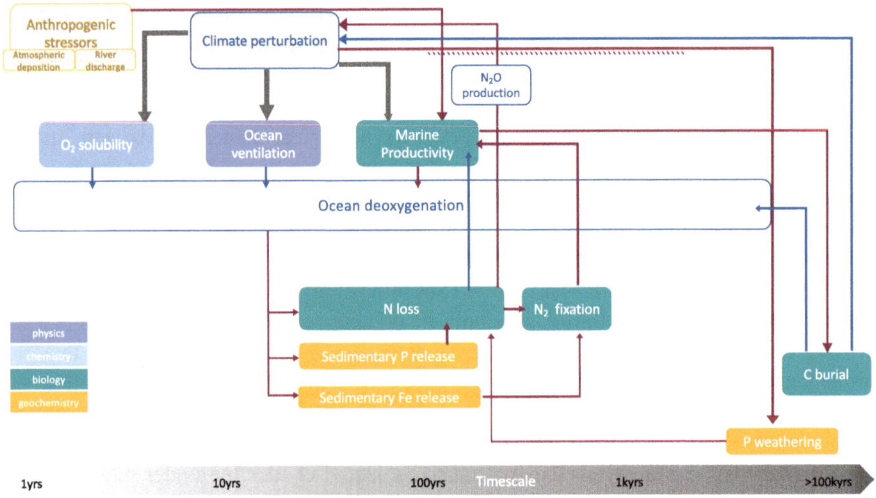

Fig. 5 Schematics of the physical, chemical and biogeochemical processes that interact and impact ocean oxygen content over multiple timescales. Red arrows represent positive feedbacks, blue arrows represent negative feedbacks. The anthropogenic climate perturbation affects O_2 solubility, ventilation and biological O_2 consumption (grey arrows) potentially with positive or negative feedbacks. With warming O_2 solubility is reduced and, via changes in stratification and mixing, ocean ventilation declines. The decline in surface nutrient supply due to enhanced stratification affects marine productivity generally with a decline in the tropics and increase in the high latitudes. The net effect is uncertain and may lead to expansion of oxygen deficient zones on decadal to centennial timescales, where anoxic microbial processes, denitrification and anammox, can lead to N loss and the production of the greenhouse gas N_2O. A negative feedback can occur if N loss is not compensated for by N_2 fixation, with the potential collapse of marine productivity O_2 can accumulate as organic matter respiration is reduced. However, if N_2 fixation is promoted by sedimentary release of P and Fe and/or P weathering, a self-sustained positive feedback can occur, potentially leading to N runaway and eventually the establishment of euxinia

of stratification and slow-down of circulation and vertical mixing, resulting in subsurface ocean deoxygenation and widespread oxygen depletion (Tsandev and Slomp 2009; Ruvalcaba Baroni et al. 2014).

2.6.1.3 Biogeochemical Feedbacks It is still a matter of debate whether the climate forcing exerted by volcanic CO_2 emissions, was sufficient to explain the aerial extent of ocean anoxia during past OAE or if additional feedbacks were required to amplify the deoxygenation trend induced by global warming (Jenkyns 2010; Wallmann et al. 2019). Phosphorus is a key nutrient for supporting biological production. This nutrient is generally provided to the ocean by continental weathering and is removed from the ocean by being incorporated in marine sediments. However, if bottom water O_2 concentrations fall below a certain threshold level this element can be released from sediments back into the water (Van Cappellen and Ingall 1994), where it can fertilise organic matter production and remineralization (Beil et al. 2020; Jenkyns et al. 2010; Watson et al. 2017). Despite large uncertainties regarding the weathering and sedimentary P fluxes model parameterizations, recent modelling studies have shown that enhanced ocean P fluxes from weathering and/or from anoxic sediments, under present, future (Niemeyer et al. 2017; Kemena, et al. 2019) and past climate scenarios (Wallmann, 2005; Monteiro et al., 2012), can promote additional ocean productivity. This in turn, exerts a positive feedback on deoxygenation, leading to a large increase in the volume of ODZ on millennial timescales. The large expansion of suboxia then causes a net loss of bioavailable N that is consumed during organic matter remineralization in ODZ in the process of denitrification. Eventually, in a negative feedback, a considerable amount of P is left unused in the ocean as N becomes the primary nutrient limiting biological productivity (and is not compensated for by the Fe-limited N_2 fixers) thus ultimately constraining the expansion of ODZ (Kemena et al. 2019). Hence, although P addition to the ocean drives additional deoxygenation, the positive redox-related feedback on benthic P release is eventually limited by the availability of N and the apparent inability of N_2 fixers to respond to the larger P inventory. Further, the inability of N_2 fixers to respond to denitrification allows, in current models that tend to suppress N_2 fixation in low temperature regions, for a net ocean O_2 accumulation in a multi-millennial Earth System Model (ESM) simulation (Oschlies et al. 2019). Models without the full representation of the N-cycle feedbacks, miss this important negative feedback limiting global deoxygenation (Wallmann 2003).

On the other hand, how a full dynamic iron cycle in an Earth system model will affect the utilisation of the added P by N_2 fixers remains to be tested. In such a model setting, primary production is not limited by N-loss and widespread deoxygenation can be self-sustained (Landolfi et al. 2013). Positive feedbacks between sedimentary P and Fe release and ODZ intensification could potentially drive a widespread and lasting expansion of oxygen-depleted ocean regions on centennial timescales, eventually leading to euxinic events (Canfield 2006). This is shown in a recent regional box-model study of the Eastern Tropical Pacific. Under current warming conditions if the Fe-demand of N_2 fixing organisms can be met by O_2-dependent sedimentary Fe release, a positive feedback loop develops on decadal timescales and amplifies O_2 decline (Wallmann et al. 2022). As N losses are not compensated for by N_2 fixation due to stoichiometric constraints (Landolfi et al. 2013), NO_3 and NO_2 can become depleted and sulfidic conditions may eventually occur (Wallmann et al. 2019, 2022). The production of hydrogen sulphide (a gas toxic to animals) may be limited by the consumption of sulphide-oxidising photoautotrophs (Meyer et al. 2008) and by the scavenging of it by organic particles, which can help to reoxygenate the ocean after an OAE and reduce atmospheric CO_2 concentrations by efficiently sequestering biological carbon and providing an alkalinity source (Hülse et al. 2019). The release of H_2S in euxinic conditions is supported by paleorecords and remains a suspected mass extinction mechanism (Meyer et al. 2008). The potential for negative H_2S feedbacks to prevent widespread deoxygenation of the water column is not well constrained.

It is unclear whether the burning of fossil fuels combined with human activities delivering additional nutrients, from atmospheric deposition and/or river-discharge and land-use changes, are sufficient to trigger such positive biogeochemical feedbacks (e.g. from increased stratification and associated reduced interior ocean ventilation, in interplay with increased phosphorus input from fertilisers-use, accelerated weathering, land-use changes, and/or recycling from sediments under anoxic conditions), if they could lead to a widespread deoxygenation and over which timescales. Given the large uncertainties over drivers, mechanisms, time-scale and critical thresholds for triggering a widespread oxygen loss, ocean deoxygenation is currently regarded as an uncertain potential tipping element and has not been included within the core tipping elements by the recent study of Armstrong McKay et al. (2022).

What is more, ESM models predict only about half of the ocean deoxygenation that is currently observed (Oschlies et al. 2018), suggesting that current models may lack key physical and biogeochemical feedback mechanisms that amplify the respiratory oxygen demand in the ocean interior (Oschlies et al. 2018). Models that include positive feedbacks between anoxia, phosphorus and/or Fe recycling from sediments and marine productivity have been shown to exhibit a non-linear behaviour, tipping into widespread deoxygenation on centennial to millennial timescales (Kemena et al. 2019; Walmann et al. 2019; Somes et al. 2017). These studies emphasise the critical role of biogeochemical feedbacks in affecting the ocean O_2 content that need to be accounted for in models for past and future projections.

2.6.1.4 Biological-Ecosystem Impacts of Tipping Ocean deoxygenation poses a serious direct threat to life in the ocean. Under low oxygen conditions (< 62 mmol m^{-3}, hypoxia) metabolic processes such as respiration, reproduction and movement may be impaired (Rabalais et al. 2010). At ecosystem level this will lead to changes in community structure towards hypoxia-tolerant species of fish and invertebrates, habitat compression, migration and mass mortality (Breitburg et al. 2018). Species extirpation (local species loss) can occur when ecophysiological tolerances of temperature and O_2 threshold are crossed. These can be quantified from the local O_2-supply to demand ratio (Penn et al. 2019). On a global scale, species extinction can occur when habitat loss exceeds a critical ocean fraction. In a model study extirpation risk has been found to be higher in the high latitude and the tropics under a high emission scenario. Extirpation, an ecological tipping point, is likely to culminate in a mass extinction similar to those in Earth's past by the end of the twenty-first century under continued global warming (Penn and Deutsch, 2022).

2.6.1.5 Biogeochemical-Climate Impacts of Tipping The expansion of ODZ (< 5 mmol m^{-3}, suboxia) will restrict aerobic remineralization and trigger anaerobic microbial processes, which affect the cycling of elements nitrogen, phosphorus, iron, sulphur and various trace metals, and the production of greenhouse gases such as nitrous oxide (N_2O) and methane (CH_4). Under low O_2 conditions, the anaerobic microbially-driven remineralization of organic matter will proceed consuming nitrate, manganese and iron oxides and sulphate sequentially. The consumption of NO_3 via the microbial processes of denitrification and anammox can increase. These microbial processes result in the loss of fixed N, that if compensated regionally for by N_2 fixation may lead to positive feedbacks and self-sustained ODZ expansion (Landolfi et al. 2013) and euxinia may develop (Canfield 2006). During denitrification and nitrification under low O_2 conditions, the production of N_2O occurs (Ward and Jansen, 2014). This potent greenhouse gas, if released to the atmosphere, has the potential to amplify global warming and deoxygenation, constituting a positive feedback. Ocean deoxygenation also leads to major changes in the redox-sensitive cycling of essential elements and metals such as P and Fe. Sediment release of these elements can occur under low bottom water oxygen concentrations (Van Cappellen and Ingall 1994; Dale et al., 2015), potentially stimulating organic matter production and remineralization, thereby also exerting positive feedback on deoxygenation (Walmann et al. 2019).

2.6.2 EO Needs and Opportunities

2.6.2.1 Monitoring Since ocean deoxygenation can cause severe ecosystem disruptions and feedbacks to other elements and climate, the availability of early warning systems, in open and coastal ocean is highly desirable. The identification of changes and possible TPs in ocean O_2 requires adequately long observational records with adequate spatial coverage. Although novel technologies have largely expanded the O_2 observing system (Grégoire et al. 2021), with O_2 sensors routinely deployed on a range of in situ platforms (such as ARGO, AUVs and gliders, moorings) that allow the assessment of intraseasonal, seasonal and interannual oxygen variability, challenges remain for identifying long-term trends. The use of EO observations such as temperature (SST), ocean productivity (Ocean Color), combined through artificial intelligence to reconstruct oxygen changes, can provide novel avenues towards monitoring and early warning systems over the satellite era. Of elevated interest for this purpose is the remotely sensed data of chlorophyll a (Chla) a phytoplankton biomass proxy, particulate back scattering (bbp) a proxy for non-algal particulate material, a the derived quantity particulate organic carbon (POC) and their 3D reconstruction, and more widespread availability of HYPER-spectral remote sensing reflectances (Rrs), complemented with field measurements and new in situ observing systems (like BGC-Argo) of key variables such as O_2 and nutrients. Of crucial importance are also observational constraints on ocean circulation and its changes over time. Abiotic transient tracers can be particularly valuable to provide information about changes in ventilation and residence times over which respiratory signals can accumulate in the ocean interior. Detection of fingerprints of ocean deoxygenation such as N_2O and H_2S fluxes at the sea surface via remote sensing (Infrared Atmospheric Sounding Interferometer, IASI) also represent potential avenues that require further development (Garcia et al. 2018).

2.6.2.2 Paleoproxies The reconstruction of regional O_2 levels of the past ocean relies on the indirect evidence of O_2 dynamics from local sedimentary proxy data such as benthic foraminifera, redox-sensitive elements, and isotopic fingerprints. Further efforts towards expanding sparse paleoproxy datasets, improving the resolution of paleo data, providing uncertainty estimates in the paleo-reconstructions are required to advance our quantitative understanding of OAE and attribution of their drivers and enhance our predictive capabilities.

2.6.2.3 Modelling ESMs are powerful tools to explore the possibility of ocean O_2 tipping over a range of timescales and forcings. The drivers, location and likelihood of tipping are sensitive to emerging physical and biogeochemical feedbacks (Fig. 5). However, biogeochemical processes are poorly constrained and are not fully represented in current state-of-the-art models. A multimodel intercomparison experiment which included key biogeochemical processes (Fig. 5) would be desirable to test mechanisms of oxygen tipping in models of different complexities. This effort could potentially build from the TipMIP programme (https://tipmip.pik-potsdam.de/), and specifically designed to tackle the following questions:

- What are the drivers, key physical processes and biogeochemical feedbacks, associated with oxygen tipping behaviour?
- What are the characteristic spatial and temporal scales of positive and negative feedbacks and what is their combined effect on ocean oxygen?
- What are the drivers' tipping points and thresholds and what are their corresponding uncertainties?
- What is the risk of crossing individual TPs, at different levels of ongoing climate and land-use change, which might trigger oxygen tipping behaviour?
- Are the respective O_2 changes abrupt and on which timescales are they reversible?
- How is the overall stability of the Earth system affected by marine deoxygenation?

Remotely sensed data can further help to constrain biogeochemical processes in climate models, thereby reducing uncertainties in the thresholds of drivers and timescales of climate tipping elements.

Overall, our current scientific knowledge on the tipping behaviour of ocean oxygen is limited, a systematic analysis, drawing from expanded current and paleoproxy observations and multimodel ensemble studies, is desirable. Further development may consider: (i) Expanding spatial and temporal coverage of multiplatform observations of O_2 and associated variables, particularly those informing about ocean ventilation and changes in ventilation (e.g. abiotic transient tracers); (ii) Testing the use of artificial intelligence (AI) tools to exploit ever growing EO-observations for O_2 studies (iii) Expanding the sedimentary record and using multivariate approaches for attribution studies (iv) Testing mechanisms of oxygen tipping in models of different complexities.

In addressing these questions, critical knowledge gaps in the oceanic O_2 controls will be reduced and understanding of Earth system homeostasis will be improved.

3 Tipping Points in Pelagic Marine Ecosystems

In the dynamic expanse of the ocean, pelagic organisms, inhabitants of the water column, exhibit continual movement—both vertically and horizontally—engaging in drifting, swimming, and migrating activities. Consequently, conventional time series measurement systems conducted over fixed grids, such as repeated transects, only offer an approximate representation of the temporal dynamics of observed species. The inherent limitation lies in our fixed spatial sampling scale, which fails to align with the fluid variability of the samples. Organisms may be missed during sampling, yet reside slightly deeper or a few metres or kilometres away in any direction, and their presence can be overestimated because of their typically patchy distribution. Moreover, the uncertainty extends to the entities being sampled—be it individuals, populations, or species—making it challenging to ascertain whether the population measured at time ’t’ is the same as that observed at time ’t + n’.

Consequently, there exists a significant contrast between time series studies of pelagic organisms freely moving in three dimensions in the water column, and benthic organisms inhabiting the seafloor and moving in two dimensions with reduced mobility. The former incorporate spatial and temporal uncertainties that are challenging to measure accurately, while the latter provides relatively precise spatial and temporal measurements, particularly for sessile species, offering a degree of population certainty.

Despite numerous studies addressing sampling design and uncertainty in each compartment (Pepin and Helbig 2012; Lin et al. 2010; van Hoey et al. 2019), to our knowledge the inherent difference between these compartments has not been thoroughly assessed.

There has been extensive research on marine benthos tipping points and phase shifts (this is the term commonly used in benthos research, while regime shifts or abrupt shift are the terms commonly used in water column research). Thus, the challenges faced by coral reefs, a global concern among ecologists, are well-documented (Hughes 1994; Matz et al. 2018). Numerous studies explore phase shift and tipping point dynamics leading to the transition of coral reefs to alternative algal states, and several provide future projections, and reef-danger warnings based on earth observations (Nim et al. 2010; Claar et al. 2020; Hobday et al. 2023; Wernberg et al. 2023; Melet et al. 2020; Heron et al. 2016). Due to their role as biodiversity hotspots, nursery grounds, nutrient cycling hubs, and CO_2 sequestering agents, coral reefs are considered a ’tipping element’ in various works (Lenton et al. 2008; Armstrong McKay et al. 2022; Wang et al. 2023).

In contrast, the pelagic ocean organisms, invisible and un-gridded, moving and mixing in three dimensions, have not yet been recognised as a tipping element. The 3D moving matrix complicates the identification of pelagic ecological shifts (often referred to as abrupt shifts), which are not marked by clear irreversibility, a crucial concept in defining tipping points. Nevertheless, shifts in these interconnected and interchanging systems can have potential global-scale consequences, and research has indicated that pelagic shifts can synchronously occur in distant oceanic basins and that the frequency and magnitude of the shifts may increase as the earth system continues to warm (Beaugrand et al 2015, 2019).

This section, therefore, concentrates on abrupt shifts and tipping points in the pelagic components of the ocean, emphasising the potential of EO to monitor their temporal and spatial dynamics and provide early warnings. All pelagic components described in this section exhibit threshold behaviour and possess the potential to impact biological elements (such as the food web) and even contribute feedback to the climate (e.g. phytoplankton), making them 'invisible' tipping elements.

The pelagic component, particularly the plankton community, plays a vital role in sustaining life on Earth. Phytoplankton, forming the foundation of marine food webs, contribute to approximately half of the global primary production. Additionally, through various processes and interactions (e.g. carbon sequestration, biological pump, aerosol/DMS production, ocean albedo, oceanic uptake of heat), they influence temperature, ocean acidity, carbon cycling, and the overall climate system (Miloslavich et al. 2018). Marine zooplankton, sustained by phytoplankton, encompass a diverse range of species, including copepods, krill, and jellyfish, along with the larval stages of most marine animals. Each plays a specific role in nutrient cycling and ecosystem stability, holding immense ecological and economic importance in marine ecosystems. Their impact extends to human societies, significantly contributing to multiple economic sectors. Fisheries and farmed seafood, sustained by plankton, provide a significant source of animal protein to more than 3 billion people globally, with the total first sale value of global production estimated at USD 406 billion in 2020 (of which USD 141 billion for capture fisheries and USD 265 billion for aquaculture, FAO 2022).

Presently, human activities exert a profound impact on marine life at scales ranging from global to local, including global warming, acidification, deoxygenation, overfishing, pollution, and notably, the burgeoning issue of plastics. If current trends persist, the weight of plastics in the world's oceans is projected to surpass that of fish by 2050 (Industry Agenda 2016). All these factors, individually or in combination, can propel marine communities towards ecological shifts.

The *pelagic tipping point dilemma* thus revolves around the vital task of identifying tipping points and ecological shifts in a vast, dynamic ecosystem. The inherent challenges stem from its vastness, spanning regional to semi-global scales, its dynamic 3D nature, and its interconnectedness with other biological and physical components of the Earth system. Thus, the potential consequences of tipping points in the pelagic ecosystem are far-reaching, surpassing the confines of specific habitats or areas. Early warnings of hard-to-identify critical changes (including dimension, intensity and pace of the changes) are consequently *crucial* and *complex* because of the inherent mismatch between sampling design and sampled population.

Satellite observations offer a promising avenue in addressing this dilemma. Their extensive scope and comprehensive coverage make a substantial contribution to this quest. These observations provide consistent, and in some cases decade-long, monitoring on a global scale. While they may be limited in the depth dimension, satellite observations excel in providing excellent horizontal coverage. This capability accommodates regions with diverse requirements, resources, and monitoring capabilities, offering valuable insights into the elusive open ocean.

3.1 Phytoplankton Productivity and DMS-Phytoplankton-Cloud Interactions

3.1.1 Nature of Tipping Point

3.1.1.1 Unpredictable Abrupt Shifts During the 21st Century Marine phytoplankton, constituting half of the global net primary production, play a pivotal role in sustaining the Earth's Biosphere (Field et al. 1998). Their productivity and diversity in the ocean control the bottom-up dynamics of crucial commercial seafood products, influencing the magnitude and pathways of energy transfer within the marine food web. In a prior review on tipping elements in marine ecosystems (Sect. 2.2 of Swingedouw et al. 2020), the remote sensing of chlorophyll-a for monitoring phytoplankton was recognised as an advanced and accessible Earth Observation tool for marine ecosystems.

Satellite ocean colour sensors have continuously provided information on the global distribution of surface phytoplankton biomass and community composition since the late 1990s (Groom et al. 2019). This information has unveiled global-scale changes, including the expansion of the oligotrophic ocean (ocean desertification) and phenological shifts in seasonal blooms which can have substantial consequences for higher trophic levels (Sect. 2.2.4 of Swingedouw et al. 2020).

A notable recent application of remote sensing to phytoplankton involves the development and use of an advanced lower trophic level marine ecosystem model. This model incorporates 35 phytoplankton groups based on various functional types and size classes (Dutkiewicz et al. 2021), offering a more comprehensive representation of phytoplankton diversity compared to current-generation Earth System models, which typically include only 1–3 phytoplankton groups (Kearney et al. 2021). Advances in ocean colour sensors and algorithms have allowed the estimation of sea surface chlorophyll-a concentration for different size classes (e.g. Ward 2015), contributing to a better understanding of the global distribution of the simulated phytoplankton community structure in recent decades (Dutkiewicz et al. 2021).

A global climate projection utilising the 35-compartment phytoplankton model under a high greenhouse gas emission scenario illustrates potential abrupt shifts in phytoplankton biomass, productivity, and community structure across various oceanic regions during the latter half of the twenty-first century (Cael et al. 2021). These shifts may exceed 50% relative to historical baselines, covering up to 25% of the tropics and subtropical gyres (Cael et al. 2021). Notably, these abrupt shifts are anticipated on species requiring specific resources like diatoms, dinoflagellates, and diazotrophs (Dutkiewicz et al. 2020), particularly in subtropical regions (Cael et al. 2021). Importantly, these shifts occur independently of gradual changes in environmental variables, such as temperature and nutrients, making them unpredictable using standard early warning signals (Cael et al. 2021).

The causes of these abrupt shifts are unclear, but their occurrence at the edges of subtropical gyres suggests a potential link to the expansion of ocean deserts (Cael et al. 2021; Karl et al. 1995). Whether these shifts are reversible or not remains an open question that requires projections beyond the twenty-first century. Nevertheless, such abrupt shifts could trigger cascading effects in higher trophic levels and biogeochemical cycles due to the strong coupling of these systems. Investigating these consequences can be achieved by incorporating outputs from phytoplankton diversity models (e.g. Cael et al. 2021) into higher trophic level marine ecosystem models (e.g. Tittensor et al. 2021). This integrative approach represents a step towards understanding ecosystem changes comprehensively.

3.1.1.2 Impacts: Implications for Dimethylsulfide Abrupt shifts in phytoplankton productivity and diversity carry significant implications for the phytoplankton-aerosol-cloud feedback mediated by dimethylsulfide (DMS), as proposed by Charlson et al. (1987). DMS, a marine biogenic source of sulphur-containing aerosols such as sulphate, plays a key role in influencing the radiative properties of the atmosphere. For instance, a projected 24% reduction in the global sea-to-air DMS flux, potentially resulting from substantial changes in phytoplankton productivity and diversity, is estimated to warm the atmosphere by up to 0.76 K globally (Six et al. 2013). In the ocean, DMS originates from dimethylsulfoniopropionate (DMSP), an organosulphur compound primarily produced by phytoplankton. The intracellular ratio of DMSP to carbon biomass varies significantly among phytoplankton functional types, ranging up to four orders of magnitude (Stefels et al. 2007), highlighting the importance of phytoplankton diversity in DMS production. The production of DMSP and DMS will change as a result of not only warming but also acidification (Schwinger et al. 2017; Six et al. 2013). However, estimates on projected DMS changes using the current-generation climate models neither fully account for these environmental effects nor these models incorporate sufficient phytoplankton diversity (Bock et al. 2021).

Using the output of the 35-compartment phytoplankton model projection (Cael et al. 2021; Henson et al. 2021), we illustrate potential regional-scale abrupt shifts in DMSP during the twenty-first century. By multiplying the modelled phytoplankton biomass of the six functional types by their respective observed intracellular ratios, we derive the annual mean sea surface DMSP distribution averaged over the present period (2005–2024; Fig. 6a) and the projected percentage changes by the end of the twenty-first century (2081–2100; Fig. 6b). The DMSP time series from 1991 to 2100 at each grid point are fitted using both linear regression (Virtanen et al. 2020) and a step function through an offline change point detection method (Truong et al. 2020). The identification of abrupt changes is based on a comparison of the root mean square error (RMSE) of the step function to that of the linear fit, considering changes as abrupt when the ratio of the step-function RMSE to the linear-fit RMSE is less than 0.99.

Projected DMSP changes are characterised by a high (close to 100%) increase in subpolar regions (beyond 40°N/S) and a decrease in the tropics and sub-tropics (Fig. 6b). Henson's et al., (2021) work analyses the association with projected changes in phytoplankton biomass and notice that in the areas above DMS and phytoplankton biomass patterns are similar. However, there are regions, such as the Bay of Bengal and the Iceland Basin, where DMSP changes are prominent despite subtle biomass and diversity changes (Fig. 1a, b of Henson et al. 2021). Regression analysis indicates gradual DMSP increases in these regions (Fig. 6b). Abrupt shifts are projected to occur in various regions without a clear pattern regarding potential drivers, such as phytoplankton biomass and community composition.

In summary, our results suggest that the contribution of marine ecosystems to radiative forcing through DMS emission is likely to change gradually at the global scale (Fig. 6c), while abrupt changes may occur at the regional scale during the twenty-first century (Fig. 6b, d). These regional abrupt changes in DMS may be crucial for new particle formation in areas with low background cloud condensation nuclei, such as the Arctic (Sharma et al. 2012). It is crucial to note that our results are based on a basic parameterisation of intracellular DMSP-to-carbon ratios, intended to demonstrate the possibility of abrupt changes in DMSP in the coming decades. In reality, these ratios vary significantly within and among species, strongly depending on environmental conditions such as nutrient stress and acidification (Six et al. 2013). A more accurate projection requires a sophisticated parameterisation (McParland and Levine 2019). Further advancements in both modelling

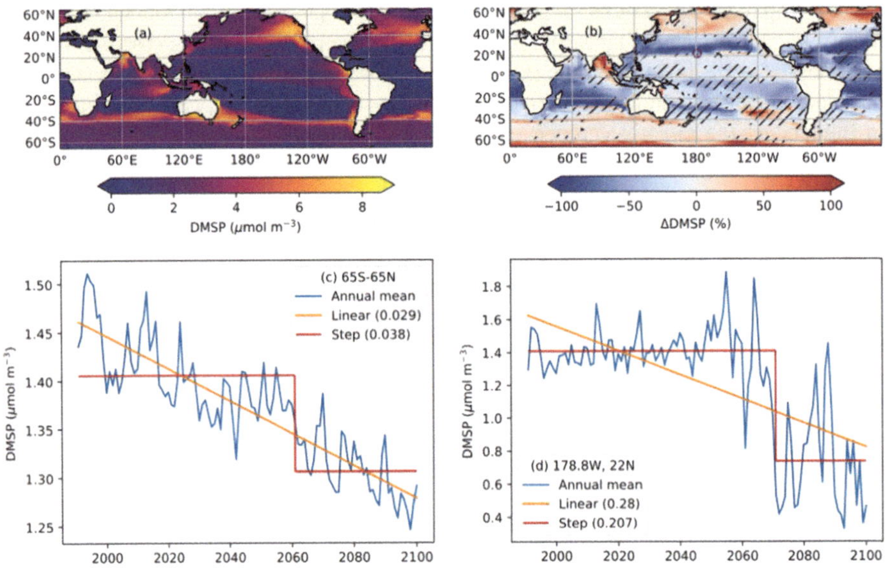

Fig. 6 Abrupt shifts in dimethylsulfoniopropionate (DMSP) in the global ocean during the twenty-first century derived from the output of a 35-compartment phytoplankton model projection based on a high greenhouse gas emission scenario. **a** Global distribution of annual mean sea surface DMSP averaged over 2005–2024, **b** projected percentage change in annual mean sea surface DMSP averaged over 2081–2100 relative to the period 2005–2024, and time series of annual mean sea surface DMSP **c** averaged over 65°S-65°N and **d** at a location in the North Pacific subtropical gyre denoted by a purple circle in **b**. DMSP is derived by multiplying the modelled carbon biomass of dinoflagellates by 22, coccolithophores by 11, diatoms by 0.86, and diazotrophs/prokaryotes/picokaryotes by 0.0015, which are the observed mean intracellular ratios for these phytoplankton functional types (Stefels et al. 2007; M. Galí and J. Stefels, personal communication, September 2023). In **b**, hatches represent areas where the step-function outperforms the linear regression in fitting the projected time series over 1991–2100 at each grid point. In **c** and **d**, orange and red lines represent the best fit based on linear regression and step function, respectively. Numbers in parentheses represent the root mean square error of each fit. See the main text for details

and remote sensing of phytoplankton functional types are essential for accurate quantification of DMSP in the ocean and monitoring the tipping point for the regional phytoplankton-aerosol-cloud feedback.

3.1.2 EO Needs and Opportunities

Detection and early warning of tipping points for these elements require further investigation into the various environmental factors influencing productivity and diversity, along with assessment through modelling. Long-term monitoring of phytoplankton productivity and diversity through remote sensing has proven effective in understanding global-scale patterns (Groom et al. 2019) and their responses to gradual climate changes (Cael et al. 2023). Continuous investment in this monitoring effort is essential to advance knowledge in the coming decades. Two ongoing satellite missions, the Global Change Observation Mission – Climate "SHIKISAI" (GCOM-C; https://global.jaxa.jp/projects/sat/gcom_c/) and the Plankton, Aerosol, Cloud, and ocean Ecosystem (PACE; https://pace.gsfc.nasa.gov/), offer prospects over the next decade, and additional missions are crucial for the future. Simultaneously, it is important to enhance the calibration of multiple missions to detect decadal changes.

Further improvements are needed in algorithms for identifying multiple phytoplankton groups, and these enhancements should occur concurrently with the parallel development of models. For instance, considering the significance of phytoplankton species composition in shaping the DMSP distribution, a top priority should be to increase the number of phytoplankton functional types in Earth System Models (ESMs), ideally to six as recommended by Stefels et al. (2007). Such an implementation would enable the assessment of the suitability of existing models with simpler complexity for projection studies (Bock et al. 2021) and open up new opportunities for further exploration.

3.2 Zooplankton

3.2.1 Nature of Tipping Point

Marine zooplankton can be considered a tipping element in marine ecosystems due to their central role in maintaining the balance and functionality of oceanic food webs and biogeochemical cycles: marine zooplankton are in fact fundamental to ocean ecosystems and global biogeochemical cycles, serving as a critical link between primary producers, like phytoplankton, and higher trophic levels, including fish, seabirds and marine mammals. Through their feeding activities, zooplankton help regulate phytoplankton populations, thus influencing primary production and the overall energy flow within marine food webs. Additionally, zooplankton play a key role in the "biological pump," a process that contributes to carbon sequestration. By consuming phytoplankton and converting this organic matter into faecal pellets, which sink to the ocean floor, zooplankton facilitate the transport of carbon from the surface to the deep ocean, effectively removing it from the atmosphere and helping mitigate climate change (Steinberg and Landry 2017; Turner 2015). Tipping points and abrupt shifts in marine zooplankton populations can therefore be extremely important because they can trigger significant and potentially irreversible changes in marine ecosystems and global biogeochemical cycles.

Abrupt community shifts (ACSs) in zooplankton have been documented in various marine ecosystems, providing valuable insights into ecological dynamics. Notable instances include occurrences in the North Pacific during 1976/1977 and 1989, as extensively studied by Ebbesmeyer et al. (1991), Francis and Hare (1994), Hare and Mantua (2000), Polovina (2005), and Field et al. (2006). Specifically, the north-east Pacific witnessed ACSs in 1976/1977 (McGowan et al. 2003; Miller et al. 2004) and 1998/1999 (Peterson and Schwing 2003; Lavaniegos and Ohman 2007), while the north-west Pacific experienced shifts in 1976/1977, 1988/1989, and 1998 (Chiba et al. 2006; Kasai and Ono 2007; Sakurai 2007). The Humboldt Current in the southern Pacific also observed ACSs in 1968/1970 and 1984/1986 (Alheit and Niquen 2004; Alheit 2009). In the north-west Atlantic, a notable event occurred in 1989/1990, documented by Frank et al. (2005), Pershing et al. (2005), Greene and Pershing (2007), Greene et al. (2012), and Greene et al. (2013).

Additionally, abrupt community shifts have been identified in European seas during the late 1980s, affecting the North Sea (Reid et al. 2001; Weijerman et al. 2005); the Baltic Sea (Alheit et al. 2005; Moellmann et al. 2009), and the north-west European shelf seas (1987/1990) (Beaugrand 2004, Heath and Beare 2008; the Black Sea (Daskalov et al. 2007, Oguz 2007) and the western and eastern (Adriatic) Mediterranean Sea (Conversi et al. 2010). This synchronicity of these shifts was investigated by Conversi et al 2010 and Beaugrand et al 2015, among others.

In the late 1990s, ACSs were observed in the north-east Atlantic and adjacent seas, the Black Sea, and San Francisco Bay (Oguz et al. 2003; Cloern et al. 2010; Luczak et al. 2011; Beaugrand et al. 2013). These shifts coincided with a global temperature shift around the same time (Reid and Beaugrand 2012). The ACS in the late 1970s originated from changes in large-scale boreal winter circulation patterns over the North Pacific in the mid-1970s, altering the environmental regime of these oceanic regions (Miller et al. 1994; Tsonis et al. 2007).

Explanations for ACSs in zooplankton often reference the theory of alternative stable states (Holling 1973; Scheffer and Carpenter 2003; Scheffer 2009). While some ecosystem shifts may result from multiple attractors, establishing their existence remains challenging (Seekell et al. 2013). Processes and alternative stable states leading to ACSs have been recognised in lakes and some marine ecosystems (Nystrom et al. 2000; Scheffer and van Nes 2004; Scheffer 2009). Utilising the METAL theory (MacroEcological Theory on the Arrangement of Life, Beaugrand and Kirby 2018; Beaugrand 2023), Beaugrand (2015) demonstrated that many ACSs in zooplankton could arise from the nonlinear interaction between the ecological niche of individual species and environmental changes without necessarily invoking the existence of a tipping point. Subsequently, Beaugrand et al. (2019) developed a global numerical model based on METAL to understand and predict long-term biological changes in oceans, including ACSs, effectively reconstructing well-observed shifts in the pelagic ocean (Fig. 7).

The framework was subsequently applied globally for the period 1960–2015 to pinpoint areas where major biological shifts are likely. The analyses revealed that significant biological shifts may occur annually but are confined to a limited part of the ocean, averaging approximately 2.8% of the total oceanic area, translating to an average area of around 10 million km^2 per year. Certain periods, such as 2005–2009, exhibited geographically restricted shifts, while others, like 2010–2014, demonstrated more extensive ones (see Fig. 8). The spatial extent of the biological shifts in 2010–2014 was particularly

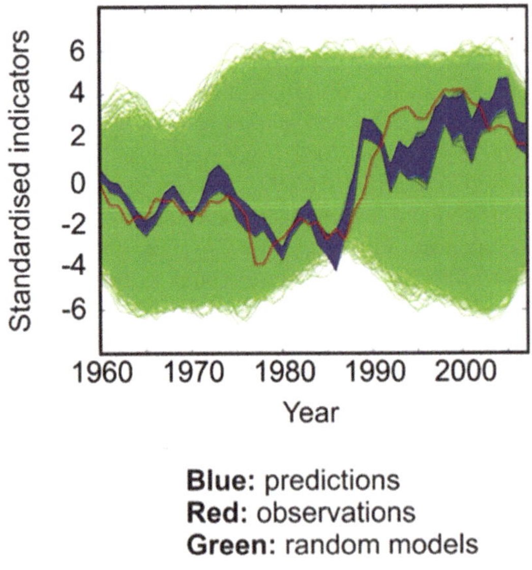

Fig. 7 Comparison of predicted (blue) and observed (red) biological shifts. Predictions were made using a model based on the MacroEcological Theory on the Arrangement of Life (METAL). Observations were synthesised using a principal component analysis combining 14 oceanic regions (Atlantic, Southern and Pacific oceans and their adjacent seas). Green curves show results using random models. The y-axis represents arbitrary standardised indicators of the biological state of the ecosystems. Negative and positive states have contrasting effects on the ecosystems

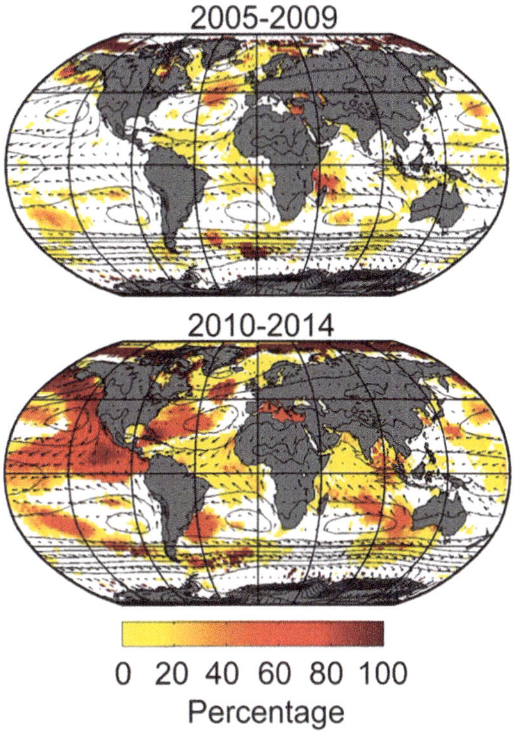

Fig. 8 Results from the global METAL model for time periods 2005–2009 (top) and 2010–2014 (bottom). The red colours denote substantial biological shifts and the yellow colours indicate minor changes. No colour denotes an absence of biological shifts

noteworthy and stemmed from the simultaneous occurrence of four phenomena: (i) a pronounced El Niño event (2015–2016),[1] (ii) the Pacific warm blob (an anomaly featuring positive temperatures in the north-eastern part of the North Pacific), (iii) the Atlantic cold blob (a cold temperature anomaly in the central part of the North Atlantic), and (iv) strong positive temperature anomalies in the Arctic Ocean (Beaugrand et al. 2019).

3.2.2 EO Needs and Opportunities

Given the crucial role that marine zooplankton plays in ocean ecosystems and global biogeochemical cycles, there is a growing need to monitor zooplankton populations on large spatial and temporal scales. Earth Observation (EO) technologies offer potential solutions, though studying zooplankton from space presents unique challenges.

Direct observation of individual zooplankton from space is not feasible due to their small size. However, proxy measurements and indirect methods, such as using ocean colour data, show promise. For instance, zooplankton abundance can be inferred from satellite-derived chlorophyll concentrations, their primary food source. Strömberg et al. (2009; https://doi.org/https://doi.org/10.1016/j.jmarsys.2009.02.004) demonstrated a relationship between chlorophyll-a concentrations and global zooplankton patterns. However, Friedland

[1] *Note that the calculation of a shift for a given year was based on a three-year period. Therefore, the time periods shown here embrace 1 year before and after in the calculation; 2005-2009 and 2010-2014 include 2004-2010 and 2009-2015.*

et al. (2018; https://doi.org/https://doi.org/10.1111/geb.12717) showed that these relationships vary significantly across regions and seasons, complicating the use of satellite data for accurate zooplankton estimates.

Isolating the optical signature of zooplankton from other particulate matter in the water remains a significant challenge. Developing algorithms to separate the zooplankton contribution to ocean colour signals is crucial for improving the accuracy of zooplankton studies using EO data. Frouin et al. (2019; https://doi.org/https://doi.org/10.3389/feart.2019.00145) reviewed advances in atmospheric correction for ocean colour remote sensing, underscoring its importance for the accurate estimation of bio-optical properties related to plankton. Persistent cloud cover in certain regions further complicates this issue, necessitating methods to work with limited or intermittent data. Moreover, managing both random and systematic errors is crucial to ensure the accuracy and reliability of the data.

Most zooplankton inhabit subsurface layers not directly observable from satellites, requiring innovative methods to infer subsurface conditions from surface observations. Techniques developed by Sauzède et al. (2015; https://doi.org/https://doi.org/10.5194/essd-7-261-2015) for estimating vertical chlorophyll profiles from satellite data could potentially be adapted for zooplankton studies. Additionally, high-frequency satellite observations, such as those from the Geostationary Ocean Color Imager (GOCI), can capture short-term variability in plankton communities (Li et al. 2022;b). Behrenfeld et al. (2019; https://doi.org/https://doi.org/10.1038/s41586-019-1796-9) utilised satellite lidar to observe diel vertical migration of zooplankton, though this technology is not yet widely available.

Distinguishing between different zooplankton species or functional groups from space is extremely challenging. Alvain et al. (2008; https://doi.org/https://doi.org/10.1029/2007GB003154) created a method to identify phytoplankton functional types from ocean colour data, which could serve as a model for similar approaches in zooplankton research.

Another challenge is the validation of zooplankton-related products derived from satellite data. Extensive in-situ sampling is necessary (Druon et al., 2019; https://doi.org/https://doi.org/10.1038/s41598-019-41212-2). However, combining satellite data with in-situ measurements involves substantial technical and methodological challenges due to differences in scale, coverage, and measurement techniques.

The examination of abrupt shifts and their attribution to environmental drivers, detailed in Sect. 3.2.1, highlights the critical significance of global-scale coverage encompassing both physical parameters like sea surface temperature (SST) and biological variables such as colour, facilitated by EO. Coordinated international efforts are essential for advancing high-resolution remote sensing and data integration capabilities to handle the large volumes of high-resolution data required to accurately represent the Earth's surface. Addressing inter-sensor biases is also critical for long-term monitoring and avoiding false trends (Balsamo et al., 2019; https://doi.org/https://doi.org/10.3390/rs11080941; Sathyendranath et al., 2023; https://doi.org/https://doi.org/10.1007/s10712-023-09805-9).

3.3 Higher Trophic Levels and Fisheries

3.3.1 Nature of Tipping Point

Pelagic ecosystems, particularly over continental shelves, do not operate in isolation, and the nonlinear interaction between plankton species and environmental changes can have cascading effects throughout the entire trophic food web. Numerous instances in the scientific literature (Lindley et al. 2010; Luczak et al. 2011, 2012) exemplify this phenomenon.

For instance, a study revealed a surge in the population of balearic shearwater (*Puffinus mauretanicus*) during the mid-1990s in the north-east Atlantic (17°W-4°E and 48°N-60°N). Concurrently, significant changes in a zooplankton composition index from the Continuous Plankton Recorder (CPR) survey were noted, along with a stepwise increase in the abundance of sardine and anchovy, the primary diet of the seabird (Luczak et al. 2011).

In regions over continental shelves, pelagic and benthic ecosystems are intricately linked, and alterations in pelagic ecosystems have pronounced impacts on benthic ecosystems (Lindley et al. 2010; Luczak et al. 2012). Profound transformations in the North Sea ecosystems, occurring at the end of the 1980s and from the middle to the end of the 1990s, were observed (Reid et al. 2001; Beaugrand et al. 2014). The abrupt rise in temperature significantly disrupted the trophodynamics of pelagic ecosystems, subsequently affecting benthic communities (Lindley et al. 2010). In the late 1990s, an increase in swimming crabs further influenced the reproductive success of lesser black-backed gull (*Larus fuscus graelsii*), establishing a connection between marine ecosystem shifts and terrestrial ecology (Luczak et al. 2012).

Alterations in the composition of North Sea zooplankton also had a considerable impact on commercially exploited fish, such as Atlantic cod (*Gadus morhua*). Post the late 1980s, changes in small planktonic crustaceans, crucial as the primary food for larval cod, occurred with a reduction in size, abundance, and biomass (Beaugrand et al. 2003). The timing of the main peak in copepod abundance shifted later in the year, coinciding with the period when fish larvae had matured and were no longer reliant on these small prey organisms. Combined with overfishing, these abrupt shifts adversely affected larval cod recruitment.

A novel model based on the METAL theory has demonstrated how overfishing and environmental changes can interact to impact cod recruitment and biomass in the North Sea (Beaugrand et al. 2022). The identification of a tipping point and the suggestion of hysteresis under certain conditions emphasis e the complex dynamics. Overexploitation can swiftly deplete fish stocks (Christensen et al. 2003; Pauly et al. 2003). Beaugrand et al. (2022) illustrated that, in favourable environmental conditions and with a hypothetical moratorium, recovery may occur, but the time required is lengthier than the duration of fishing-induced depletion. The recovery duration is influenced by the interaction between the low stock level and the initially slow exponential population growth phase, dependent on the stock's growth coefficient. However, when environmental conditions become unfavourable, recovery may become exceedingly challenging, spanning decades or even be impossible if adverse conditions persist, indicating irreversibility.

This example also elucidates why, despite a fishing moratorium near Newfoundland, recovery, albeit partial, took more than two decades (DFO 2018). Notably, recovery proves problematic for numerous other fish stocks (e.g. haddock, flatfish, Hutchings 2000). Hence, preventing collapse emerges as a more attainable goal than attempting to reverse it. Rebuilding a stock remains achievable only if environmental conditions remain conducive (Möllmann et al. 2021).

3.3.2 EO Needs and Opportunities

In this section, we review the significant challenges and promising opportunities associated with using Earth Observation (EO) to study higher trophic levels and fisheries.

Challenges:

1. Indirect Observations: One major obstacle is the inability to directly observe higher trophic species from space, which necessitates the use of proxy measurements and habitat modelling. This indirect approach is complicated by the complex, nonlinear relationships between environmental conditions and higher trophic levels, as well as time lags between observable environmental changes and their effects on species. These factors make it difficult to establish clear cause-and-effect relationships and accurately interpret EO data (Grémillet and Boulinier, 2009; https://doi.org/https://doi.org/10.3354/meps08212). The temporal mismatch between environmental changes and biological responses can obscure true relationships between oceanographic conditions and species distributions or abundances (Visser and Both, 2005; https://doi.org/https://doi.org/10.1098/rspb.2005.3356). This issue is particularly pronounced for long-lived species, where environmental effects may only become apparent after significant delays (Durant et al., 2007; https://doi.org/:https://doi.org/10.3354/cr033271). Additionally, the limited historical record of EO data (Swingedouw et al. 2020) and methodological consistency problems when reconstructing long time series from different sensors (Sathyendranath et al., 2023; https://doi.org/https://doi.org/10.1007/s10712-023-09805-9) further complicate the analysis.
2. Data integration Issues: Similar to the other biological components, Integrating EO data with in-situ observations, catch data, and ecosystem models is crucial for a comprehensive understanding of marine ecosystems but faces several challenges. Differences in data resolution, scale, and format can hinder this integration (Rose et al., 2015; https://doi.org/https://doi.org/10.1111/cobi.12397). The processes affecting fisheries and higher trophic levels often occur at scales that do not align with the resolution of available EO data, potentially missing fine-scale movements and leading to challenges in representing ecosystem dynamics (Rose et al., 2015; https://doi.org/https://doi.org/10.1111/cobi.12397; Kavanaugh et al., 2016; https://doi.org/https://doi.org/10.1093/icesjms/fsw086).

Opportunities:

1. Monitoring Fishing Activities: EO has proven valuable in monitoring fishing activities. Kroodsma et al. (2018; https://doi.org/https://doi.org/10.1126/science.aao5646) utilised AIS data and satellite imagery to create a global map of fishing effort, providing insights into the spatial and temporal patterns of fishing activities. This information is essential for enforcing fishing regulations and designing effective marine protected areas.
2. Habitat Modelling: Significant progress has been made in using satellite-derived oceanographic data to model and map potential habitats for species such as the blue shark (Druon et al., 2022; https://doi.org/https://doi.org/10.3389/fmars.2022.828412). Muller-Karger et al. (2018; https://doi.org/https://doi.org/10.3389/fmars.2018.00211) proposed a framework for integrating satellite observations with in-situ data and models to support ecosystem-based management of fisheries and marine resources. However, these approaches still require refinement.
3. Data Integration Advances: A strategic combination of remote sensing, in situ observations, and appropriate modelling is essential (Stuart et al. 2011). EO technology addresses critical gaps, providing an opportunity for all countries and individuals, even those without the capacity to collect essential data, to contribute to local solutions for global challenges (e.g. Almar et al. 2022). Recent advances in data assimilation techniques have shown promise in bridging the gap between satellite observations and ecosystem models (de Souza et al., 2016; https://doi.org/https://doi.org/10.1371/journ

al.pone.0163760; Schwartz-Belkin & Portman, 2023; https://doi.org/https://doi.org/10.1016/j.ocecoaman.2022.106280). By implementing models trained on a combination of EO and in situ data, Hazen et al. (2018) introduce an innovative multispecies and dynamic approach that utilises daily satellite data to monitor ocean features, aligning the scales of management, species movement, and fisheries.

4. Policy advances: Policy makers must understand the significance of satellite EO data in evidence-based decision-making. For example, NASA's Applied Sciences Programme within the Earth Science Division offers free training on the practical application of satellite data, covering objectives such as fisheries management (Andries et al. 2022). Global Fishing Watch builds geospatial datasets and hosts an online mapping platform that allows anyone with internet access to monitor various types of ocean-going vessels, sharing data and map products with scientists and practitioners, notwithstanding challenges discussed in (Drakopulos et al. 2023).

Future efforts should focus on improving our ability to detect and predict tipping points in marine ecosystems. This can be achieved through higher resolution satellite data, improved algorithms for detecting subtle changes in ecosystem indicators, and better methods for integrating multiple data sources to identify early warning signs of ecological shifts.

Integrating remote sensing EO technology with environmental decision-making enhances the efficiency and sustainability of solutions. Therefore, the broader scientific community must consistently support programs that enable the application of satellite EO data for global societal benefit and the stewardship of our planet (Kansakar & Hossain 2016).

3.4 Marine Biodiversity and Biomass

3.4.1 Nature of Tipping Point

While human activities, such as fishing, pollution, and species invasion, exert a significant impact on pelagic marine biodiversity, global climate change also plays a substantial role. However, assessing the extent of this stressor and the presence of tipping points remains challenging due to data and theoretical approach limitations.

A theoretical framework based on METAL estimated the vulnerability of global ocean biodiversity to climate change (Beaugrand et al. 2015). This framework suggested that climate change could swiftly reorganise marine biodiversity across extensive oceanic regions. Observations have since confirmed an increase in marine biodiversity in extratropical regions Hiddink and Hofstedeter 2008; Gordó-Vilaseca et al. 2023), aligning with the model's predictions. Rapid and substantial changes in marine biodiversity occurred in the north-east Atlantic and its adjacent seas in the late 1980s, influencing the entire trophodynamics of the North Sea across phytoplankton, fish, pelagic and benthic realms, and extending to terrestrial ecosystems (Beaugrand 2009). Although no clear tipping point has been identified, these changes are believed to result from the nonlinear interaction between species' ecological niches and shifts in the environmental regime (Beaugrand and Kirby 2018).

Comparisons of biodiversity changes expected by the end of this century with those observed between the last glacial maximum (LGM; about 20,000 years ago) and today, or between the mid-Pliocene (2.3 million years ago) and today, offer insights. The study highlights that the intensity of biodiversity reorganisation depends on the magnitude of warming (Beaugrand et al. 2015). Under scenarios of small global warming (RCP2.6), biological changes would be relatively benign, reflecting a modest percentage of historical changes. However, for moderate warming (RCP4.5), changes in marine biodiversity are considerably more extensive and robust than those observed over the past 50 years. In the case of severe global warming (RCP6.0 and 8.5), a substantial portion of the global ocean experiences a change in marine biodiversity equivalent to or exceeding that observed between the LGM/mid-Pliocene and today. This underscores the significant impact of climate warming on marine biodiversity. Any reorganisation of marine biodiversity is bound to influence species interactions and, consequently, ecosystem functioning, provisioning, and regulating services (Cheung et al. 2009), emphasising the importance of understanding the effects of climate change on biodiversity.

3.4.2 EO Needs and Opportunities

Earth Observation (EO) is crucial for species distribution mapping, particularly through habitat mapping. Satellites use spectral bands to identify marine habitats like coral reefs, seagrass beds, and kelp forests, with changes in these habitats indicating shifts in dependent species. For example, Landsat satellites have been instrumental in mapping coral reefs globally. (Li et al. 2021) used Landsat to create a global coral reef map, essential for understanding coral-dependent species distributions and detecting ecosystem tipping points. EO data combined with species distribution models helps predict suitable habitats and guide conservation efforts, as demonstrated by Yati et al. (2024; https://doi.org/https://doi.org/10.1016/j.marenvres.2024.106540), who used this approach to propose marine protected areas. Monitoring changes in predicted habitats over time allows detection of species distribution shifts, potentially indicating broader ecosystem changes.

EO also facilitates direct observation of larger marine species or large multi-individual aggregations. High-resolution satellite imagery, such as WorldView-3, has been used to detect and count whales, as shown by Cubaynes et al. (2019; https://doi.org/https://doi.org/10.1111/mms.12544). Similarly, satellites equipped with ocean colour sensors, such as MODIS, can map harmful algal blooms, which impact marine ecosystems, as demonstrated in studies by Zhao et al. (2020; https://doi.org/https://doi.org/10.1016/j.envadv.2020.100008) and others.

Satellite imagery is also useful in mapping seabird colonies, such as Lynch et al.'s (2012; https://doi.org/https://doi.org/10.1007/s00300-011-1138-3) survey of Adélie penguins in Antarctica.

Evaluating and monitoring future biodiversity requires robust models, similar to those used in generating Fig. 9. These models should be calibrated and evaluated based on recent biodiversity changes, using environmental parameters often derived from remote sensing. Additionally, empirical and data-driven early warning approaches, as demonstrated by studies like Dakos et al. (2010) and Verbesselt et al. (2016), present promising avenues. These approaches can be directly fuelled by Earth Observation (EO) data or utilise the outputs of ecosystem models.

Future EO needs and opportunities include:

- Hyperspectral sensors for improved habitat discrimination
- Higher temporal resolution data from geostationary satellites
- Improved atmospheric correction algorithms for coastal waters

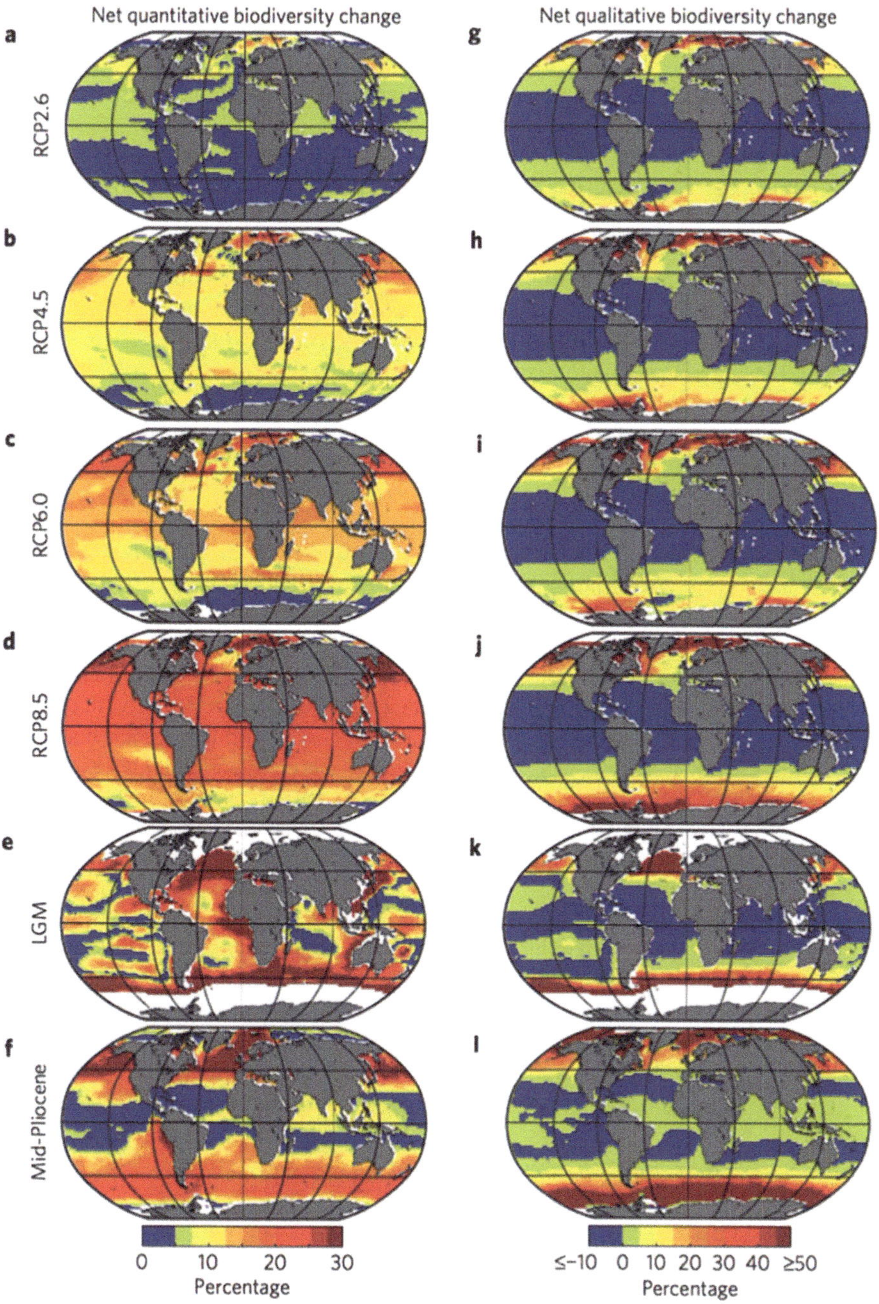

Fig. 9 Past and future quantitative (left-hand maps) and qualitative (right-hand maps) change in marine biodiversity. Maps on top four rows **a–d** and **g–j** show the difference between projected future (2081–2100) and present (2006–13) under the four emissions scenarios. Maps on bottom two rows show difference between the last glacial maximum (LGM, maps **e** and **k**) and 1960–69, and the Mid-Pliocene (maps **f** and **l**) and 2006–13. Regions of largest percentage change shown as oranges and reds, and smaller changes as blues and greens. From Beaugrand and colleagues (Beaugrand et al. 2015)

- LiDAR technology for detailed coastal topography and bathymetry
- Advanced SAR systems for all-weather monitoring of marine environments

The development of data fusion techniques and integrated satellite platforms, which combine data from optical, radar, and thermal sensors, holds significant promise, e.g. Moroni et al. (2016; https://doi.org/https://doi.org/10.1016/j.marpolbul.2015.07.045). This multi-sensor approach, coupled with advanced algorithms and AI, as shown by Guirado et al. (2019; https://doi.org/https://doi.org/10.1038/s41598-019-50795-9) for detecting whales, is revolutionising ecological research.

Advancements in EO technology, like higher resolution sensors optimised for coastal and inland waters, offer new opportunities for monitoring complex environments. Kutser et al. (2016; https://doi.org/https://doi.org/10.1080/01431161.2016.1186852) reviewed these methods, emphasising their importance for fine-scale ecological monitoring. Developing new biodiversity metrics based on EO data is crucial for conservation, as discussed by Pettorelli et al. (2016; https://doi.org/https://doi.org/10.1002/rse2.15), while long-term data consistency, as stressed by Mélin et al. (2017; https://doi.org/https://doi.org/10.1016/j.rse.2017.03.039), is essential for tracking ecological changes.

Combining EO with in-situ data enhances the accuracy of biodiversity assessments. Muller-Karger et al. (2018; https://doi.org/https://doi.org/10.3389/fmars.2018.00211) proposed a framework for such integration, which is vital for a holistic understanding of ecosystems. Advances in satellite-derived bathymetry and seafloor classification, as reviewed by Hedley et al. (2016; https://doi.org/https://doi.org/10.3390/rs8020118), hold promise for better mapping of underwater habitats. Integrating satellite observations with emerging techniques like environmental DNA (eDNA) sampling, or data from biogeochemical-Argo floats could provide a more comprehensive picture of marine biodiversity, as explored by Yamasaki et al. (2017; https://doi.org/https://doi.org/10.1016/j.cosust.2018.03.005) and Sauzède et al. (2016; https://doi.org/https://doi.org/10.1002/2015JC011408).

Cloud-based platforms and standardised data products are key for timely biodiversity assessments. Duffy et al. (2019; https://doi.org/https://doi.org/10.3389/fmars.2019.00317) emphasised the importance of near-real-time EO data for conservation.

Looking ahead, satellite-based acoustic sensors could monitor underwater noise levels, potentially affecting marine life. While still in its early stages, this innovative approach, discussed by Sun et al. (2021; https://doi.org/https://doi.org/10.3390/s21237849), represents the kind of innovative thinking needed to fully leverage EO for marine biodiversity monitoring. These advancements will significantly enhance our ability to monitor and understand marine ecosystems using Earth Observation technologies.

4 Synthesis and Prospects

Specific needs for Earth Observation (EO) depend strongly on the detail of the Tipping Element under consideration, and the reader is referred to the individual TE discussions in Sects. 2 and 3 of this paper. However, some common themes emerge. These themes are not all unique to ocean tipping points, but the nature of the ocean and of abrupt changes places particular emphasis on some of them. References are given beside each point below to relevant TE-specific discussion in earlier sections:

- The value of long-term, uninterrupted and homogeneous timeseries is clear in several areas, and is particularly important when studying abrupt changes that span decadal timescales (e.g. AMOC §2.2.2, ecosystem dynamics, especially involving long-lived species, §3.2.2, 3.3.2). Issues of continuity and cross-calibration should therefore be high on the agenda of future mission planning, as well as resolving apparent differences between alternative instruments or products (e.g. polar SSS, §2.2.2)
- An important role for EO emerges in monitoring and early detection of changes in critical fine scale processes (e.g. transports through Arctic gateways §2.3.2, 2.4.2, sub-mesoscale dynamics §2.5.2, patchy and moving planktonic systems §3.1.2, 3.2.2). In many cases the value of EO for this purpose can only be realised through combination with in situ, and model data (especially for processes involving the sub-surface ocean, which is not directly accessible to EO). Machine Learning, perhaps trained on high-resolution process-resolving models, provides particular new opportunities to integrate multiple datasets, augmenting the spatial and temporal resolution, and vertical coverage, of EO.
- Another important role for EO is in providing 'truth' data to assess and develop parametrisations of fine scale processes for use in coarser models (e.g. air-sea-ice interactions §2.2.2, 2.3.2). Many models used for large-scale climate assessments lack resolution in key areas (e.g. in space, time or ecosystem complexity).
- In other cases, state-of-the-art models (especially regional or subsystem-specific models) are ahead of current EO capabilities in resolution or complexity (e.g. multi-species ecosystem models, §3.3, 3.4). This creates a demand for higher resolution or accuracy than is possible with current EO systems. As well as driving future technological improvements, the existence of complex models provides opportunities to develop improved algorithms to maximise the information extracted from existing instruments, including EO-derived proxies for variables that cannot be observed directly. Again, Machine Learning is opening up new possibilities in this area.
- Combination or joint analysis of several data types is challenging, and in some cases methods of consistently assimilating multiple data types into (inevitably biased) models are a limiting factor in obtaining full value from existing EO datasets (e.g. for decadal prediction systems, §2.2.2). Developments in modelling and multivariate data assimilation must therefore proceed in tandem with future observational developments if the full potential of EO is to be realised.
- While data assimilating models, incorporating multiple data streams, offer a long-term path to monitoring and early warning, alternative methods of analysis based directly on EO data can provide valuable insights. For example, empirical methods for identifying approaching tipping through linking the changing properties of timeseries to simple conceptual models of tipping dynamics (e.g. critical slowing down §2.1, or space-for-time, §2.2), are well suited to the high spatial or temporal resolution of many EO datasets. The advent of high-resolution models and machine learning techniques opens up many new possibilities for training and evaluating such methods.
- It will never be possible to observe everything we would like to in sufficient detail. So inevitably, indirect or proxy relationships will need to be used between observed variables and the actual system variable of interest (e.g. various AMOC proxies §2.1.2, key elements of the oxygen cycle §2.6.2, indicators of ecosystem structure §3.2.2, 3.3.2). When developing empirical early warning indicators and proxies, it will be important to build confidence through model studies that such methods can detect the changes that may occur in future, as well as what has happened in the past (e.g. for AMOC §2.1.1). For long timescale processes palaeoclimatic data/model studies may be an

important source of calibration (e.g. biogeochemical processes §2.6). It will also be important to establish that relationships developed for one region are also applicable to other regions (e.g. for planktonic ecosystems §3.2.2) Specific attention should be paid to the combination of different observables, both EO and in situ, to increase the robustness of warning signals (e.g. a combination of temperature and salinity, both surface and subsurface and from different locations, to derive an early warning indicator for AMOC or SPG, §2.1.2, 2.2.2).

- The multivariate approaches outlined above imply an increasing need for improvements in data standardisation and accessibility, despite the challenges of differing space–time scales, timeliness, quality assurance and error characteristics (see e.g. discussion in §3.3.2). This presents a particular challenge to the international scientific community, as solutions developed for a particular problem may not be useful in the context of another.

As the focus of climate science moves from establishing the challenges of climate change to supporting societal responses to those challenges, so the demand for detail in monitoring changes in the climate system increases. The likelihood of crossing climate tipping points is expected to increase as the climate warms (IPCC 2021), and there is an urgent need for a joined-up approach, integrating EO and in situ observations with climate modelling, downstream impacts assessment and societal decision-making and policy (Wood et al. 2023). Earth Observation has a key role to play in avoiding the worst risks of climate tipping, and in building resilience to those risks that cannot be avoided.

Acknowledgements This paper is an outcome of the Workshop on "Tipping Points and Understanding EO data needs for a Tipping Element Model Intercomparison Project" held at the International Space Science Institute (ISSI) in Bern, Switzerland (10-14 October 2022). R. Wood and J. Baker were supported by the Met Office Hadley Centre Climate Programme funded by BEIS. R. Wood, J. Baker and T. Stocker were supported by the European Union's Horizon 2020 research and innovation programme under grant agreement no. 820970 (TiPES project). This paper is TiPES contribution number 274. T. Stocker acknowledges support by SNF #200492. J. Boutin acknowledges support from the European Space Agency (ESA) CCI+SSS phase 2 project. I. Frenger acknowledges support from German Federal Ministry of Education and Research (BMBF) funded project Humboldt Tipping (HTP; IF, grant no. 01LC2323B) K. Kvale acknowledges support from the New Zealand Ministry of Business, Innovation and Employment within the Antarctic Science Platform, grant ANTA1801. K.K. also acknowledges support from the New Zealand Ministry of Business, Innovation and Employment through the Global Change through Time programme (Strategic Science Investment Fund, contract C05X1702). A. Landolfi acknowledges support from the European Space Agency (ESA) Ocean Science Cluster CAREHeat (Contract No. 4000137121/21/I-DT) and MiTHO (Contract No. 4000142100/23/I-DT) projects. W. Maslowski acknowledges support from the U.S. Department of Energy (DOE) Regional and Modelling Analysis (RGMA) and Office of Naval Research (ONR) Arctic and Global Prediction (AGP) programs. A.Romanou was supported by the NASA-Modeling, Analysis and Prediction program (MAP) (grant number NNX16AC93 G, PI Romanou). Christopher J. Somes was supported by the Deutsche Forschungsgemeinschaft (project no. 445549720). D. Swingedouw received financial support from the French government in the framework of the University of Bordeaux's IdEx "Investments for the Future" program / RRI Tackling Global Change and from the UKRI DECADAL project and from TipESM project funded by the European Union's Horizon Europe research and innovation programme under grant agreement No 101137673.

Author contributions Authors Conversi, Donner, Landolfi, Maslowski, Romanou, Stocker, Swingedouw and Wood contributed to the conceptualisation of the paper, initiated at the ISSI Workshop on "Tipping Points and Understanding EO data needs for a Tipping Element Model Intercomparison Project" in October 2022.

Funding Open access funding provided by Consiglio Nazionale Delle Ricerche (CNR) within the CRUI-CARE Agreement.

Declarations

Conflict of interest There are no conflict of interests.

References

Aagaard K, Coachman LK, Carmack E (1981) On the halocline of the Arctic Ocean. Deep Sea Res Part A 28(6):529–545. https://doi.org/10.1016/0198-0149(81)90115-1

de Abreu L, Shackleton NJ, Schönfeld J, Hall MA, Chapman MR (2003) Millenial-scale oceanic climate variability off the Western Iberian margin during the last two glacial periods. Mar Geol 196(1–2):1–20. https://doi.org/10.1016/S0025-3227(03)00046-X

Alheit J (2009) Consequences of regime shifts for marine food webs. Int J Earth Sci 98:261–268

Alheit J, Mollmann C, Dutz J, Kornilovs G, Loewe P, Mohrholz V, Wasmund N (2005) Synchronous ecological regime shifts in the central Baltic and the North Sea in the late 1980s. ICES J Mar Sci 62:1205–1215

Alheit J, Niquen M (2004) Regime shifts in the Humboldt current ecosystem. Prog Oceanogr 60:201–222

Alkhayuon H, Ashwin P, Jackson LC, Quinn C, Wood RA (2019) Basin bifurcations, oscillatory instability and rate-induced thresholds for AMOC in a global oceanic box model. Proc r Soc A 475:20190051. https://doi.org/10.1098/rspa.2019.0051

Alley RB, Ágústsdóttir AM (2005) The 8k event: cause and consequences of a major Holocene abrupt climate change. Quatern Sci Rev 24(10–11):1123–1149. https://doi.org/10.1016/j.quascirev.2004.12.004

Almar R, Kestenare E, Duong D, Bergsma E, Louvat D et al (2022) Remote sensing to contribute to local solutions for global challenges. J Coastal Res 38:188–201. https://doi.org/10.2112/JCOASTRES-D-21-00023.1

Alvain S, Moulin C, Dandonneau Y, Loisel H (2008) Seasonal distribution and succession of dominant phytoplankton groups in the global ocean: a satellite view. Glob Biogeochem Cycles 22:GB3001. https://doi.org/10.1029/2007GB003154

Andries A, Morse S, Murphy RJ, Lynch J, Woolliams ER (2022) Using data from earth observation to support sustainable development indicators: an analysis of the literature and challenges for the future. Sustainability 14:1191

Baker J, Renshaw R, Jackson LC, Dubois C,Iovino D, Zuo H (2022a) Overturning variations in the subpolar North Atlantic in an ocean reanalyses ensemble. Copernicus Marine service ocean state report, Issue 6, Journal of Operational Oceanography

Baker JA, Renshaw R, Jackson LC, Dubois C, Iovino D, Zuo H, Perez RC, Dong S, Kersalé M, Mayer M, Mayer J, Speich S, Lamont T (2023) Overturning and heat transport variations in the South Atlantic in an ocean reanalysis ensemble, State of the Planet, 1-osr7, 4 pp., https://doi.org/10.5194/sp-1-osr7-4-2023

Bakker P, Schmittner A, Lenaerts JTM, Abe-Ouchi A, Bi D, van den Broeke MR, Chan WL, Hu A, Beadling RL, Marsland SJ, Mernild SH, Saenko OA, Swingedouw D, Sullivan A, Yin J (2016) Fate of the Atlantic Meridional overturning circulation: strong decline under continued warming and Greenland melting. Geophys Res Lett 43(23):12252–12260. https://doi.org/10.1002/2016GL070457

Balsamo G, Agusti-Panareda A, Albergel C, Arduini G, Beljaars A, Bidlot J, Blyth E, Bousserez N, Boussetta S, Brown A et al (2019) Correction: satellite and in situ observations for advancing global earth surface modelling: a review. Remote Sensing 10:2038. https://doi.org/10.3390/rs11080941

Bamber JL et al (2018) Land ice freshwater budget of the Arctic and North Atlantic Oceans: 1. Data, methods, and results. J Geophys Res Oceans 123(3):1827–1837

Barthélémy S, Brajard J, Bertino L et al (2022) Super-resolution data assimilation. Ocean Dyn 72:661–678. https://doi.org/10.1007/s10236-022-01523-x

Beaugrand G (2004) The North Sea regime shift: evidence, causes, mechanisms and consequences. Prog Oceanogr 60:245–262
Beaugrand G (2009) Decadal changes in climate and ecosystems in the North Atlantic Ocean and adjacent seas. Deep-Sea Res II 56:656–673
Beaugrand G (2015) Theoretical basis for predicting climate-induced abrupt shifts in the oceans. Philos Trans R Soc b: Biol Sci 370:20130264
Beaugrand G (2023) Towards an understanding of large-scale biodiversity patterns on land and in the sea. Biology 12:339
Beaugrand G, Balembois A, Kléparski L, Kirby RR (2022) Addressing the dichotomy of fishing and climate in fishery management with the FishClim model. Commun Biol 5:1146
Beaugrand G, Brander KM, Lindley JA, Souissi S, Reid PC (2003) Plankton effect on cod recruitment in the North Sea. Nature 426:661–664
Beaugrand G, Conversi A, Atkinson A, Cloern J, Chiba S, Fonda-Unami S, Kirby RR, Greene CG, Goberville E, Otto SA, Reid PC, Stemmann L, Edwards M (2019) Prediction of unprecedented biological shifts in the global ocean. Nat Clim Chang 9:237–243
Beaugrand G, Conversi A, Chiba S, Edwards M, Fonda-Umani S, Greene C, Mantua N, Otto SA, Reid PC, Stachura M, Stemmann L, Sugisaki H (2015) Synchronous marine pelagic regime shifts in the Northern Hemisphere. Philos Trans R Soc b: Biol Sci 370:20130272
Beaugrand G, Edwards M, Raybaud V, Goberville E, Kirby RR (2015) Future vulnerability of marine biodiversity compared with contemporary and past changes. Nat Clim Chang 5:695–701
Beaugrand G, Harlay X, Edwards M (2014) Detecting plankton shifts in the North Sea: a new abrupt ecosystem shift between 1996 and 2003. Mar Ecol Prog Ser 502:85–104
Beaugrand G, Kirby RR (2018) How do marine pelagic species respond to climate change? Theories and observations. Ann Rev Mar Sci 10:169–197
Beaugrand G, McQuatters-Gollop A, Edwards M, Goberville E (2013) Long-term responses of North Atlantic calcifying plankton to climate change. Nature Clim Change 3:263–267
Beer E, Eisenman I, Wagner T, Fine E (2023) A Possible Hysteresis in the Arctic Ocean due to Release of Subsurface Heat During Sea Ice Retreat. J Phys Oceanogr 53:1323–1335. https://doi.org/10.1175/JPO-D-22-0131
Behrenfeld MJ, Gaube P, Della Penna A et al (2019) Global satellite-observed daily vertical migrations of ocean animals. Nature 576:257–261. https://doi.org/10.1038/s41586-019-1796-9
Beil S, Kuhnt W, Holbourn A, Scholz F, Oxmann J, Wallmann K et al (2020) Cretaceous oceanic anoxic events prolonged by phosphorus cycle feedbacks. Clim past 16:757–782. https://doi.org/10.5194/cp-16-757-2020
Belkin IM (2004) Propagation of the "Great Salinity Anomaly" of the 1990s around the northern North Atlantic. Geophys Res Lett 31:L08306. https://doi.org/10.1029/2003GL019334
Bellomo K, Meccia VL, D'Agostino R et al (2023) Impacts of a weakened AMOC on precipitation over the Euro-Atlantic region in the EC-Earth3 climate model. Clim Dyn 61:3397–3416. https://doi.org/10.1007/s00382-023-06754-2
Benedetti-Cecchi L (2021) Complex networks of marine heatwaves reveal abrupt transitions in the global ocean. Sci Rep 11:1739. https://doi.org/10.1038/s41598-021-81369-3
Bilbao R, Wild S, Ortega P, Acosta-Navarro J, Arsouze T, Bretonnière P-A et al (2021) Assessment of a full-field initialized decadal climate prediction system with the CMIP6 version of EC-Earth. Earth Syst Dyn 12:173–196. https://doi.org/10.5194/esd-12-173-2021
Bilo TC, Straneo F, Holte J, leBras IA-A (2022) Arrival of new Great Salinity Anomaly weakens convection in the Irminger Sea. Geophys Res Lett. https://doi.org/10.1029/2022GL098857
Blunier T, Chappellaz J, Schwander J et al (1998) Asynchrony of Antarctic and Greenland climate change during the last glacial period. Nature 394:739–743. https://doi.org/10.1038/29447
Bock J, Michou M, Nabat P, Abe M, Mulcahy JP, Olivié DJL, Schwinger J, Suntharalingam P, Tjiputra J, van Hulten M, Watanabe M, Yool A, Séférian R (2021) Evaluation of ocean dimethylsulfide concentration and emission in CMIP6 models. Biogeosciences 18:3823–3860. https://doi.org/10.5194/bg-18-3823-2021
Boers N (2021) Observation-based early-warning signals for a collapse of the Atlantic Meridional overturning circulation. Nat Clim Change 11(August):680–688. https://doi.org/10.1038/s41558-021-01097-4
Boers N, Ghil M, Stocker TF (2022) Theoretical and paleoclimatic evidence for abrupt transitions in the Earth system. Environ Res Lett 17(9):093006. https://doi.org/10.1088/1748-9326/ac8944
Bonjean F, Boutin J, Vergely J-L, Richaume P and Sabia R (2024) Recovery of SMOS salinity variability in RFI-contaminated regions. In IEEE Transactions on Geoscience and Remote Sensing, https://doi.org/10.1109/TGRS.2024.3408049

Bonnet R, Swingedouw D, Gastineau G et al (2021) Increased risk of near term global warming due to a recent AMOC weakening. Nat Commun 12:6108. https://doi.org/10.1038/s41467-021-26370-0
Bopp L, Resplandy L, Untersee A, Le Mezo P, Kageyama M (2017) Ocean (de)oxygenation from the last glacial maximum to the twenty-first century: insights from earth system models. Phil Trans R Soc A 375:20160323. https://doi.org/10.1098/rsta.2016.0323
Born A, Stocker TF, Raible CC, Levermann A (2013) Is the Atlantic subpolar gyre bistable in comprehensive cou- pled climate models? Clim Dyn 40:2993–3007
Boulton CA, Allison LC, Lenton TM (2014) Early warning signals of atlantic meridional overturning circulation collapse in a fully coupled climate model. Nat Commun 5(7):5752. https://doi.org/10.1038/ncomms6752
Boutin J et al (2021) Correcting sea surface temperature spurious effects in salinity retrieved from spaceborne L-band radiometer measurements. IEEE Trans Geosci Remote Sens 59(9):7256–7269. https://doi.org/10.1109/tgrs.2020.3030488
Boutin J, Reul N, Koehler J, Martin A, Catany R, Guimbard S et al (2021) Satellite-based sea surface salinity designed for ocean and climate studies. J Geophys Res (Oceans). https://doi.org/10.1029/2021JC017676
Boutin J, Vergely JL, Bonjean F, Perrot X, Zhou Y, Dinnat E, Lang R, Levine D, Sabia R (2023) New Seawater Dielectric constant parametrization and application to smos retrieved salinity. IEEE Transactions of Geoscience and Remote Sensing 61:1–13. https://doi.org/10.1109/TGRS.2023.3257923
Breitburg D, Levin LA, Oschlies A, Grégoire M, Chavez FP, Conley DJ et al (2018) Declining oxygen in the global ocean and coastal waters. Science 359:eaam240. https://doi.org/10.1126/science.aam7240
Broecker WS (1987) Unpleasant surprises in the greenhouse? Nature 328(6126):123
Broecker WS (1998) Paleocean circulation during the Last Deglaciation: A bipolar seesaw? Paleoceangr Paleoclimatology 13(2):119–121. https://doi.org/10.1029/97PA03707
Broecker WS, Peteet DM, Rind D (1985) Does the ocean–atmosphere system have more than one stable mode of operation? Nature 315(6014):21
Broeker WS (1991) The great ocean conveyor. Oceanography 4(2):79–89. https://doi.org/10.5670/oceanog.1991.07
Brown N, Galbraith ED (2016) Hosed vs. unhosed: interruptions of the Atlantic Meridional Overturning Circulation in a global coupled model, with and without freshwater forcing. Clim past 12:1663–1679. https://doi.org/10.5194/cp-12-1663-2016
Bryan F (1986) High-latitude salinity effects and interhemispheric thermohaline circulations. Nature 323:301–304
Buckley MW, Marshall J (2016) Observations, inferences, and mechanisms of the Atlantic Meridional overturning circulation: a review. Rev Geophys 54(1):5–63. https://doi.org/10.1002/2015RG000493
Cabré A, Marinov I, Bernardello R, Bianchi D (2015) Oxygen minimum zones in the tropical Pacific across CMIP5 models: mean state differences and climate change trends. Biogeosciences 12:5429–5454. https://doi.org/10.5194/bg-12-5429-2015
Cael BB, Bisson K, Boss E, Dutkiewicz S, Henson S (2023) Global climate-change trends detected in indicators of ocean ecology. Nature 619(7970):7970. https://doi.org/10.1038/s41586-023-06321-z
Cael BB, Dutkiewicz S, Henson S (2021) Abrupt shifts in 21st-century plankton communities. Sci Adv 7(44):eabf8593. https://doi.org/10.1126/sciadv.abf8593
Caesar L, Rahmstorf S, Robinson A, Feulner G, Saba V (2018) Observed fingerprint of a weakening Atlantic Ocean overturning circulation. Nature 556(7700):191–196. https://doi.org/10.1038/s41586-018-0006-5
Canfield DE (2006) Models of oxic respiration, denitrification and sulfate reduction in zones of coastal upwelling. Geochim Cosmochim Acta 70:5753–5765. https://doi.org/10.1016/j.gca.2006.07.023
Canfield DE, Glazer A, Falkowski PG (2010) The evolution and future of earth's nitrogen cycle. Science 330:192–196. https://doi.org/10.1126/science.1186120
Van Cappellen P, Ingall ED (1994) Benthic phosphorus regeneration, net primary production, and ocean anoxia: a model of the coupled marine biogeochemical cycles of carbon and phosphorus. Paleoceanography 9:677–692. https://doi.org/10.1029/94PA01455
Carpenter SR, Brock WA (2006) Rising variance: a leading indicator of ecological transition. Ecology Lett 9:311–318
Castelao RM, Oliver H, Medeiros PM (2023) Satellite-derived Lagrangian transport 693 pathways in the Labrador Sea. Rem Sens 15(23):5545. https://doi.org/10.3390/rs15235545
Castellana D, Dijkstra HA (2020) Noise-induced transitions of the Atlantic Meridional overturning circulation in CMIP5 models. Sci Rep 10(1):1–9. https://doi.org/10.1038/s41598-020-76930-5

Chang Y-L, Miyazawa Y, Miller ML, Tsukamoto K (2019) Influence of ocean circulation and the Kuroshio large meander on the 2018 Japanese eel recruitment season. PLoS ONE 14(9):e0223262. https://doi.org/10.1371/journal.pone.0223262
Charlson RJ, Lovelock JE, Andreae MO, Warren SG (1987) Oceanic phytoplankton, atmospheric sulphur, cloud albedo and climate. Nature 326(6114):655–661. https://doi.org/10.1038/326655a0
Chen X, Tung K-K (2018) Global surface warming enhanced by weak Atlantic overturning circulation. Nature 559:387–391. https://doi.org/10.1038/s41586-018-0320-y
Cheung WWL, Lam VWY, Sarmiento JL, Kearney K, Watson R, Pauly D (2009) Projecting global marine biodiversity impacts under climate change scenarios. Fish Fish 10:235–251
Chiba S, Tadokoro K, Sugisaki H, Saino T (2006) Effects of decadal climate change on zooplankton over the last 50 years in the western subarctic North Pacific. Glob Change Biol 12:907–920
Christensen V, Guénette S, Heymans JJ, Walthers CJ, Watson R, Zeller D, Pauly D (2003) Hundred-year decline of North Atlantic predatory fishes. Fish Fish 4:1–2
Claar DC, Starko S, Tietjen KL et al (2020) Dynamic symbioses reveal pathways to coral survival through prolonged heatwaves. Nat Commun 11:6097. https://doi.org/10.1038/s41467-020-19169-y
Clement Kinney J, Assmann KM, Maslowski W, Björk G, Jakobsson M, Jutterström S, Lee YJ, Osinski R, Semiletov I, Ulfsbo A, Wåhlström I, Anderson LG (2022) On the circulation, water mass distribution, and nutrient concentrations of the western Chukchi Sea. Ocean Sci 18:29–49. https://doi.org/10.5194/os-18-29-2022
Cloern JE, Hieb KA, Jacobson T, Sanso B, Di Lorenzo E, Stacey MT, Largier JL, Meiring W, Peterson WT, Powell TM, Winder M, Jassby AD (2010) Biological communities in San Francisco Bay track large-scale climate forcing over the North Pacific. Geophys Res Lett 37:L21602
Collins M, Sutherland M, Bouwer L, Cheong S-M, Frolicher T, Jacot Des Combes H, et al. (2019) IPCC Special report on the ocean and cryosphere in a changing climate. Chapter 6: Extremes, abrupt changes and managing risks. Cambridge, NY: Cambridge University Press
Conversi A, Fonda-Umani S, Peluso T, Molinero JC, Santojanni A, Edwards M (2010) The mediterranean sea regime shift at the end of the 1980s, and intriguing parallelisms with other european basins. PLoS ONE 5:e10633
Le Corre M, Gula J, Tréguier A-M (2020) Barotropic vor ticity balance of the North Atlantic subpolar gyre in an eddy- resolving model. Ocean Sci 16:451–468
Crichton KA, Wilson JD, Ridgwell A et al (2023) What the geological past can tell us about the future of the ocean's twilight zone. Nat Commun 14:2376. https://doi.org/10.1038/s41467-023-37781-6
Crichton KA, Wilson JD, Ridgwell A, Pearson PN (2021) Calibration of temperature-dependent ocean microbial processes in the cGENIE.muffin (v0.9.13) Earth system model. Geosci Model Dev 14:125–149. https://doi.org/10.5194/gmd-14-125-2021
Cubaynes HC, Fretwell PT, Bamford C, Gerrish L, Jackson JA (2019) Whales from space: four mysticete species described using new VHR satellite imagery. Mar Mam Sci 35:466–491. https://doi.org/10.1111/mms.12544
DFO (2018) Stock assessment of northern cod (NAFO Divisions 2J3KL) in 2018. In. DFO Can. Sci. Advis. Sec. Sci. Advis. Rep., St. John's
Dakos V et al (2010) Spatial correlation as leading indicator of catastrophic shifts. Theor Ecol 3:163–174. https://doi.org/10.1007/s12080-009-0060-6
Dansgaard W, Johnsen S, Clausen H et al (1993) Evidence for general instability of past climate from a 250-kyr ice-core record. Nature 364:218–220. https://doi.org/10.1038/364218a0
Daskalov GM, Grishin AN, Rodionov S, Mihneva V (2007) Trophic cascades triggered by overfishing reveal possible mechanisms of ecosystem regime shifts. Proc Natl Acad Sci 104:10518–10523
Defrance D, Ramstein G, Charbit S, Vrac M, Sultan B, Swingedouw D, Dumas C, Gemenne F, Alvarez-Solas J, Vanderlinden JP (2017) Consequences of rapid ice-sheet melting on the Sahelian population vulnerability. PNAS 114(25):6533–6538. https://doi.org/10.1073/pnas.1619358114
DelSole T, Nedza D (2022) Reconstructing the Atlantic overturning circulation Using linear machine learning techniques. Atmos Ocean 60:541–553. https://doi.org/10.1080/07055900.2021.1947181
Desbruyères DG, Mercier H, Maze G, Daniault N (2019) Surface predictor of overturning circulation and heat content change in the subpolar North Atlantic. Ocean Sci 15:809–817. https://doi.org/10.5194/os-15-809-2019
Deshayes J, Tréguier AM, Barnier B, Lecointre A, le Sommer J, Molines JM et al (2013) Oceanic hindcast simulations at high resolution suggest that the Atlantic MOC is bistable. Geophys Res Lett 40:3069–3073. https://doi.org/10.1002/grl.50534
Dickson RR, Meincke J, Malmberg S-A, Lee AJ (1988) The "Great Salinity Anomaly" in the northern North Atlantic, 1968–1982. Prog Oceanogr 20:103–151

Dima M, Lohmann G (2010) Evidence for two distinct modes of large-scale ocean circulation changes over the last century. J Clim 23:5–16
Ding Q, Schweiger A, L'Heureux M et al (2017) Influence of high-latitude atmospheric circulation changes on summertime Arctic sea ice. Nature Clim Change 7:289–295. https://doi.org/10.1038/nclimate3241
Dong S, Goni G, Domingues R, Bringas F, Goes M, Christophersen J, Baringer M (2021) Synergy of In Situ and satellite ocean observations in determining meridional heat transport in the Atlantic Ocean. J Geophys Res Oceans 126:e2020JC017073. https://doi.org/10.1029/2020JC017073
Dosser HV, Rainville L (2016) Dynamics of the changing nearinertial internal wave field in the Arctic Ocean. J Phys Oceanogr 46:395–415. https://doi.org/10.1175/JPO-D-15-0056.1
Drakopulos L, Silver JJ, Nost E, Gray N, Hawkins R (2023) Making global oceans governance in/visible with Smart Earth: the case of global fishing watch. Environ Plan E: Nat Space 6:1098–1113
Druckenmiller ML, Thoman RL, Moon TA, Eds. (2022) Arctic Report Card 2022, https://doi.org/10.25923/yjx6-r184
Druon J-N et al (2022) Global-Scale environmental niche and habitat of blue shark (Prionace glauca) by size and sex: a pivotal step to improving stock management. Front Mar Sci 9:828412. https://doi.org/10.3389/fmars.2022.828412
Druon J-N, Hélaouët P, Beaugrand G et al (2019) Satellite-based indicator of zooplankton distribution for global monitoring. Sci Rep 9:4732. https://doi.org/10.1038/s41598-019-41212-2
Duffy JE et al (2019) Toward a coordinated global observing system for seagrasses and marine macroalgae. Front Mar Sci 6:317. https://doi.org/10.3389/fmars.2019.00317
Durant JM et al (2007) Climate and the match or mismatch between predator requirements and resource availability. Clim Res 33:271–283. https://doi.org/10.3354/cr033271
Dutkiewicz S, Boyd PW, Riebesell U (2021) Exploring biogeochemical and ecological redundancy in phytoplankton communities in the global ocean. Glob Change Biol 27(6):1196–1213. https://doi.org/10.1111/gcb.15493
Dutkiewicz S, Cermeno P, Jahn O, Follows MJ, Hickman AE, Taniguchi DAA, Ward BA (2020) Dimensions of marine phytoplankton diversity. Biogeosciences 17(3):609–634. https://doi.org/10.5194/bg-17-609-2020
Ebbesmeyer CC, Cayan DR, McLain DR, Nichols FH, Peterson DH, Redmond KT (1991) 1976 step in the Pacific climate: forty environmental changes between 1968–1975 and 1977–1984.
Estella-Perez V, Mignot J, Guilyardi E et al (2020) Advances in reconstructing the AMOC using sea surface observations of salinity. Climate Dyn 55:975–992. https://doi.org/10.1007/s00382-020-05304-4
FAO (2022) The State of World Fisheries and Aquaculture 2022 Towards Blue Transformation. FAO, Rome
Field DB, Baumgartner TR, Charles CD, Ferreira-Bartrina V, Ohman MD (2006) Planktonic foraminifera of the California Current reflect 20th-century warming. Science 311:63–66
Field CB, Behrenfeld MJ, Randerson JT, Falkowski P (1998) Primary production of the Biosphere: integrating terrestrial and oceanic components. Science 281(5374):237–240. https://doi.org/10.1126/science.281.5374.237
Fine EC, Cole ST (2022) Decadal observations of internal wave energy, shear, and mixing in the western Arctic Ocean. J Geophys Res Oceans 127:e2021JC018056. https://doi.org/10.1029/2021JC018056
Frajka-Williams E (2015) Estimating the Atlantic overturning at 26°N using satellite altimetry and cable measurements. Geophys Res Lett 42:3458–3464. https://doi.org/10.1002/2015GL063220
Frajka-Williams E, Ansorge IJ, Baehr J, Bryden HL, Chidichimo MP, Cunningham SA, Danabasoglu G, Dong S, Donohue KA, Elipot S, Heimbach P, Holliday NP, Hummels R, Jackson LC, Karstensen J, Lankhorst M, Le Bras IA, Lozier MS, McDonagh EL, Wilson C (2019) Atlantic meridional overturning circulation: observed transport and variability. Front Mar Sci 6:1–18. https://doi.org/10.3389/fmars.2019.00260
Francis RC, Hare SR (1994) Decadal-scale regime shifts in the large marine ecosystems of the north-east Pacific: a case for historical science. Fish Oceanogr 3:279–291
Frank KT, Petrie B, Choi JS, Leggett WC (2005) Trophic cascades in a formerly cod-dominated ecosystem. Science 308:1621–1623
Fraser NJ, Cunningham SA (2021) 120 Years of AMOC variability reconstructed from observations using the Bernoulli inverse. Geophys Res Lett 48(18):e2021GL093893. https://doi.org/10.1029/2021GL093893
Friedland KD, Mouw CB, Asch RG et al (2018) Phenology and time series trends of the dominant seasonal phytoplankton bloom across global scales. Glob Ecol Biogeogr 27:551–569. https://doi.org/10.1111/geb.12717
Frouin RJ, Franz BA, Ibrahim A et al (2019) Advances in atmospheric correction for ocean colour remote sensing. Front Earth Sci 7:145. https://doi.org/10.3389/feart.2019.00145

Frölicher TL, Aschwanden MT, Gruber N, Jaccard SL, Dunne JP, Paynter D (2020) Contrasting upper and deep ocean oxygen response to protracted global warming. Global Biogeochem Cycles. https://doi.org/10.1029/2020GB006601
Frölicher TL, Joos F, Plattner G-K, Steinacher M, Doney SC (2009) Natural variability and anthropogenic trends in oceanic oxygen in a coupled carbon cycle-climate model ensemble. Global Biogeochem Cycles. https://doi.org/10.1029/2008GB003316
Fu Y, Lozier MS, Biló TC et al (2023b) Seasonality of the Meridional Overturning Circulation in the subpolar North Atlantic. Commun Earth Environ 4:181. https://doi.org/10.1038/s43247-023-00848-9
Fu Y, Lozier MS, Biló TC, Bower AS, Cunningham SA, Cyr F, de Jong MF, deYoung B, Drysdale L, Fraser N, Fried N, Furey HH, Han G, Handmann P, Holliday NP, Holte J, Inall ME, Johns WE, Jones S, Karstensen J, Li F, Pacini A, Pickart RS, Rayner D, Straneo F, Yashayaev I (2023a) Meridional overturning circulation observed by the overturning in the Subpolar North Atlantic Program (OSNAP). Commun Earth Environ 4(1):181. https://doi.org/10.35090/gatech/70342
Galaasen EV et al (2014) Rapid reductions in North Atlantic deep water during the peak of the last interglacial period. Science 343:1129–1132. https://doi.org/10.1126/science.1248667
Galbraith E, Kienast M, The NICOPP working group members (2013) The acceleration of oceanic denitrification during deglacial warming. Nature Geosci 6:579–584. https://doi.org/10.1038/ngeo1832
García OE, Schneider M, Ertl B, Sepúlveda E, Borger C, Diekmann C, Wiegele A, Hase F, Barthlott S, Blumenstock T, Raffalski U, Gómez-Peláez A, Steinbacher M, Ries L, de Frutos AM (2018) The MUSICA IASI CH4 and N2O products and their comparison to HIPPO, GAW and NDACC FTIR references. Atmos Meas Tech 11:4171–4215. https://doi.org/10.5194/amt-11-4171-2018
Gary SF, Fox AD, Biastoch A, Roberts JM, Cunningham SA (2020) Larval behavior, dispersal and population connectivity in the deep sea. Sci Rep 10(1):10675. https://doi.org/10.1038/s41598-020-67503-7
Gascard JC, Festy J, le Goff H, Weber M, Bruemmer B, Offermann M et al (2008) Exploring Arctic transpolar drift during dramatic sea ice retreat. EOS Trans Am Geophys Union 89(3):21–22
Gent PR (2018) A commentary on the Atlantic meridional overturning circulation stability in climate models. Ocean Model 122:57–66. https://doi.org/10.1016/j.ocemod.2017.12.006
Gordó-Vilaseca C, Stephenson F, Colld M, Lavina C, Costello MJ (2023) Three decades of increasing fish biodiversity across the northeast Atlantic and the Arctic Ocean. Proc Natl Acad Sci USA 120:2120869120
Greene CH, Meyer-Gutbrod E, Monger BC, McGarry LP, Pershing AJ, Belkin IM, Fratantoni PS, Mountain DG, Pickart RS, Proshutinsky A, Conversi A (2013) Remote climate forcing of decadal-scale regime shifts in Northwest Atlantic shelf ecosystems. Limnol Oceanogr 58:803–816
Greene CH, Monger BC, McGarry LP, Connelly MD, Schnepf NR, Pershing AJ, Belkin IM, Fratantoni PS, Mountain DG, Pickart RS, Ji RB, Bisagni JJ, Chen CS, Hakkinen SMA, Haidvogel DB, Wang J, Head E, Smith P, Conversi A (2012) Recent Arctic climate change and its remote forcing of Northwest Atlantic shelf ecosystems. Oceanography 25:208–213
Greene CH, Pershing AJ (2007) Climate drives sea change. Science 315:1084–1085
Groom S, Sathyendranath S, Ban Y, Bernard S, Brewin R, Brotas V, Brockmann C, Chauhan P, Choi J, Chuprin A, Ciavatta S, Cipollini P, Donlon C, Franz B, He X, Hirata T, Jackson T, Kampel M, Krasemann H et al (2019) Satellite ocean colour: current status and future perspective. Front Mar Sci 6:485. https://doi.org/10.3389/fmars.2019.00485
Grégoire M, Garçon V, Garcia H, Breitburg D, Isensee K, Oschlies A et al (2021) A global ocean oxygen database and atlas for assessing and predicting deoxygenation and ocean health in the open and coastal ocean. Front Mar Sci 8:724913. https://doi.org/10.3389/fmars.2021.724913
Grémillet D, Boulinier T (2009) Spatial ecology and conservation of seabirds facing global climate change: a review. Mar Ecol Prog Series 391:121–138
Guirado E, Tabik S, Rivas ML et al (2019) Whale counting in satellite and aerial images with deep learning. Sci Rep 9:14259. https://doi.org/10.1038/s41598-019-50795-9
Haine TWN, Curry B, Gerdes R, Hansen E, Karcher M, Lee C, Rudels B, Spreen G, de Steur L, Stewart KD, Woodgate R (2015) Arctic freshwater export: status, mechanisms, and prospects. Glob Planet Change 125:13–35. https://doi.org/10.1016/j.gloplacha.2014.11.013
Handoh IC, Lenton TM (2003) Periodic mid-Cretaceous oceanic anoxic events linked by oscillations of the phosphorus and oxygen biogeochemical cycles. Glob Biogeochem Cycles 17:2003GB002039. https://doi.org/10.1029/2003GB002039
Hare SR, Mantua NJ (2000) Empirical evidence for North Pacific regime shifts in 1977 and 1989. Prog Oceanogr 47:103–145
Hayasaki M, Kawamura R, Mori M, Watanabe M (2013) Response of extratropical cyclone activity to the Kuroshio large meander in northern winter. Geophys Res Lett 40:2851–2855

Hayashida H, Kiss AE, Miyama T, Miyazawa Y, Yasunaka S (2023) Anomalous nutricline drives marked biogeochemical contrasts during the Kuroshio large meander. J Geophys Res (Oceans) 128:e2023JC019697. https://doi.org/10.1029/2023JC019697
Hazen EL, Scales KL, Maxwell SM, Briscoe DK, Welch H, Bograd SJ, Bailey H, Benson SR, Eguchi T, Dewar H, Kohin S, Costa DP, Crowder LB, Lewison RL (2018) A dynamic ocean management tool to reduce bycatch and support sustainable fisheries. Sci Adv 4:eaar3001
Heath MR, Beare DJ (2008) New primary production in northwest European shelf seas, 1960–2003. Mar Ecol Prog Ser 363:183–203
Hedley JD, Roelfsema CM, Chollett I, Harborne AR, Heron SF, Weeks S, Skirving WJ, Strong AE, Eakin CM, Christensen TRL et al (2016) Remote sensing of coral reefs for monitoring and management: a review. Remote Sens 8(2):118. https://doi.org/10.3390/rs8020118
Heinze C et al (2023) Reviews and syntheses: abrupt ocean biogeochemical change under human-made climatic forcing—warming, acidification, and deoxygenation. Biogeosciences. https://doi.org/10.5194/bg-2023-182
Heinze C, Blenckner T, Martins H, Rusiecka D, Doscher R, Gehlen M, Gruber N, Holland E, Hov O, Joos F, Matthews JBR, Rodven R, Wilson S (2021) The quiet crossing of ocean tipping points. Proc Natl Acad Sci 118:e2008478118. https://doi.org/10.1073/pnas.2008478118
Held H, Kleinen T (2004) Detection of climate system bifurcations by degenerate fingerprinting. Geophys Res Lett 31:L23207. https://doi.org/10.1029/2004GL020972
Helm KP, Bindoff NL, Church JA (2011) Observed decreases in oxygen content of the global ocean. Geophys Res Lett 38:L23602. https://doi.org/10.1029/2011GL049513
Henson SA, Cael BB, Allen SR, Dutkiewicz S (2021) Future phytoplankton diversity in a changing climate. Nat Commun 12(1):5372. https://doi.org/10.1038/s41467-021-25699-w
Heron SF, Maynard JA, van Hooidonk R, Eakin CM (2016) Warming trends and bleaching stress of the world's coral reefs 1985–2012. Sci Rep 6:38402. https://doi.org/10.1038/srep38402
Hewitt HT, Roberts M, Mathiot P, Biastoch A, Blockley E, Chassignet EP et al (2020) Resolving and parameterising the ocean mesoscale in earth system models. Curr Clim Change Rep 6:137–152. https://doi.org/10.1007/s40641-020-00164-w
Hibler WD (1979) A dynamic thermodynamic sea ice model. J Phys Oceanog 9(4):815–846
Hiddink JG, ter Hofstede R (2008) Climate induced increases in species richness of marine fishes. Glob Change Biol 14:453–460
Hirschi JJ et al (2020) The Atlantic meridional overturning circulation in high-resolution Models. J Geophys Res: Oceans 125(4):1–35. https://doi.org/10.1029/2019JC015522
Hobday AJ et al (2023) With the arrival of El Niño, prepare for stronger marine heatwaves. Nature 621(7977):38–41
Van Hoey G, Wischnewski J, Craeymeersch J et al (2019) Methodological elements for optimising the spatial monitoring design to support regional benthic ecosystem assessments. Environ Monit Assess 191:423. https://doi.org/10.1007/s10661-019-7550-9
Hofmann M, Rahmstorf S (2009) On the stability of the Atlantic meridional overturning circulation. Proc Natl Acad Sci 106(49):20584–20589. https://doi.org/10.1007/s00382-014-2153-1
Holliday NP, Bersch M, Berx B, Chafik L, Cunningham S, Florindo- López C et al (2020) Ocean circulation causes the largest freshening event for 120 years in eastern subpolar North Atlantic. Nat Commun 11:585. https://doi.org/10.1038/s41467-020-14474-y
Holling CS (1973) Resilience and stability of ecological systems. Annu Rev Ecol Syst 4:1–23
Holt J, Polton J, Huthnance J, Wakelin S, O'Dea E, Harle J, Yool A, Artioli Y, Blackford J, Siddorn J, Inall M (2018) Climate-driven change in the North Atlantic and Arctic Oceans can greatly reduce the circulation of the North Sea. Geophys Res Lett 45:11827–11836. https://doi.org/10.1029/2018GL078878
Hughes TP (1994) Catastrophes, phase shifts, and large-scale degradation of a Caribbean coral reef. Science 265:1547–1551
Hutchings JA (2000) Collapse and recovery of marine fishes. Nature 406:882–885
Hülse D, Arndt S, Ridgwell A (2019) Mitigation of extreme Ocean Anoxic Event conditions by organic matter sulfurization. Paleoceanogr Paleoclimatology 34(4):476–489. https://doi.org/10.1029/2018PA003470
Hülse D, Lau KV, van de Velde SJ et al (2021) End-Permian marine extinction due to temperature-driven nutrient recycling and euxinia. Nat Geosci 14:862–867. https://doi.org/10.1038/s41561-021-00829-7
IPCC, Summary for Policymakers. In: Climate Change 2021: The Physical Science Basis. Contribution of Working Group I to the Sixth Assessment Report of the Intergovernmental Panel on Climate Change [Masson-Delmotte V, Zhai P, Pirani A, Connors SL, Péan C, Berger S, Caud N, Chen Y, Goldfarb L,

Gomis MI, Huang M, Leitzell K, Lonnoy E, Matthews JBR, Maycock TK, Waterfield T, Yelekçi O, Yu R, and Zhou B (eds.)]. Cambridge University Press, Cambridge, pp. 3–32
IPCC, (2019) Collins M, Sutherland M, Bouwer L, Cheong S-M, Frölicher TL, Jacot Des Combes H, Roxy MK, Losada I, McInnes KL, Ratter B, Rivera-Arriaga E, Susanto RD, Swingedouw D, Tibig L, Bakker P, Eakin CM, Emanuel K, Grose M, Hemer M, Jackson L, Kääb A, Kajtar JB, Knutson T, Laufkötter C, Noy I, Payne M, Ranasinghe R, Sgubin G, Timmermans M-L, IPCC Special Report on the Ocean and Cryosphere in a Changing Climate
Industry Agenda (2016) The new plastics economy rethinking the future of plastics. The World economic forum: Geneva, Switzerland. Vol. 36
Ito T, Minobe S, Long MC, Deutsch C (2017) Upper ocean O_2 trends: 1958–2015. Geophys Res Lett 44:4214–4223. https://doi.org/10.1002/2017GL073613
Jaccard SL, Galbraith ED (2012) Large climate-driven changes of oceanic oxygen concentrations during the last deglaciation. Nature Geosci 5:151–156. https://doi.org/10.1038/ngeo1352
Jackson LC, Dubois C, Forget G, Haines K, Harrison M, Iovino D, Köhl A, Mignac D, Masina S, Peterson KA, Piecuch CG, Roberts C, Robson J, Storto A, Toyoda T, Valdivieso M, Wilson C, Wang Y, Zuo H (2019) The mean state and variability of the North Atlantic circulation: a perspective from ocean reanalyses. J Geophys Res: Oceans 124(12):9141–9170. https://doi.org/10.1029/2019JC015210
Jackson LC, Kahana R, Graham T, Ringer MA, Woollings T, Mecking JV, Wood RA (2015) Global and European climate impacts of a slowdown of the AMOC in a high resolution GCM. Climate Dyn 45:3299–3316. https://doi.org/10.1007/s00382-015-2540-2
Jackson LC, Peterson KA, Roberts CD, Wood RA (2016) Recent slowing of Atlantic overturning circulation as a recovery from earlier strengthening. Nat Geosci 9:518–521. https://doi.org/10.1038/NGEO2715
Jackson LC, Smith RS, Wood RA (2017) Ocean and atmosphere feedbacks affecting AMOC hysteresis in a GCM. Clim Dyn 49:173–219. https://doi.org/10.1007/s00382-016-3336-8
Jackson LC, Wood RA (2017) Timescales of AMOC decline in response to freshwater forcing. Clim Dyn. https://doi.org/10.1007/s00382-017-3957-6
Jackson LC, Wood RA (2018) Hysteresis and resilience of the AMOC in an eddy-permitting GCM. Geophys Res Lett. https://doi.org/10.1029/2018GL078104
Jackson LC, Wood RA (2020) Fingerprints for early detection of changes in the AMOC. J Clim 33(16):7027–7044. https://doi.org/10.1175/JCLI-D-20-0034.1
Jackson LC, de Asenjo EA, Bellomo K, Danabasoglu G, Haak H, Hu A, Jungclaus J, Lee W, Meccia VL, Saenko O, Shao A, Swingedouw D (2023) Understanding AMOC stability: the North Atlantic Hosing Model Intercomparison Project. Geosci. Model Dev. 16. https://doi.org/10.5194/gmd-16-1975-2023
Jenkyns HC (2010) Geochemistry of oceanic anoxic events: review. Geochem Geophys Geosyst. https://doi.org/10.1029/2009GC002788
Jeong H, Turner AK, Roberts AF, Veneziani M, Price SF, Asay-Davis XS, Van Roekel LP, Lin W, Caldwell PM, Park H-S, Wolfe JD, Mametjanov A (2023) Southern Ocean polynyas and dense water formation in a high-resolution, coupled earth system model. Cryosphere 17:2681–2700. https://doi.org/10.5194/tc-17-2681-2023
Jiménez C, Tenerelli J, Prigent C, Kilic L, Lavergne T, Skarpalezos S et al (2021) Ocean and sea ice retrievals from an end-to-end simulation of the Copernicus imaging microwave radiometer (CIMR) 1.4–36.5 GHz measurements. J Geophys Res: Oceans 126:e2021JC017610. https://doi.org/10.1029/2021JC017610
Johnson JT et al (2021) (2021) Microwave radiometry at frequencies from 500 to 1400 MHz: an emerging technology for earth observations. IEEE J Sel Top Appl Earth Obs Remote Sens 14:4894–4914. https://doi.org/10.1109/jstars.2021.3073286
Kacimi S, Kwok R (2022) Arctic snow depth, ice thickness, and volume from ICESat-2 and CryoSat-2: 2018–2021. Geophys Res Lett 49:e2021GL097448. https://doi.org/10.1029/2021GL097448
Kansakar P, Hossain F (2016) A review of applications of satellite earth observation data for global societal benefit and stewardship of planet earth. Space Policy 36:46–54
Karl D, Letelier R, Hebel D et al (1995) Ecosystem changes in the North Pacific subtropical gyre attributed to the 1991–92 El Niño. Nature 373:230–234. https://doi.org/10.1038/373230a0
Kasai H, Ono T (2007) Has the 1998 regime shift also occurred in the oceanographic conditions and lower trophic ecosystem of the Oyashio region? J Oceanogr 63:661–669
Kavanaugh MT et al (2016) Seascapes as a new vernacular for pelagic ocean monitoring, management and conservation. ICES J Mar Sci 73:1839–1850. https://doi.org/10.1093/icesjms/fsw086
Kawabe M (1995) Variations of current path, velocity, and volume transport of the Kuroshio in relation with the large meander. J Phys Oceanograph 25(12):3103–3117

Kearney KA, Bograd SJ, Drenkard E, Gomez FA, Haltuch M, Hermann AJ, Jacox MG, Kaplan IC, Koenigstein S, Luo JY, Masi M, Muhling B, Pozo Buil M, Woodworth-Jefcoats PA (2021) Using global-scale earth system models for regional fisheries applications. Front Mar Sci 8:622206. https://doi.org/10.3389/fmars.2021.622206

Keil P, Mauritsen T, Jungclaus J, Hedemann C, Olonscheck D, Ghosh R (2020) Multiple drivers of the North Atlantic warming hole. Nat Clim Chang 10(7):667–671. https://doi.org/10.1038/s41558-020-0819-8

Kemena TP, Landolfi A, Oschlies A, Wallmann K, Dale AW (2019) Ocean phosphorus inventory: large uncertainties in future projections on millennial timescales and their consequences for ocean deoxygenation. Earth Syst Dynam 10:539–553. https://doi.org/10.5194/esd-10-539-2019

Kersalé M, Meinen CS, Perez RC, le Hénaff M, Valla D, Lamont T, Sato OT, Dong S, Terre T, van Caspel M, Chidichimo MP, van den Berg M, Speich S, Piola AR, Campos EJD, Ansorge I, Volkov DL, Lumpkin R, Garzoli SL (2020) Highly variable upper and abyssal overturning cells in the South Atlantic. Sci Adv 6:eaba7573. https://doi.org/10.1126/sciadv.aba7573

Kersalé M, Meinen CS, Perez RC, Piola AR, Speich S, Campos EJD, Garzoli SL, Ansorge I, Volkov DL, le Hénaff M, Dong S, Lamont T, Sato OT, van den Berg M (2021) Multi-year estimates of daily heat transport by the Atlantic meridional overturning circulation at 34.5 S. J Geophys Res-Oceans 126:e2020JC016947. https://doi.org/10.1029/2020JC016947

Kilbourne KH, Wanamaker AD, Moffa-Sanchez P, Reynolds DJ, Amrhein DE, Butler PG, Gebbie G, Goes M, Jansen MF, Little CM, Mette M, Moreno-Chamarro E, Ortega P, Otto-Bliesner BL, Rossby T, Scourse J, Whitney NM (2022) Atlantic circulation change still uncertain. Nat Geosci 15(3):165–167. https://doi.org/10.1038/s41561-022-00896-4

Kim HJ, An S, Kim SK, Park JH (2021) Feedback processes modulating the sensitivity of atlantic thermohaline circulation to freshwater forcing timescales. J Clim 34(12):5081–5092. https://doi.org/10.1175/JCLI-D-20-0897.1

Klus A, Prange M, Varma V, Bruno Tremblay L, Schulz M (2018) Abrupt cold events in the North Atlantic Ocean in a transient Holocene simulation. Clim past 14(8):1165–1178. https://doi.org/10.5194/cp-14-1165-2018

Kroodsma DA et al (2018) Tracking the global footprint of fisheries. Science 359:904–908. https://doi.org/10.1126/science.aao5646

Kutser T, Hedley JD, Giardino C, Roelfsema C, Brando VE (2016) Remote sensing for inland and coastal water monitoring. Int J Remote Sens 37:626–638. https://doi.org/10.1080/01431161.2016.1186852

Kvale KF, Turner KE, Landolfi A, Meissner KJ (2019) Phytoplankton calcifiers control nitrate cycling and the pace of transition in warming icehouse and cooling greenhouse climates. Biogeosciences 16:1019–1034. https://doi.org/10.5194/bg-16-1019-2019

Kwok R (2018) Arctic sea ice thickness, volume, and multiyear ice coverage: losses and coupled variability (1958–2018). Environ Res Lett 13:105005. https://doi.org/10.1088/1748-9326/aae3ec

Landerer FW, Wiese DN, Bentel K, Boening C, Watkins MM (2015) North Atlantic meridional overturning circulation variations from GRACE ocean bottom pressure anomalies. Geophys Res Lett 42:8114–8121. https://doi.org/10.1002/2015GL065730

Landolfi A, Dietze H, Koeve W, Oschlies A (2013) Overlooked runaway feedback in the marine nitrogen cycle: the vicious cycle. Biogeosciences 10:1351–1363. https://doi.org/10.5194/bg-10-1351-2013

Lapointe F, Bradley RS (2021) Little ice age abruptly triggered by intrusion of Atlantic waters into the Nordic Seas. Sci Adv 7:eabi8230. https://doi.org/10.1126/sciadv.abi8230

Laufkötter C, Vogt M, Gruber N, Aita-Noguchi M, Aumont O, Bopp L, Buitenhuis E, Doney SC, Dunne J, Hashioka T, Hauck J, Hirata T, John J, Le Quéré C, Lima ID, Nakano H, Seferian R, Totterdell I, Vichi M, Völker C (2015) Drivers and uncertainties of future global marine primary production in marine ecosystem models. Biogeosciences 12:69556984. https://doi.org/10.5194/bg-12-6955-2015

Laufkötter C, Vogt M, Gruber N, Aumont O, Bopp L, Doney SC et al (2016) Projected decreases in future marine export production: the role of the carbon flux through the upper ocean ecosystem. Biogeosciences 13:4023–4047. https://doi.org/10.5194/bg-13-4023-2016

Lavaniegos BE, Ohman MD (2007) Coherence of long-term variations of zooplankton in two sectors of the California current system. Prog Oceanogr 75:42–69

Lee YJ, Maslowski W, Cassano JJ, Clement Kinney J, Craig AP, Kamal S, Osinski R, Seefeldt MW, Stroeve J, Wang H (2023a) Causes and evolution of winter polynyas north of Greenland. Cryosphere 17:233–253. https://doi.org/10.5194/tc-17-233-2023

Lee YJ, Watts M, Maslowski W, Kinney JC, Osinski R (2023b) Assessment of the pan-arctic accelerated rate of sea ice decline in CMIP6 historical simulations. J Climate 36:6069–6089. https://doi.org/10.1175/JCLI-D-21-0539.1

Lehman SJ, Keigwin LD (1992) Sudden changes in North Atlantic circulation during the last deglaciation. Nature 356:757–762

Lenton TM, Held H, Kriegler E, Hall JW, Lucht W, Rahmstorf S et al (2008) Tipping elements in the Earth's climate system. Proc Natl Acad Sci USA 105:1786–1793. https://doi.org/10.1073/pnas.0705414105

Lenton TM, Rockström J, Gaffney O, Rahmstorf S, Richardson K, Steffen W, Schellnhuber HJ (2019) Climate tipping points—too risky to bet against. Nature 575(7784):592–595

Li H, Fedorov A, Liu W (2021) AMOC stability and diverging response to arc- tic sea ice decline in two climate models. J Clim 34(13):5443–5460. https://doi.org/10.1175/JCLI-D-20-0572.1

Li J, Liu X, Zhang W, Huang Y, Shen H (2021) Mapping global coral reefs using landsat satellites. Coral Reefs 40:389–399. https://doi.org/10.1007/s00338-020-02005-6

Li L, Lozier MS, Li F (2022) Century-long cooling trend in subpolar North Atlantic forced by atmosphere: an alternative explanation. Clim Dyn 58(9–10):2249–2267. https://doi.org/10.1007/s00382-021-06003-4

Li X, Wang X, Huang W, Xu Z, Hu L et al (2022) High-frequency satellite observations using GOCI to capture plankton variability. J Environ Manage 320:115311. https://doi.org/10.1016/j.jenvman.2022.115311

Lin P, Ji R, Davis CS, McGillicuddy DJ Jr (2010) Optimizing plankton survey strategies using observing system simulation experiments. J Mar Syst 82:187–194. https://doi.org/10.1016/j.jmarsys.2010.05.005

Lin P, Pickart RS, Heorton H et al (2023) (2023) Recent state transition of the Arctic Ocean's Beaufort Gyre. Nat Geosci 16:485–491. https://doi.org/10.1038/s41561-023-01184-5

Lindley JA, Beaugrand G, Luczak C, Dewarumez J-M, Kirby RR (2010) Warm-water decapods and the trophic amplification of climate in the North Sea. Biol Let 6:773–776

Little CM, Zhao M, Buckley MW (2020) Do surface temperature indices reflect centennial-timescale trends in Atlantic meridional overturning circulation strength? Geophys Res Lett 47:e2020GL090888

Liu Q, Babanin AV, Zieger S, Young IR, Guan C (2016) Wind and wave climate in the Arctic Ocean as observed by altimeters. J Climate 29:7957–7975. https://doi.org/10.1175/JCLI-D-16-0219.1

Liu W, Xie SP, Liu Z, Zhu J (2017) Overlooked possibility of a collapsed Atlantic meridional overturning circulation in warming climate. Sci Adv 3(1):1–8. https://doi.org/10.1126/sciadv.1601666

Lizarbe Barreto DA, Chevarria Saravia R, Nagai T, Hirata T (2021) Phytoplankton increase along the Kuroshio due to the large meander. Front Mar Sci 8:677632. https://doi.org/10.3389/fmars.2021.677632

Lobelle D, Sévellec F, Beaulieu C, Livina V, Frajka-Williams E et al. (2022) AMOC Trends From 1850 to 2100 At Interannual To Multi-Decadal Time Scales Corroborated By Changes In Salinity Budget. Res Sq. https://doi.org/10.21203/rs.3.rs-1231837/v1

Long MC, Deutsch C, Ito T (2016) Finding forced trends in oceanic oxygen. Global Biogeochem Cycles 30:381–397. https://doi.org/10.1002/2015GB005310

Lu W, Ridgwell A, Thomas E et al (2018) Late inception of a resiliently oxygenated upper ocean. Science 361:174–177. https://doi.org/10.1126/science.aar5372

Luczak C, Beaugrand G, Jaffré M, Lenoir S (2011) Climate change impact on Balearic Shearwater through a trophic cascade. Biol Let 7:702–705

Luczak C, Beaugrand G, Lindley JA, Dewarumez J-M, Dubois PJ, Kirby RR (2012) North Sea ecosystem changes from swimming crabs to seagulls. Biol Let 8:821–824

Lux K, Ashwin P, Wood R, Kuehn C (2022) Assessing the impact of parametric uncertainty on tipping points of the Atlantic meridional overturning circulation. Environ Res Lett 17(7):075002. https://doi.org/10.1088/1748-9326/ac7602

Lynch HJ, White R, Black AD et al (2012) Detection, differentiation, and abundance estimation of penguin species by high-resolution satellite imagery. Polar Biol 35:963–968. https://doi.org/10.1007/s00300-011-1138-3

Lynch-Stieglitz J (2017) The Atlantic meridional overturning circulation and abrupt Climate change. Ann Rev Mar Sci 9(1):83–104

Ma L (2019) Response of ocean dynamics to multiple equilibria of the Kuroshio path south of Japan. Dyn Atmosph Ocean 85:57–71

Manucharyan GE, Lopez-Acosta R, Wilhelmus MM (2022) Spinning ice floes reveal intensification of mesoscale eddies in the western Arctic Ocean. Sci Rep 12:7070. https://doi.org/10.1038/s41598-022-10712-z

Marchau VAWJ, Walker WE, Bloemen PJTM, Popper SW (2019) Decision making under deep uncertainty: from theory to practice. Springer, Amsterdam

Martin T, Biastoch A (2023) On the ocean's response to enhanced Greenland runoff in model experiments: relevance of mesoscale dynamics and atmospheric coupling. Ocean Sci 19:141–167. https://doi.org/10.5194/os-19-141-2023

Maslowski W, Clement Kinney J, Higgins M, Roberts A (2012) The future of Arctic sea ice. Annu Rev Earth Planet Sci 40:625–654

Maslowski W, Clement Kinney J, Okkonen S, Osinski R, Roberts A, Williams W (2014) The Large scale ocean circulation and physical processes controlling Pacific-Arctic interaction. In: Grebmeier J, Maslowski W (eds) The Pacific Arctic Region: ecosystem status and trends in a rapidly changing environment. Springer, Germany
Maslowski W, Marble D, Walczowski W, Semtner A (2001) On large-scale shifts in the Arctic Ocean and sea-ice conditions during 1979–98. Ann Glaciol 33:545–550. https://doi.org/10.3189/172756401781818978
Matz MV, Treml EA, Aglyamova GV, Bay LK (2018) Potential and limits for rapid genetic adaptation to warming in a great barrier reef coral. PLoS Genet 14:e1007220. https://doi.org/10.1371/journal.pgen.1007220
Maykut GA, McPhee MG (1995) Solar heating of the Arctic mixed layer. J Geophys Res: Oceans 100(C12):24691–24703
McGowan JA, Bograd SJ, Lynn RJ, Miller AJ (2003) The biological response to the 1977 regime shift in the California current. Deep Sea Res Part II 50:2567–2582
McParland EL, Levine NM (2019) The role of differential DMSP production and community composition in predicting variability of global surface DMSP concentrations. Limnol Oceanogr 64(2):757–773. https://doi.org/10.1002/lno.11076
McPhee MG (2008) Air-ice-ocean interaction: turbulent ocean boundary layer exchange processes. Springer, New York, p 215
Mckay DA et al (2022) Exceeding 15°C global warming could trigger multiple climate tipping points. Science 377(6611):abn750. https://doi.org/10.1126/science.abn7950
Mecking JV, Drijfhout SS, Jackson LC, Graham T (2016) Stable AMOC off state in an eddy-permitting coupled climate model. Clim Dyn 47(7–8):2455–2470. https://doi.org/10.1007/s00382-016-2975-0
Melet A, Teatini P, Le Cozannet G, Jamet C, Conversi A, Benveniste J, Almar R (2020) Earth observations for monitoring marine coastal hazards and their drivers. Surv Geophys 41:1489–1534. https://doi.org/10.1007/s10712-020-09594-5
Mercier H, Lherminier P, Sarafanov A, Gaillard F, Daniault N, Desbruyères D, Falina A, Ferron B, Gourcuff C, Huck T, Thierry V (2015) Variability of the meridional overturning circulation at the Greenland-Portugal OVIDE section from 1993 to 2010. Progr Oceanogr 132:250–261. https://doi.org/10.1016/j.pocean.2013.11.001
Meyer KM, Kump LR (2008) Oceanic euxinia in earth history: causes and consequences. Annu Rev Earth Planet Sci 36:251–288
Michel SLL, Swingedouw D, Ortega P, Gastineau G, Mignot J, McCarthy G, Khodri M (2022) Early warning signal for a tipping point suggested by a millennial Atlantic multidecadal variability reconstruction. Nat Commun 13(1):2–4. https://doi.org/10.1038/s41467-022-32704-3
Michel SLL, Dijkstra HA, Guardamagna F, Jacques-Dumas V, van Westen RM and von der Heydt AM (2023) Deep Learning reconstruction of Atlantic Meridional Overturning Circulation 1 strength (submitted)
Miller AJ, Cayan DR, Barnett TP, Graham NE, Oberhuber JM (1994) The 1976–77 climate shift of the Pacific Ocean. Oceanography 7:21–26
Miller AJ, Chai F, Chiba S, Moisan JR, Neilson DJ (2004) Decadal-scale climate and ecosystem interactions in the North Pacific Ocean. J Oceanogr 60:163–188
Miloslavich P, Bax NJ, Simmons SE et al (2018) Essential ocean variables for global sustained observations of biodiversity and ecosystem changes. Glob Change Biol 24:2416–2433. https://doi.org/10.1111/gcb.14108
Moat BI, Smeed DA, Rayner D, Johns WE, Smith R, Volkov D, Baringer MO, Collins J (2023) Atlantic meridional overturning circulation observed by the RAPID-MOCHA-WBTS (RAPID-Meridional Overturning Circulation and Heatflux Array-Western Boundary Time Series) array at 26N from 2004 to 2022 (v2022.1), British Oceanographic Data Centre - Natural Environment Research Council, UK. https://doi.org/10.5285/04c79ece-3186-349a-e063-6c86abc0158c
Moffa-Sánchez P, Moreno-Chamarro E, Reynolds DJ, Ortega P, Cunningham L, Swingedouw D, Amrhein DE, Halfar J, Jonkers L, Jungclaus JH, Perner K, Wanamaker A, Yeager S (2019) Variability in the Northern North Atlantic and Arctic Oceans Across the last two millennia: a review. Paleoceanogr Paleoclimatology 34:1399–1436. https://doi.org/10.1029/2018PA003508
Monteiro FM, Pancost RD, Ridgwell A, Donnadieu Y (2012) Nutrients as the dominant control on the spread of anoxia and euxinia across the Cenomanian-Turonian oceanic anoxic event (OAE2): model-data comparison. Paleoceanography. https://doi.org/10.1029/2012PA002351
Moore GWK, Schweiger A, Zhang J, Steele M (2018) Collapse of the 2017 winter Beaufort high: a response to thinning sea ice? Geophys Res Lett 45:2860–2869. https://doi.org/10.1002/2017GL076446

Moroni D, Pieri G, Tampucci M, Salvetti O (2016) A proactive system for maritime environment monitoring. Marit Pollut Bull 102:316–322. https://doi.org/10.1016/j.marpolbul.2015.07.045
Muilwijk M, Nummelin A, Heuzé C, Polyakov IV, Zanowski H, Smedsrud LH (2023) Divergence in climate model projections of future arctic atlantification. J Climate 36:1727–1748. https://doi.org/10.1175/JCLI-D-22-0349.1
Muller-Karger FE, Miloslavich P, Bax NJ, Simmons SE, Costello MJ, Sousa-Pinto I et al (2018) Integrating satellite observations with in-situ data and models for ecosystem management. Front Mar Sci 5:211. https://doi.org/10.3389/fmars.2018.00211
Murazaki K, Tsujino H, Motoi T, Kurihara K (2015) Influence of the Kuroshio large meander on the climate around Japan based on a regional climate model. J Met Soc Japan Ser II 93(2):161–179
Mélin F, Sathyendranath S, Brewin R, Jackson T et al (2017) Evaluating long-term consistency of satellite ocean-colour data for tracking ecosystem changes. Remote Sens Environ 203:39–52. https://doi.org/10.1016/j.rse.2017.03.039
Möllmann C, Cormon X, Funk S et al (2021) Tipping point realized in cod fishery. Sci Rep 11:14259. https://doi.org/10.1038/s41598-021-93843-z
Möllmann C, Diekmann R, Müller-Karulis B, Kornilovs G, Plikshs M, Axe P (2009) Reorganization of a large marine ecosystem due to atmospheric and anthropogenic pressure: a discontinuous regime shift in the Central Baltic Sea. Glob Change Biol 15:1377–1393
Nakamura H, Nishina A, Minobe S (2012) Response of storm tracks to bimodal Kuroshio path states south of Japan. J Clim 25(21):7772–7779
Niemeyer D, Kemena TP, Meissner KJ, Oschlies A (2017) A model study of warming-induced phosphorus–oxygen feedbacks in open-ocean oxygen minimum zones on millennial timescales. Earth Syst Dyn 8:357–367. https://doi.org/10.5194/esd-8-357-2017
Nim CJ, Skirving WJ, Eakin CM (2010) Satellite monitoring of reef vulnerability in a changing climate. NOAA Technical Report CRCP 1. NOAA Coral Reef Conservation Program. Silver Spring, MD. https://doi.org/10.1126/scien ce.aaf99 33
Nystrom M, Folke C, Moberg F (2000) Coral reef disturbance and resilience in a human-dominated environment. Trends Ecol Evol 15:413–417
Oguz T (2007) Nonlinear response of Black Sea pelagic fish stocks to over-exploitation. Mar Ecol Prog Ser 34:211–228
Oguz T, Cokacar T, Malanotte-Rizzoli P, Ducklow HW (2003) Climatic warming and accompanying changes in the ecological regime of the Black Sea during 1990s. Global Biogeochem Cycles 17:1088
Oltmanns M, Karstensen J, Moore GWK, Josey SA (2020) Rapid cooling and increased storminess triggered by freshwater in the North Atlantic. Geophys Res Lett 47:e2020GL087207. https://doi.org/10.1029/2020GL087207
Orihuela-Pinto B, England MH, Taschetto AS (2022) Interbasin and interhemispheric impacts of a collapsed Atlantic overturning circulation. Nat Clim Chang 12(6):558–565. https://doi.org/10.1038/s41558-022-01380-y
Oschlies A (2021) A committed fourfold increase in ocean oxygen loss. Nat Commun 12:2307. https://doi.org/10.1038/s41467-021-22584-4
Oschlies A, Brandt P, Stramma L, Schmidtko S (2018) Drivers and mechanisms of ocean deoxygenation. Nature Geosci 11:467–473. https://doi.org/10.1038/s41561-018-0152-2
Oschlies A, Koeve W, Landolfi A, Kähler P (2019) Loss of fixed nitrogen causes net oxygen gain in a warmer future ocean. Nat Commun 10:2805. https://doi.org/10.1038/s41467-019-10813-w
Oschlies, A (2019). Ocean deoxygenation from climate change, in IUCN Ocean Deoxygenation: Everyone's Problem, eds. D. Laffoley and J.M. Baxter, p. 107–118
O'Hare G (2011) Updating our understanding of climate change in the North Atlantic: the role of global warming and the Gulf Stream. Geography 96(1):5–15. https://doi.org/10.1080/00167487.2011.12094303
Pauly D, Alder J, Bennett E, Christensen V, Tyedmers P, Watson R (2003) The future for fisheries. Science 302:1359–1361
Pedro JB, van Ommen TD, Rasmussen SO, Morgan VI, Chappellaz J, Moy AD, Masson-Delmotte V, Delmotte M (2011) The last deglaciation: timing the bipolar seesaw. Clim past 7:671–683. https://doi.org/10.5194/cp-7-671-2011
Pepin P, Helbig JA (2012) Sampling variability of ichthyoplankton surveys–exploring the roles of scale and resolution on uncertainty. Fish Res 117–118:137–145
Pershing AJ, Greene CH, Jossi JW, O'Brien L, Brodziak JKT, Bailey BA (2005) Interdecadal variability in the Gulf of Maine zooplankton community, with potential impacts on fish recruitment. ICES J Mar Sci 62:1511–1523

Peterson WT, Schwing FB (2003) A new climate regime in northeast pacific ecosystems. Geophys Res Lett. https://doi.org/10.1029/2003GL017528

Pettorelli N, To Bühne HS, Tulloch A, Dubois G, Macinnis-Ng C, Queirós AM, Keith DA, Wegmann M, Schrodt F, Stellmes M, Sonnenschein R, Geller GN, Roy S, Somers B, Murray N, Bland L, Geijzendorffer I, Kerr JT, Broszeit S, Leitão PJ, Duncan C, El Serafy G, He KS, Blanchard JL, Lucas R, Mairota P, Webb TJ, Nicholson E (2018) Satellite remote sensing of ecosystem functions: opportunities, challenges and way forward. Remote Sens Ecol Conserv 4:71–93. https://doi.org/10.1002/rse2.59

Pohl A, Lu Z, Lu W et al (2021) Vertical decoupling in Late Ordovician anoxia due to reorganization of ocean circulation. Nat Geosci 14:868–873. https://doi.org/10.1038/s41561-021-00843-9

Pohl A, Ridgwell A, Stockey RG et al (2022) Continental configuration controls ocean oxygenation during the Phanerozoic. Nature 608:523–527. https://doi.org/10.1038/s41586-022-05018-z

Polkova J et al (2023) Initialization shock in the ocean circulation reduces skill in decadal predictions of the Atlantic subpolar gyre. Front Clim. https://doi.org/10.3389/fclim.2023.1273770/abstract

Polkova J, Brune S, Kadow C, Romanova V, Gollan G, Baehr J, Glowienka-Hense R, Greatbatch RJ, Hense A, Illing S, Köhl A, Kröger J, Müller WA, Pankatz K, Stammer D (2019) Initialization and ensemble generation for decadal climate predictions: a comparison of different methods. J Adv Model Earth Syst 11:149–172. https://doi.org/10.1029/2018MS001439

Polovina JJ (2005) Climate variation, regime shifts, and implications for sustainable fisheries. Bull Mar Sci 76:233–244

Polyakov IV et al (2020) Weakening of cold halocline layer exposes sea ice to oceanic heat in the Eastern Arctic Ocean. J Climate 33:8107–8123. https://doi.org/10.1175/JCLI-D-19-0976.1

Proshutinsky A, Krishfield RA, Timmermans ML, Toole JM, Carmack E et al (2009) Beaufort Gyre freshwater reservoir: state and variability from observations. J Geophys Res 114:C00A10

Proshutinsky A, Krishfield R, Toole JM, Timmermans M-L, Williams W, Zimmermann S et al (2019) Analysis of the Beaufort Gyre freshwater content in 2003–2018. J Geophys Res: Oceans 124:9658–9689. https://doi.org/10.1029/2019JC015281

Qiu B, Chen S (2021) Revisit of the occurrence of the Kuroshio large meander south of Japan. J Phys Oceanograph 51(12):3679–3694

Qiu B, Chen S, Oka E (2023) Why did the 2017 Kuroshio large meander event become the longest in the past 70 years? Geophys Res Lett 50(10):e2020GL103548

Qiu B, Chen S, Schneider N, Oka E, Sugimoto S (2020) On reset of the wind-forced decadal Kuroshio Extension variability in late 2017. J Clim 33(24):10813–10828. https://doi.org/10.1175/jcli-d-20-0237.1

Rabalais NN, Díaz RJ, Levin LA, Turner RE, Gilbert D, Zhang J (2010) Dynamics and distribution of natural and human-caused hypoxia. Biogeosciences 7:585–619. https://doi.org/10.5194/bg-7-585-2010

Rahmstorf S (2002) Ocean circulation and climate during the past 120,000 years. Nature 419:207–214. https://doi.org/10.1038/nature01090

Rampal P, Weiss J, Marsan D (2009) Positive trend in the mean speed and deformation rate of Arctic sea ice, 1979–2007. J Geophys Res: Oceans. https://doi.org/10.1029/2008JC005066

Reid PC, Beaugrand G (2012) Global synchrony of an accelerating rise in sea surface temperature. J Mar Biol Assoc UK 92:1435–1450

Reid PC, Borges M, Svenden E (2001) A regime shift in the North Sea circa 1988 linked to changes in the North Sea horse mackerel fishery. Fish Res 50:163–171

Renssen H, Goosse H, Fichefet T, Campin J-M (2001) The 8.2 kyr BP event simulated by a global atmosphere—sea-ice—ocean model. Geophys Res Lett 28:1567–1570. https://doi.org/10.1029/2000GL012602

Rind D, Schmidt GA, Jonas J, Miller R, Nazarenko L, Kelley M, Romanski J (2018) Multicentury instability of the atlantic meridional circulation in rapid warming simulations With GISS modele2. J Geophys Res (Atmos) 123:6331–6355. https://doi.org/10.1029/2017JD027149

Rippeth TP, Lincoln BJ, Lenn Y-D, Green JAM, Sundfjord A, Bacon S (2015) Tide-mediated warming of Arctic halocline by Atlantic heat fluxes over rough topography. Nat Geosci 8:191–194. https://doi.org/10.1038/ngeo2350

Ritchie PDL, Smith GS, Davis KJ et al (2020) Shifts in national land use and food production in Great Britain after a climate tipping point. Nat Food 1:76–83. https://doi.org/10.1038/s43016-019-0011-3

Ritchie P, Alkhayuon H, Cox P, Wieczorek S (2022). Rate-induced tipping in natural and human systems. November, 1–19

Rockström J, Steffen W, Noone K, Persson Å, Chapin FSI, Lambin E et al (2009) Plan etary boundaries: exploring the safe operating space for humanity. Ecol Soc. https://doi.org/10.5751/ES-03180-140232

Romanou A, Rind D, Jonas J, Miller R, Kelley M, Russell G, Orbe C, Nazarenko L, Latto R, Schmidt GA (2023) Stochastic Bifurcation of the North Atlantic Circulation under a Midrange Future

Climate Scenario with the NASA-GISS ModelE. J Clim 36(18):6141–6161. https://doi.org/10.1175/jcli-d-22-0536.1
Rooth C (1982) Hydrology and ocean circulation. Prog Oceanogr 11(2):131–149
Rose RA, Byler D, Eastman JR, Fleishman E, Geller G, Goetz S, Guild L, Hamilton H, Hansen M, Headley R, Hewson J, Horning N, Kaplin BA, Laporte N, Leidner A, Leimgruber P, Morisette J, Musinsky J, Pintea L, Prados A, Radeloff VC, Rowen M, Saatchi S, Schill S, Tabor K, Turner W, Vodacek A, Vogelmann J, Wegmann M, Wilkie D, Wilson C (2015) Ten ways remote sensing can contribute to conservation. Conserv Biol 29:350–359. https://doi.org/10.1111/cobi.12397
Rothrock DA (1975) The energetics of the plastic deformation of pack ice by ridging. J Geophys Res 80(33):4514–4519
Ruvalcaba Baroni I, Topper RPM, Van Helmond NAGM, Brinkhuis H, Slomp CP (2014) Biogeochemistry of the North Atlantic during oceanic anoxic event 2: role of changes in ocean circulation and phosphorus input. Biogeosciences 11:977–993. https://doi.org/10.5194/bg-11-977-2014
Sakurai Y (2007) An overview of the Oyashio ecosystem. Deep-Sea Res Part Ii-Topical Stud Oceanogr 54:2526–2542
Sathyendranath S, Brewin RJW, Ciavatta S, Jackson T, Kul G, Jönsson B et al (2023) Ocean biology studied from space. Surv Geophys 44:1287–1308. https://doi.org/10.1007/s10712-023-09805-9
Sauzède R, Lavigne H, Claustre H, Uitz J, Schmechtig C, D'Ortenzio F, Guinet C, Pesant S (2015) Vertical distribution of chlorophyll a concentration and phytoplankton community composition from in situ fluorescence profiles: a first database for the global ocean. Earth Syst Sci Data 7:261–273. https://doi.org/10.5194/essd-7-261-2015
Scheffer M (2009) Critical transitions in nature and society. Princeton University Press, Princeton
Scheffer M, Carpenter SR (2003) Catastrophic regime shifts in ecosystems: linking theory to observation. Trends Ecol Evol 18:648–656
Scheffer M, van Nes EH (2004) Mechanisms for marine regime shifts: can we use lakes as microcosms for oceans? Prog Oceanogr 60:303–319
Schiller A, Mikolajewicz U, Voss R (1997) The stability of the North Atlantic thermohaline circulation in a coupled ocean-atmosphere general circulation model. Clim Dyn 13:325–347. https://doi.org/10.1007/s003820050169
Schmidtko S, Stramma L, Visbeck M (2017) Decline in global oceanic oxygen content during the past five decades. Nature 542:335–339. https://doi.org/10.1038/nature21399
Schmittner A, Galbraith ED, Hostetler SW, Pedersen TF, Zhang R (2007) Large fluctuations of dissolved oxygen in the Indian and Pacific oceans during Dansgaard-Oeschger oscillations caused by variations of North Atlantic deep water subduction. Paleoceanography. https://doi.org/10.1029/2006PA001384
Schneider A, Tanhua T, Roether W, Steinfeldt R (2014) Changes in ventilation of the Mediterranean Sea during the past 25 years. Ocean Sci 10:1–16. https://doi.org/10.5194/os-10-1-2014
Schwartz-Belkin I, Portman E (2023) A review of geospatial technologies for improving marine spatial planning: challenges and opportunities. Ocean Coast Manag 231:106280. https://doi.org/10.1016/j.ocecoaman.2022.106280
Schwinger J, Asaadi A, Goris N, Lee H (2022) Possibility for strong northern hemisphere high-latitude cooling under negative emissions. Nat Commun 13(1):1095. https://doi.org/10.1038/s41467-022-28573-5
Schwinger J, Tjiputra J, Goris N, Six KD, Kirkevåg A, Seland Ø, Ilyina T (2017) Amplification of global warming through pH dependence of DMS production simulated with a fully coupled Earth system model. Biogeosciences 14(15):3633–3648
Seekell DA, Cline TJ, Carpenter KR, Pace ML (2013) Evidence of alternate attractors from a whole-ecosystem regime shift experiment. Thyroid Res 6:385–394
Segschneider J, Schneider B, Khon V (2018) Climate and marine biogeochemistry during the Holocene from transient model simulations. Biogeosciences 15:3243–3266. https://doi.org/10.5194/bg-15-3243-2018
Serreze MC, Barry RG (2005) The Arctic climate system. Cambridge University Press, Cambridge. https://doi.org/10.1017/CBO9780511535888
Sgubin G, Swingedouw D, de Cortázar-Atauri IG et al (2019) The impact of possible decadal-scale cold waves on viticulture over Europe in a context of global warming. Agronomy 9:397
Sgubin G, Swingedouw D, Drijfhout S et al (2017) Abrupt cooling over the North Atlantic in modern climate models. Nat Commun 8:14375
Sharma S, Chan E, Ishizawa M, Toom-Sauntry D, Gong SL, Li SM, Tarasick DW, Leaitch WR, Norman A, Quinn PK, Bates TS, Levasseur M, Barrie LA, Maenhaut W (2012) Influence of transport and ocean ice extent on biogenic aerosol sulfur in the Arctic atmosphere. J Geophys Res: Atmos 117(D12):D12209. https://doi.org/10.1029/2011JD017074

Shu Q, Wang Q, Guo C, Song Z, Wang S, He Y, Qiao F (2023) Arctic ocean simulations in the CMIP6 Ocean Model Intercomparison Project (OMIP). Geosci Model Dev 16:2539–2563. https://doi.org/10.5194/gmd-16-2539-2023

Sinet S, Von Der Heydt AS, Dijkstra HA (2022) AMOC stabilization under the interaction with tipping polar ice sheets. Geophys Res Lett 50:e2022GL100305

Six KD, Kloster S, Ilyina T, Archer SD, Zhang K, Maier-Reimer E (2013) Global warming amplified by reduced sulphur fluxes as a result of ocean acidification. Nat Clim Chang 3(11):975–978. https://doi.org/10.1038/nclimate1981

Skinner LC, Elderfield H (2007) Rapid fluctuations in the deep North Atlantic heat budget during the last glaciation. Paleoceanography. https://doi.org/10.1029/2006PA001338

Solodoch A, Stewart AL, McC HA, Manucharyan E (2023) Machine learning-derived inference of the meridional overturning circulation from satellite-observable variables in an ocean state estimate. J Adv Mod Earth Sys. https://doi.org/10.1029/2022MS003370

Somes CJ, Schmittner A, Muglia J, Oschlies A (2017) A three-dimensional model of the marine nitrogen cycle during the last glacial maximum constrained by sedimentary isotopes. Front Mar Sci 4:108. https://doi.org/10.3389/fmars.2017.00108

de Souza EN, Boerder K, Matwin S, Worm B (2016) Correction: improving fishing pattern detection from satellite ais using data mining and machine learning. PlosOne. https://doi.org/10.1371/journal.pone.0163760

Stefels J, Steinke M, Turner S, Malin G, Belviso S (2007) Environmental constraints on the production and removal of the climatically active gas dimethylsulphide (DMS) and implications for ecosystem modelling. Biogeochemistry 83(1–3):245–275. https://doi.org/10.1007/s10533-007-9091-5

Steinacher M, Joos F, Frölicher TL, Bopp L, Cadule P, Cocco V et al (2010) Projected 21st century decrease in marine productivity: a multi-model analysis. Biogeosciences 7:979–1005. https://doi.org/10.5194/bg-7-979-2010

Steinberg DK, Landry MR (2017) Zooplankton and the ocean carbon cycle. Ann Rev Mar Sci 9:413–444

Stocker TF (1998) The seesaw effect. Science 282:61–62

Stocker TF, Schmittner A (1997) Influence of CO_2 emission rates on the stability of the thermohaline circulation. Nature 388:862–865

Stocker TF, Jones RG, Hegglin MI, Lenton TM, Hegerl GC, Seneviratne SI, v.d Wel N, Wood RA (2024) Reflecting on the science of climate tipping points to inform and assist policy making and address the risks they pose to society. Surv Geophys. https://doi.org/10.1007/s10712-024-09844-w

Stommel H (1961) Thermohaline convection with two stable regimes of flow. Tellus 13:224–230

Storf SR, Box JE, Feulner G, Mann ME, Robinson A, Rutherford S, Schaffernicht EJ (2015) Exceptional twentieth-century slowdown in Atlantic Ocean overturning circulation. Nat Clim Change 5:475–480. https://doi.org/10.1038/nclimate2554

Stramma L, Johnson GC, Sprintall J, Mohrholz V (2008) Expanding oxygen-minimum zones in the tropical oceans. Science 320:655–658. https://doi.org/10.1126/science.1153847

Stramma L, Schmidtko S, Bograd SJ, Ono T, Ross T, Sasano D et al (2020) Trends and decadal oscillations of oxygen and nutrients at 50–300 m depth in the equatorial and North Pacific. Biogeosciences 17:813–831. https://doi.org/10.5194/bg-17-813-2020

Strömberg KHPTJ, Smyth JI, Allen S, Pitois TDO (2009) Estimation of global zooplankton biomass from satellite ocean colour. J Mar Sys 78:18–27. https://doi.org/10.1016/j.jmarsys.2009.02.004

Stuart V, Platt T, Sathyendranath S (2011) The future of fisheries science in management: a remote-sensing perspective. ICES J Mar Sci 68:644–650

Sugimoto S, Qiu B, Kojima A (2019) Marked coastal warming off Tokai attributable to Kuroshio large meander. J Oceanograph 76:141–154

Sugimoto S, Qiu B, Schneider N (2021) Local atmospheric response of the Kuroshio large meander path in summer and its remote influence on the climate of Japan. J Clim 34(9):3571–3589

Sun C, Liu Z, Du Q, Fang X et al (2021) Satellite-based acoustic sensors for monitoring underwater noise levels. Sensors 21(23):7849. https://doi.org/10.3390/s21237849

Swingedouw D, Bily A, Esquerdo C, Borchert LF, Sgubin G, Mignot J et al (2021) On the risk of abrupt changes in the North Atlantic subpolar gyre in CMIP6 models. Ann N Y Acad Sci 1504:187–201. https://doi.org/10.1111/nyas.14659

Swingedouw D, Braconnot P, Delecluse P, Guilyardi E, Marti O (2007) Quantifying the AMOC feedbacks during a 2xCO2 stabilization experiment with land-ice melting. Clim Dyn 29:521–534

Swingedouw D, Speranza CI, Bartsch A, Durand G, Jamet C, Beaugrand G, Conversi A (2020) Early warning from space for a few key tipping points in physical, biological, and social-ecological systems. Surv Geophys 41:1237–1284. https://doi.org/10.1007/s10712-020-09604-6

Thomas MD, Fedorov AV (2019) Mechanisms and Impacts of a partial AMOC recovery under enhanced freshwater forcing. Geophys Res Lett 46(6):3308–3316. https://doi.org/10.1029/2018GL080442
Timmermans ML, Marshall J (2020) Understanding Arctic Ocean circulation: a review of ocean dynamics in a changing climate. J Geophys Res Oceans 125:e2018JC014378
Timmermans ML, Toole JM (2023) The Arctic Ocean's Beaufort Gyre. Ann Rev Mar Sci 15(1):223–248
Tittensor DP, Novaglio C, Harrison CS et al (2021) Next-generation ensemble projections reveal higher climate risks for marine ecosystems. Nat Clim Change 11(11):11. https://doi.org/10.1038/s41558-021-01173-9
Truong C, Oudre L, Vayatis N (2020) Selective review of offline change point detection methods. Signal Process 167:107299. https://doi.org/10.1016/j.sigpro.2019.107299
Tsandev I, Slomp CP (2009) Modeling phosphorus cycling and carbon burial during cretaceous oceanic anoxic events. Earth Planet Sci Lett 286:71–79. https://doi.org/10.1016/j.epsl.2009.06.016
Tsonis AA, Swanson K, Kravtsov S (2007) A new dynamical mechanism for major climate shifts. Geophys Res Lett. https://doi.org/10.1029/2007GL030288
Turner JT (2015) Zooplankton fecal pellets, marine snow, phytodetritus and the ocean's biological pump. Prog Oceanogr 130:205–248
Vellinga M, Wood RA (2002) Global climatic impacts of a collapse of the Atlantic thermohaline circulation. Clim Change 54:251–267
Veneziani M, Maslowski W, Lee YJ, D'Angelo G, Osinski R, Petersen MR, Weijer W, Craig AP, Wolfe JD, Comeau D, Turner AK (2022) An evaluation of the E3SMv1 Arctic ocean and sea-ice regionally refined model. Geosci Model Dev 15:3133–3160. https://doi.org/10.5194/gmd-15-3133-2022
Verbesselt J, Umlauf N, Hirota M et al (2016) Remotely sensed resilience of tropical forests. Nature Clim Change 6:1028–1031. https://doi.org/10.1038/nclimate3108
Vinogradova N, Lee T, Boutin J et al (2019) Satellite salinity observing system: recent discoveries and the way forward. Front Mar Sci 6(May):1–23. https://doi.org/10.3389/fmars.2019.00243
Virtanen P, Gommers R, Oliphant TE et al (2020) SciPy 1.0: Fundamental algorithms for scientific computing in python. Nat Method 17(3):3. https://doi.org/10.1038/s41592-019-0686-2
Visser ME, Both C (2005) Temporal mismatch between environmental changes and biological responses. Proc R Soc B: Biol Sci 272:2561–2569. https://doi.org/10.1098/rspb.2005.3356
De Vries P, Weber SL (2005) The Atlantic freshwater budget as a diagnostic for the existence of a stable shut down of the meridional overturning circulation. Geophys Res Lett 32:L09606. https://doi.org/10.1029/2004GL021450
Wallmann K (2003) Feedbacks between oceanic redox states and marine productivity: a model perspective focused on benthic phosphorus cycling. Glob Biogeochem Cycles. https://doi.org/10.1029/2002GB001968
Wallmann K, Flögel S, Scholz F, Dale AW, Kemena TP, Steinig S et al (2019) Periodic changes in the Cretaceous ocean and climate caused by marine redox see-saw. Nat Geosci 12:456–461. https://doi.org/10.1038/s41561-019-0359-x
Wallmann K, José YS, Hopwood MJ, Somes CJ, Dale AW, Scholz F et al (2022) Biogeochemical feedbacks may amplify ongoing and future ocean deoxygenation: a case study from the Peruvian oxygen minimum zone. Biogeochemistry 159:45–67. https://doi.org/10.1007/s10533-022-00908-w
Wang Q (2024) On the release and renewal of freshwater in the Beaufort Gyre of the Arctic Ocean. J Phys Oceanogr 54:1213–1227. https://doi.org/10.1175/JPO-D-23-0184.1
Wang S, Foster A, Lenz EA et al (2023) Mechanisms and impacts of earth system tipping elements. Rev Geophys 61(1):e2021RG000757
Wang Q, Koldunov NV, Danilov S, Sidorenko D, Wekerle C, Scholz P et al (2020) Eddy kinetic energy in the Arctic Ocean from a global simulation with a 1-km Arctic. Geophys Res Lett 47:e2020GL088550. https://doi.org/10.1029/2020GL088550
Ward BA (2015) Temperature-Correlated Changes in Phytoplankton Community Structure Are Restricted to Polar Waters. PLoS ONE 10(8):e0135581. https://doi.org/10.1371/journal.pone.0135581
Wassenburg JA, Vonhof HB, Cheng H et al (2021) Penultimate deglaciation Asian monsoon response to North Atlantic circulation collapse. Nat Geosci 14(12):937–941. https://doi.org/10.1038/s41561-021-00851-9
Watson AJ (2016) Oceans on the edge of anoxia. Science 354:1529–1530. https://doi.org/10.1126/science.aaj2321
Watson AJ, Lenton TM, Mills BJW (2017) Ocean deoxygenation, the global phosphorus cycle and the possibility of human-caused large-scale ocean anoxia. Phil Trans R Soc A 375:20160318. https://doi.org/10.1098/rsta.2016.0318

Watts M, Maslowski W, Lee YJ, Kinney JC, Osinski R (2021) A spatial evaluation of Arctic sea ice and regional limitations in CMIP6 historical simulations. J Climate 34:6399–6420. https://doi.org/10.1175/JCLI-D-20-0491.1

Weijer W, Cheng W, Drijfhout SS, Fedorov AV, Hu A, Jackson LC, Liu W, McDonagh EL, Mecking JV, Zhang J (2019) Stability of the Atlantic meridional overturning circulation: a review and synthesis. J Geopys Res (Oceans) 124:5336–5375. https://doi.org/10.1029/2019JC015083

Weijerman M, Lindeboom H, Zuur AF (2005) Regime shifts in marine ecosystems of the North Sea and Wadden Sea. Mar Ecol Prog Ser 298:21–39

Wernberg T et al (2023) Impacts of climate change on marine foundation species. Annl Rev Mar Sci 16:247–282

Williamson MS, Collins M, Drijfhout SS, Kahana R, Mecking JV, Lenton TM (2018) Effect of AMOC collapse on ENSO in a high resolution general circulation model. Clim Dyn 50(7–8):2537–2552. https://doi.org/10.1007/s00382-017-3756-0

Willis JK (2010) Can in situ floats and satellite altimeters detect long-term change in Atlantic Ocean overturning? Geophys Res Lett. https://doi.org/10.1029/2010GL042372

Wood RA, Crucifix M, Lenton TM, Mach KJ, Moore C, New M, Sharpe S, Stocker TF, Sutton RT (2023) A climate science toolkit for high impact-low likelihood climate risks. Earth's Future 11:e2022EF003369. https://doi.org/10.1029/2022EF003369

Wood RA, Rodríguez JM, Smith RS, Jackson LC, Hawkins E (2019) Observable, low-order dynamical controls on thresholds of the Atlantic meridional overturning circulation. Clim Dyn 53(11):6815–6834. https://doi.org/10.1007/s00382-019-04956-1

Xu H, Tokinaga H, Xie S-P (2010) Atmospheric effects of the Kuroshio large meander during 2004–05. J Clim 23(17):4704–4715

Yamamoto A, Abe-Ouchi A, Shigemitsu M, Oka A, Takahashi K, Ohgaito R, Yamanaka Y (2015) Global deep ocean oxygenation by enhanced ventilation in the Southern Ocean under long-term global warming, global biogeochem. Cycles 29:1801–1815. https://doi.org/10.1002/2015GB005181

Yamasaki Y, Sugimura T, Noda T (2017) Environmental DNA sampling for marine biodiversity assessments. Curr Opin Environ Sustain 29:113–118. https://doi.org/10.1016/j.cosust.2018.03.005

Yang Y, Liang XS (2019) New perspectives on the generation and maintenance of the Kuroshio large meander. J Phys Oceanogr 49:2095–2113

Yashayev I, Loder JA (2016) Recurrent replenishment of Labrador Sea Water and associated decadal-scale variability. J Geophys Res (Oceans) 121:8095–8114. https://doi.org/10.1002/2016JC012046

Yasheyev I, Seidov D, Demirov E (2015) A new collective view of oceanography of the Arctic and North Atlantic basins. Prog Oceanogr 132:1–21. https://doi.org/10.1016/j.pocean.2014.12.012

Yati E, Sadiyah L, Satria F, Alabia ID, Sulma S, Prayogo T et al (2024) Spatial distribution models for the four commercial tuna in the sea of maritime continent using multi-sensor remote sensing and maximum entropy. Mar Environ Res 198:106540. https://doi.org/10.1016/j.marenvres.2024.106540

Yin J et al (2006) Is a shutdown of the thermohaline circulation irreversible? Clim Dyn. https://doi.org/10.1029/2005JD006562

Zanna L, Bolton T (2020) Data-driven equation discovery of ocean mesoscale closures. Geophys Res Lett 47:e2020GL088376. https://doi.org/10.1029/2020GL088376

Zhang J et al (2021) Labrador Sea freshening linked to Beaufort Gyre freshwater release. Nat Commun 12:1229

Zhao Y, Liu D, Wei X (2020) Monitoring harmful algal blooms using ocean colour sensors. Environ Adv 2:100008. https://doi.org/10.1016/j.envadv.2020.100008

Publisher's Note Springer Nature remains neutral with regard to jurisdictional claims in published maps and institutional affiliations.

Authors and Affiliations

Richard A. Wood[1] · Jonathan A. Baker[1] · Grégory Beaugrand[2] · Jacqueline Boutin[3] · Alessandra Conversi[4] · Reik V. Donner[5,6] · Ivy Frenger[7] · Eric Goberville[8] · Hakase Hayashida[9] · Wolfgang Koeve[7] · Karin Kvale[10] · Angela Landolfi[11] · Wieslaw Maslowski[12] · Andreas Oschlies[7] · Anastasia Romanou[13] · Christopher J. Somes[7] · Thomas F. Stocker[14] · Didier Swingedouw[15]

✉ Angela Landolfi
angela.landolfi@cnr.it

Richard A. Wood
richard.wood@metoffice.gov.uk

[1] Met Office Hadley Centre, Exeter EX1 3PB, UK

[2] Laboratoire d'Océanologie Et de Géosciences UMR 8187, LOG, CNRS, University of Lille, University of Littoral Côte d'Opale, 62930 Wimereux, France

[3] Sorbonne Université, CNRS, IRD, MNHN, Laboratoire d'Océanographie et du Climat: Expérimentations et Approches Numériques, LOCEAN/IPSL, Paris, France F-75005

[4] National Research Council of Italy, CNR–ISMAR, Forte Santa Teresa, Loc. Pozzuolo, 19032 Lerici, SP, Italy

[5] Magdeburg-Stendal University of Applied Sciences, 39114 Magdeburg, Germany

[6] Potsdam Institute for Climate Impact Research (PIK)–Member of the Leibniz Association, 14473 Potsdam, Germany

[7] GEOMAR Helmholtz Centre for Ocean Research Kiel, 24148 Kiel, Germany

[8] Unité Biologie Des Organismes Et Ecosystèmes Aquatiques (BOREA), Muséu Unité Biologie des Organismes et Ecosystèmes Aquatiques (BOREA), Muséum National d'Histoire Naturelle, CNRS, IRD, Sorbonne Université, Université de Caen Normandie, Université des Antilles, Sorbonne Université, 75006 Paris, France

[9] Application Laboratory, Japan Agency for Marine-Earth Science and Technology, Yokohama 236-0001, Japan

[10] GNS Science, 1 Fairway Ave, Lower Hutt 5013, New Zealand

[11] National Research Council of Italy, CNR–ISMAR, 00133 Rome, Italy

[12] Naval Postgraduate School, Monterey, CA 93943, USA

[13] Dept of Applied Physics and Applied Mathematics, NASA Goddard Institute for Space Studies, Columbia University, New York, NY 10027, USA

[14] Climate and Environmental Physics and Oeschger Centre for Climate Change Research, University of Bern, 3012 Bern, Switzerland

[15] EPOC, UMR 5805, Bordeaux INP, University of Bordeaux, CNRS, 33600 Pessac, France

Surveys in Geophysics (2025) 46:375–420
https://doi.org/10.1007/s10712-024-09863-7

Extreme Events Contributing to Tipping Elements and Tipping Points

A. Romanou[1] · G. C. Hegerl[2] · S. I. Seneviratne[3] · B. Abis[4] · A. Bastos[5,6] · A. Conversi[7] · A. Landolfi[8] · H. Kim[9] · P. E. Lerner[10] · J. Mekus[1,11] · B. L. Otto-Bliesner[12] · F. S. R. Pausata[13] · I. Pinto[14] · L. Suarez-Guiterrez[3,15]

Received: 15 January 2024 / Accepted: 3 September 2024 / Published online: 16 November 2024

Abstract
This review article provides a synthesis and perspective on how weather and climate extreme events can play a role in influencing tipping elements and triggering tipping points in the Earth System. An example of a potential critical global tipping point, induced by climate extremes in an increasingly warmer climate, is Amazon rainforest dieback that could be driven by regional increases in droughts and exacerbated by fires, in addition to deforestation. A tipping element associated with the boreal forest might also be vulnerable to heat, drought and fire. An oceanic example is the potential collapse of the Atlantic meridional overturning circulation due to extreme variability in freshwater inputs, while marine heatwaves and high acidity extremes can lead to coral reef collapse. Extreme heat events may furthermore play an important role in ice sheet, glacier and permafrost stability. Regional severe extreme events could also lead to tipping in ecosystems, as well as in human systems, in response to climate drivers. However, substantial scientific uncertainty remains on mechanistic links between extreme events and tipping points. Earth observations are of high relevance to evaluate and constrain those links between extreme events and tipping elements, by determining conditions leading to delayed recovery with a potential for tipping in the atmosphere, on land, in vegetation, and in the ocean. In the subsurface ocean, there is a lack of consistent, synoptic and high frequency observations of changes in both ocean physics and biogeochemistry. This review article shows the importance of considering the interface between extreme events and tipping points, two topics usually addressed in isolation, and the need for continued monitoring to observe early warning signs and to evaluate Earth system response to extreme events as well as improving model skill in simulating extremes, compound extremes and tipping elements.

Keywords Extreme events · Tipping elements · Tipping points · Compound extreme events · Climate change · Climate risk

A. Romanou, G. Hegerl, S. I. Seneviratne have contributed equally to this work.

Extended author information available on the last page of the article

Article Highlights

- Drought, heat, and other climate extreme events can trigger land tipping points
- Marine heatwaves, high acidity and low oxygen extremes and possibly intrinsic variability can trigger ocean tipping points
- Close monitoring, at high frequencies, including from space, is required to assess occurrence, response and resilience to extremes including as early warning signs

1 Introduction

1.1 Research Context and Scope of Article

In recent years, climate research on both extreme events and tipping points has steadily increased to respectively become (Fig. 1):

a. *(Climate extreme events)* a mainstream research field addressed in several thousands of articles per year. This topic has been highlighted in the cross-Working Group Special report on Extremes (IPCC 2012), with ocean extremes discussed in the special report on ocean and cryosphere (SROCC 2019; Collins et al. 2019). More recently, extreme events were summarized in a full chapter of the latest 6th Assessment Report of the Intergovernmental Panel on Climate Change (IPCC 2021; Seneviratne et al. 2021).
b. *(Tipping points)* a research topic of growing importance in climate science which was addressed in several key articles in recent years (e.g., Lenton et al. 2008, 2023; Armstrong McKay et al. 2022; Wang et al. 2023a, b). The increasing possibility of crossing of tipping points is highlighted as a key climate risk, for example, in the World Climate Research programme lighthouse activity 'safe landing climates' (https://www.wcrp-climate.org/safe-landing-climates; accessed 7/2024)

Despite the prominence of both research topics, to our knowledge the links between extreme events and tipping points have not been explicitly addressed so far. In this review article, we highlight a) how extreme events in both the land and ocean systems can affect global and regional tipping points, b) the links between several types of extreme events in the context of tipping point research, c) the potential for monitoring changes in extreme events and tipping points, in particular through Earth Observations (EO) to provide early warning, and d) further avenues for research at the interface between climate extremes and tipping point research.

1.2 Definitions

Following the terminology of Lenton et al. (2008), we refer to a **tipping point** as a critical threshold beyond which a system reorganizes, often abruptly and/or irreversibly (see Chen et al. 2021), or in other words, the point at which a small perturbation can yield large changes ("little things can make a big difference", Gladwell (2000)). However, we do not limit our assessment to global Earth System tipping points (e.g., Armstrong McKay et al. 2022) but we also consider regional tipping points and critical thresholds in the Earth System, as well as tipping points and critical thresholds associated with society or ecosystem impacts (e.g., breadbasket failures, regional forest dieback) resulting from human-induced climate change.

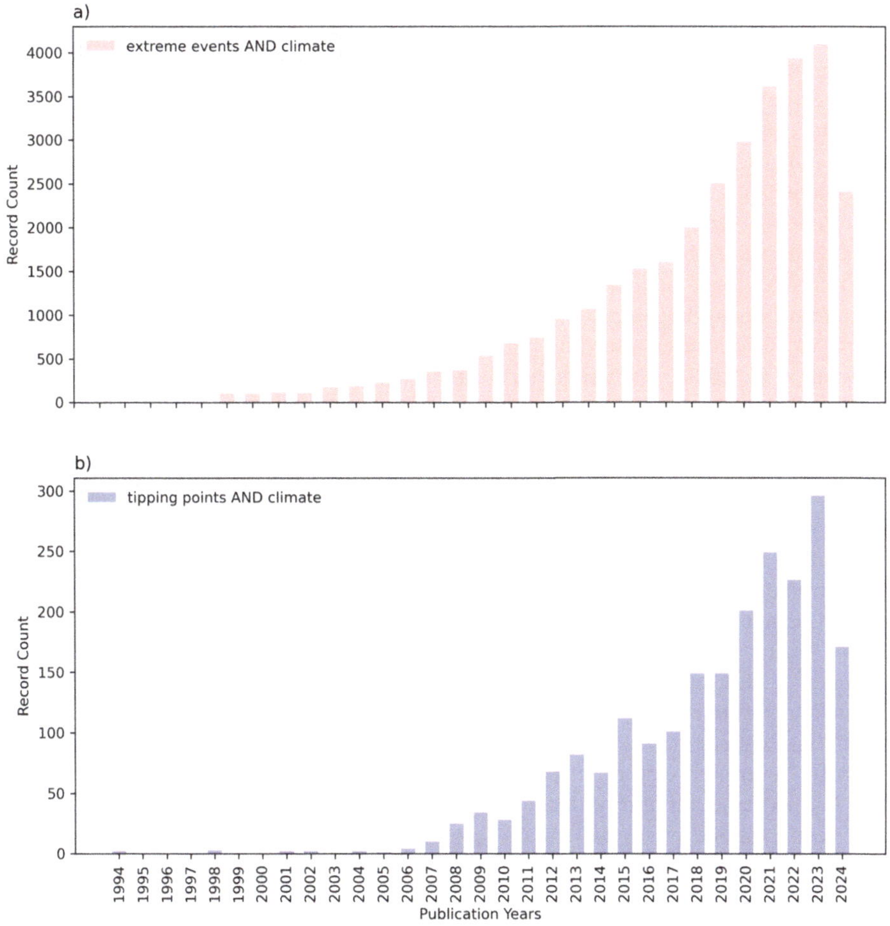

Fig. 1 Number of publications in Web of Science Core Collection on the topic: **a** "Extreme events AND climate"; and Number of publications on the topic: **b** "Tipping points AND climate" (search: 23.07.2024)

Earth System **tipping elements** are at a minimum subcontinental-scale components of the Earth System that could potentially undergo state shifts triggered at some level of climate forcing by modest levels of additional forcing that exceed critical thresholds or **tipping points** (Armstrong McKay et al. 2022).

Once tipping points have been crossed, changes are not easily reversible, at least not under the same forcing conditions that triggered them, or they take significantly longer to recover than the time it took to reach them (Lenton et al. 2023).

Extreme events belong to the most visible and impactful consequences of climate change (Seneviratne et al. 2021). Here, we discuss the link between such tipping points/ and tipping elements and extreme events, including compound extreme events.

Extreme weather and climate events are defined as follows in the 6th Assessment Report of the Intergovernmental Panel on Climate Change (IPCC; see also Seneviratne et al. 2021): An **extreme weather event** is 'an event that is rare at a particular place and time of year', and an **extreme climate event** is 'a pattern of extreme weather that persists for some time, such as a season'. **Rare events** are typically in the extremely low or high range of a given variable, and thus extreme events are often also defined in terms of their probability of occurrence, such as exceeding the 90th or 99th percentile of temperature of a climatological baseline, e.g., a preindustrial, mid-twentieth century, or late-twentieth century baseline, or a moving-mean baseline, but it depends on the study.

In the context of tipping points, compound events, a subcategory of extreme events, can be of particularly high relevance. **Compound weather and climate events** describe combinations of multiple climate drivers and/or hazards that contribute to societal or environmental risk (Zscheischler et al. 2018; Bustamante et al 2023). Examples of compound events that may lead to long-lasting impacts include drought with heat or fire, multiple storm hits (these are temporally compound), and extremes of ocean temperature occurring together with high acidity. Extreme events, particularly compound events, can also lead to cascading impacts and chains of subsequent events (e.g., AghaKouchak et al. 2020). As compound weather and climate events may occur more frequently in many regions, especially in relation to high emission scenarios (Ridder et al 2022), their potential triggering of tipping points needs to be monitored.

Figure 2 displays global and regional tipping elements identified in the study of Armstrong McKay (2022) and their links to some weather and climate extreme events. We note that not all tipping points and elements are necessarily affected by extreme events, and that even for the tipping elements which are influenced by extreme events, other non-extreme conditions and non-climatic processes can play a role as well. We focus both on changes with global and large-scale ramifications, and also refer to events where tipping occurs on the interface between ecosystems and society.

Figure 3 illustrates how extreme events can push a given system beyond a critical threshold determining a tipping point, related to the concept of a "noise-induced tipping point" (Ashwin et al. 2012; Boers et al. 2022). Here we focus on cases where the amplitude of a severe extreme event pushes the system further toward the alternative state than less extreme variability does. For example, noise-induced transitions in AMOC have been explored in several studies (e.g., Castellana and Dijkstra 2020). Long-term impacts of some extremes, such as vegetation loss to fire or drought, and degradation may also lead to a persistent alteration of forest properties and thus diminish the viability of the previous state (e.g., Saatchi et al. 2021).

In this article, we review the literature on weather and climate extreme events, including compound events, which might be relevant in the context of tipping points or tipping elements. We discuss the extent to which extreme events can:

a. Contribute to reported global Earth System tipping points, such as Amazon rainforest loss, the Atlantic Meridional Overturning Circulation (AMOC) shutdown, or loss of major ice sheets (see Armstrong McKay et al. 2022; Wang et al. 2023a, b).
b. Contribute to regional climate tipping points of the Earth system, both those highlighted in review papers (e.g., Boreal forest change, see Wang et al. 2023a, b) and other cases where extreme events may lead to long-lasting or irreversible changes, such as long-lasting vegetation change.
c. Contribute to global and regional society and ecosystem tipping points, such as migration in the dustbowl region in response to drought, degradation of ecosystem services, food insecurity and other economic pressures due to coral and fisheries collapse and loss of regional biodiversity; humanitarian crises in response to extremes.
d. We also identify evidence of processes that increase Earth System resilience, i.e. processes which lead to increased carbon uptake following some extremes.

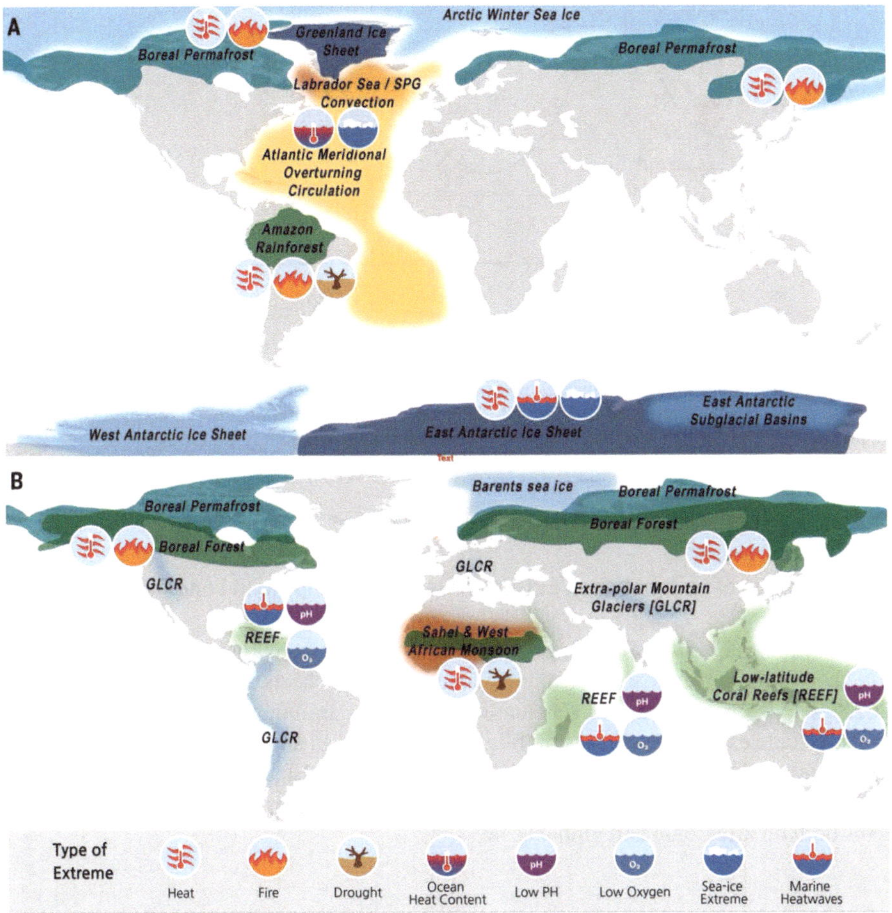

Fig. 2 Schematic of extremes and compound extremes on land and ocean and their link to tipping elements of global (**A**) and of regional (**B**) impact, drawing on Fig. 1 from Armstrong McKay et al (2022). Symbols indicate links to extreme events. For land: Boreal forest responds to drought, fire, heat; Boreal permafrost to heat; Amazon rainforest to drought and fire. Mountain glaciers (GLCR) respond to heat. For ocean: AMOC responds to marine heatwaves and low salinity extremes, coral reefs respond to marine heatwaves, high acidity, and low oxygen concentration extremes

This paper is organized as follows: We first show from the perspective of both observations and process studies how extreme and compound events on land (Sect. 2) and in the ocean (Sect. 3) contribute to tipping points. This is followed by a discussion of the state of understanding based on paleoclimate data (Sect. 4), observations and climate modeling (Sect. 5), and finally recommendations and opportunities to use Earth observations (EO) for early warning of extreme events that can lead to tipping.

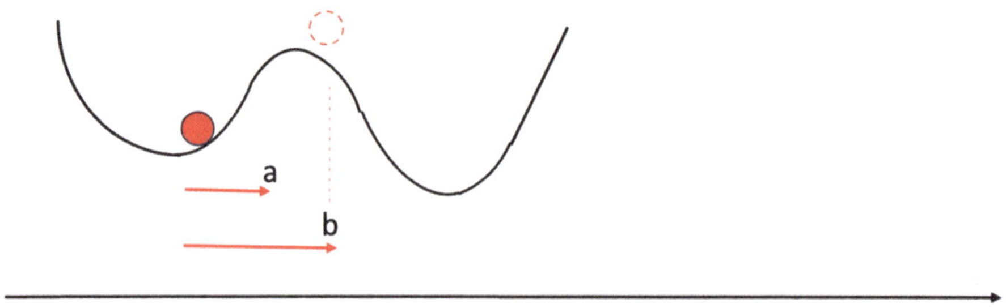

Fig. 3 Links between extreme events (arrows) and the crossing of a critical threshold associated with a tipping point (dashed line circle). Compared to a standard perturbation ("**a**"), an extreme event ("**b**") can push the system sufficiently away from its base state (filled circle) to induce a tipping point. The extreme event could also be a compound event, or a sequence of multiple events (temporally compound)

2 Land Extreme Events from a Tipping Point Perspective

2.1 Overview of Link between Extreme Events and Tipping in Land Systems

Global climate tipping elements arising from land systems include the Amazon rainforest, the boreal permafrost, and major glaciers and ice sheets (Armstrong-McKay et al. 2022, see also new combined schematic Fig. 2), and arguably all of these are sensitive to large-scale extreme events, as highlighted hereafter. In addition, changes in the boreal forest have been identified as a tipping element and are also sensitive to extreme events. Finally, extreme events could also trigger impact-related tipping points in the land systems both on a regional and global scale.

2.1.1 Amazon Rainforest (Global Tipping Element)

The Amazon rainforest shows well-documented vulnerability to extremes, particularly increasing drought (combined with extreme temperatures) under increasing global warming (Seneviratne et al. 2021), and enhanced fire activity, which is amplified by human disturbance (Hubau et al. 2020; Silva Junior et al. 2021; Rosan et al. 2022; Lapola et al. 2023). The combination of these changes could lead to irreversible degradation of the forest, and with it its tipping toward a degraded forest or savanna state (see review in Armstrong McKay et al. 2022; Wang et al. 2023a, b; Nobre et al. 2016; see Fig. 2).

Due to its role in water recycling, Amazon rainforest losses are expected to be self-amplifying (Zemp et al. 2017) and to have global ramifications (Lenton et al. 2008), such as disruption of water transport to important agricultural regions of the world, with irreversible effects for at least decades. The main drivers of Amazon forest loss in models and observations are deforestation and the effects of climate change, such as increased soil moisture drought associated both with increased evaporation and decreased precipitation as well as the combination of dry and hot conditions leading to a stronger incidence of fire. Analyses have already identified the Amazon as a sensitive system in a degraded state (Saatchi et al. 2021; Boulton et al. 2022; Lapola et al. 2023). Armstrong McKay et al. (2022) provide a best estimate that this tipping point may be triggered around 3 °C of global mean warming since the preindustrial period (with minimum value of 2 °C).

However, uncertainties about the future response of the Amazon forest to climate extremes are substantial, especially about the impact of their interactions. For example, recent research suggests that deforestation impacts water recycling and can increase the possibility of forest loss (e.g., Bochow et al. 2023); but also impacts the precipitation regime in the dry season in complex ways, which might lead to opposing effects (Khanna et al. 2017). The realism of processes linked to the land–atmosphere coupling under drought in climate models is also uncertain. For example, soil moisture and rainfall may be coupled too strongly in low resolution models (Lee and Hohenegger 2024). There is also high uncertainty about the ability of state-of-the-art Earth System Models (ESMs) to simulate forest resilience to climate change. Some studies indicate that ESMs underestimate the sensitivity of ecosystem carbon uptake to drought variability under the current climate (Humphrey et al. 2018), and others report an observed increase in this sensitivity in the recent decades (Liu et al. 2023a, b), thus challenging the assessment of the exact conditions under which the Amazon rainforest dieback as global tipping point could be triggered. Also, fires at the edges of humid forests could play an important role in this context (Silva Junior et al. 2020). Furthermore, Earth System models lack the representation of processes that could confer increased resilience of the Amazon rainforest to climate extremes, such as vegetation demographics and biodiversity which can contribute to buffer impacts and enhancing ecosystem resilience (Sakschewski et al. 2016; Fisher et al. 2018). On the other hand, they generally also do not represent drought-induced forest mortality adequately, and there is evidence that they may not be able to represent fires as large as some recently observed, at least not in Australia (Sanderson and Fisher 2020). Therefore, while some processes not represented in Earth System Models could imply a higher resilience, the balance of evidence suggests that the feedbacks triggering this tipping point may be underestimated in current models and thus this tipping element could be triggered at a lower level of global warming than estimated in Armstrong McKay et al. (2022), which would have substantial policy implications. More research is necessary to clarify this point.

2.1.2 Boreal Permafrost (Global Tipping Element)

Boreal permafrost could thaw under rising temperatures, affecting the natural and built environments and water cycle balance of the region and further releasing currently trapped methane into the atmosphere (Smith et al. 2022; see companion paper by Brovkin et al., 2024). This is first a regional tipping element due to its importance to the regional water cycle, and second permafrost thaw can also have global implications due to possible carbon and methane release. Permafrost change has been linked to vegetation changes and extreme events, such as extreme heat, wildfires, and precipitation shifts (see Wang et al. 2023a, b).

2.1.3 Glaciers and Ice Sheets (Global Tipping Element)

As temperatures continue to rise, glaciers and ice sheets will also continue to melt, leading to potentially irreversible ice mass loss (Hanna et al. 2020) and, in turn, to impacts ranging from irreversible sea level rise (Frederikse et al. 2020) to changes to the Earth's albedo that could further exacerbate both warming and ice melt (Box et al. 2012). The melting of large ice sheets is interlinked with complex processes on the interface between ocean and ice creating substantial uncertainty in the dynamics and future evolution of major ice sheets. Furthermore, ice-sheet melting can be exacerbated by hot spells, as exemplified by the 2010 record melting in the Greenland Ice Sheet (Tedesco et al. 2011). In 2010, many locations in Greenland experienced record-breaking temperatures in winter, spring, summer and the year as a whole. The combination of abnormally warm conditions both in the cold and in the warm seasons lead to a melting period 50 days longer in 2010 compared to the 1979–2009 average, which started exceptionally early and ended exceptionally late (Tedesco et al. 2011).

2.1.4 Boreal Forest

Beyond the above-mentioned global tipping elements, another tipping element in the land systems includes shifts (Northward expansion and Southward retreat) in the Boreal forest. It is listed as a regional tipping element in Fig. 2, although it may have global consequences on Earth's albedo and is classified as a global tipping element in another recent review paper (Wang et al. 2023a, b). Warming, enhanced by Arctic amplification, leads to growing season shrub encroachment in the high Arctic (e.g., Myers-Smith et al. 2015), causing an expansion of tree cover into currently treeless tundra ecosystems, at rates of up to 4 km per decade (Dial et al. 2022). However, the Southern parts of the boreal forest are in retreat attributed to fire and increased logging (Rotbarth et al. 2023; see also Beck et al. 2011; Carpino et al. 2018; Wang et al. 2023a, b; and references therein). Empirical evidence suggests that disturbances are responsible for switches between alternative states rather than gradual changes (Scheffer et al. 2012; Abis and Brovkin 2017; Lenton et al. 2023), for instance due to shift from evergreen to deciduous species (Mack et al. 2021; Massey et al. 2023). In many events (see discussion below), extreme heat, fire and ecological changes are tightly linked in high latitudes, given that vegetation and climate can feed back on each other (Bastos et al. 2023). For example, compositional changes due to more frequent fires in Alaska have been reported to result in biomass gains (Mack et al. 2021) due to feedbacks with nutrients and other ecological processes. Boreal forest change can be globally important because it has the potential to influence warming by changing albedo (Massey et al. 2023), yet at present the impacts of forest changes are uncertain, and their potential to lead to tipping are underexplored.

2.1.5 Further Land-Related Tipping Elements, Including in Ecosystems and Society

Further tipping elements are linked to the monsoon systems, particularly the Sahel and West African monsoon system (see schematic in Fig. 2; Armstrong McKay et al 2022), which may not be per se irreversible (Wang et al. 2023a, b), but can have huge societal consequences. However, the mechanisms are complex, such as changes in the onset and retreat of monsoons and thus not clearly linked to individual extreme events. Other changes related to tipping in the literature include long-term change in heat/moisture climate due to soil-moisture changes in heatwaves, such as in Central Europe (Seneviratne et al. 2006) and in a case in central East Asia (Zhang et al. 2020).

Furthermore, extreme events may lead to cascading effects and socioeconomic tipping in the human system, for example through heat and drought-induced failure of breadbasket regions, and long-lasting impacts or transitions in societies affected by them; thus, leading to global and regional tipping elements affecting society and ecosystems.

In the following, we discuss the mechanisms by which specific types of extreme events may lead to long-lasting changes, threshold behavior and with it, tipping points. We focus on drought and heat given the larger body of literature focusing on these events, but long-lasting impacts are not limited to them. For example, high-impact compound events can also arise from severe storm hits, particularly in sequence. An example is multiple or a single large tropical cyclone landfall, which creates potential for severe and long-lasting flooding, and impacts the viability of infrastructure. Multiple storm hits can lead to very slow recovery times and impact habitability, with a negative impact on community resilience (e.g., Binte and Hasan 2023).

2.2 Drought-Heat-Fire Events and Their Potential Link to Tipping

Headline messages from the recent Intergovernmental Report on Climate Change (IPCC 2021) indicate that extreme high-temperature events that used to recur every 50 years in a preindustrial climate, now are likely to occur about every decade, and will potentially increase to every 3–4 years at 2 °C of global warming, and such heatwaves are also expected to be more intense. The frequency of drought has also already increased with a previous 10-year event expected to be more than twice as frequent, and more intense, in a 2 °C warmer world (Seneviratne et al. 2021).

As discussed above, both Amazon rainforest dieback and boreal forest changes have the potential to interact with severe and large-scale extreme events. However, here we argue that other regions might also be sensitive to long-lasting, near-term irreversible impacts from large-scale severe events, which can lead to long-term consequences, such as vegetation loss or change (e.g., Seidl and Turner 2022). We discuss mechanisms by which extremes could create long-lasting transitions in the Earth system. For example, an increase in compound hot-dry conditions is also a key characteristic of projected climate change in the Mediterranean region (Seneviratne et al. 2021), where it could potentially lead to a marked loss of local vegetation beyond 2 °C of global warming (Guiot and Cramer 2016).

Co-occurring or 'compound' drought and heat events amplify each other, both in terms of intensity and in terms of impacts (Seneviratne et al. 2010; Allen et al. 2015; Zscheischler and Seneviratne 2017; Zscheischler et al. 2018). In particular, under soil-moisture limitation and increased atmospheric aridity, vegetation productivity and stomatal conductance decline (Fu et al. 2022), reducing surface cooling, with resulting increased heat enhancing evaporative demand and creating a positive feedback (e.g., Seneviratne et al. 2021). Where physiological or ecological thresholds are crossed (e.g., Lawrence et al. 2022; Allen et al. 2015; Hartmann et al. 2022), dieback can occur; increased wildfires result in carbon stocks that are rapidly released into the atmosphere, followed by increased carbon sequestration as vegetation regrows (Byrne et al. 2021; Qin et al. 2022). Furthermore, after severe compound drought and heat stress, vegetation recovery can be delayed, for example due to reduced growth or non-reversible damages (Ruehr et al. 2019), and potential impacts may be worsened by increasingly frequent successive or multi-year extremes (Suarez-Gutierrez et al. 2023). This may lead to legacy effects (Ruehr et al. 2019; Kannenberg et al. 2019; Yu et al. 2022a, b) that increase vulnerability to further compound drought and heat stress. If subsequent events occur before complete vegetation recovery, they can lead to larger impacts, for example as occurred in the 2018–2019 successive extreme hot and dry summers in central Europe (Bastos et al. 2021a, b). Widespread tree mortality in the aftermath of these events (Hlàsny et al. 2021) have turned the forests in the Czech Republic into a carbon source (EUCRA 2024). Thus, crossing a threshold for the coupled vegetation/

climate system, or cumulative effects of repeated extremes, can create a regional 'tipping point' where a small perturbation can cause a rapid, nonlinear change into a different state.

Fire is endemic to some forest or wooded ecosystems, which are in fact often dependent on fire (e.g., Bowman et al. 2009), but these are adapted to their corresponding fire disturbance regime (return period, magnitude, severity; Turner 2010). Changes in fire regimes such as increased fire return frequencies and fire intensity, or an expansion of fires into less frequently burnt areas, can lead to vegetation loss, productivity changes, or shifts to a different vegetation state altogether (Turner and Seidl 2023). For instance, a decrease in coniferous forests, potentially replaced by early successional broadleaved tree species, or a decrease in fire-resisting tree species (Anoszko et al. 2022; Liu et al. 2022; Abis and Brovkin 2019). Wildfires can also affect the regional water cycle (e.g., Zemp et al. 2017).

The following two examples highlight the potential for extreme events to lead to severe, long-lasting and sometimes surprising consequences:

The first such example is the extreme US 'Dust Bowl' drought (Fig. 4), and resulting severe land surface changes. It was initiated by long-term (Schubert et al. 2004) but not unique drought (Cook et al. 2022a, b). The 'Dust Bowl droughts' contributed to extreme heat (Cowan et al. 2017), crop failure and dust storms in some cases reaching the US East Coast, which in turn amplified drought and heat (Cook et al. 2008) with serious consequences for human health and migration (Worster et al., 1979; see review in Hegerl et al. 2018). Some ESMs simulate dust resulting from vegetation loss (see Cowan et al. 2020a,b). This event was caused by a combination of extreme events and societal factors (rapid transition of the great plains to agriculture, which was vulnerable to drought) and led to a tipping point in the human system, but also to very long-lasting (and in some locations still not reversed) changes in the land surface, hence breached regional tipping points.

A second example that illustrates potential for severe and possibly irreversible change is the occurrence of recent extreme fire seasons in high latitudes, including in the summers of 2019–2021 in Siberia (Scholten et al. 2022), where many strong fire events in recent decades were preceded by severe heat (Hegedűs et al. 2024). Recent fires and droughts in Siberia have decreased its carbon sink (Fan et al. 2023). The 2020 Siberian heat wave is considered to have been almost impossible without human influence on climate (Ciavarella et al. 2021) and led to particularly strong fire activity. However, the timing of the events is also important, for example the 2020 Siberian heat wave led to an increase in spring vegetation productivity, which

Fig. 4 A farmer and his sons were caught in a dust storm in Cimarron County, Oklahoma, April 1936. NPx 66-174(32) from Forward with Roosevelt; F.D. Roosevelt presidential library and museum

offset later losses in summer and fall (Kwon et al. 2021). Northern Canada has seen an intense fire season in 2023, with evacuation of the town of Yellowknife and smoke emissions from the fires affecting air quality across a broad swath of North America (Wang et al. 2023a, b). These fires also released a record amount of carbon into the atmosphere, with possible implications for carbon budgets (Zheng et al. 2023), although recovery needs to be assessed. Nevertheless, the increased rates of change and unprecedented magnitude of many recent extremes highlight risk and change in the boreal forest.

Several other recent extreme events caused severe and slow-to-reverse impacts regionally, some with global ramifications that illustrate the potential for even more severe extreme events to cause tipping, particularly in ecosystems and human systems. For instance, the 2003 European heat wave was a record-shattering extreme event (Schär et al. 2004) that led to tens of thousands of excess deaths, severe drought stress on vegetation, loss of agricultural production, and multiple other consequences. The 2021 Pacific Northwest Heat wave saw record-shattering heat (White et al. 2023)—where the highest temperature measured ever above 50° of latitude (49.6 °C) occurred—an intense fire season, and vast agricultural and ecological losses, including the deaths of more than 10 million marine animals such as barnacles and sea stars due to the co-occurrence of low tide in South facing shores with peak daily temperatures (White et al. 2023). Severe vegetation impacts resulted from the hot and dry European summer of 2019. The Horn of Africa has recently experienced severe drought exacerbated by heat (Kimutai et al. 2023) with devastating consequences for populations living there. In the Southern Hemisphere, recent Australian summers have broken records for heat and wildfires and have been linked to greenhouse gas increases (van Oldenborgh et al. 2021). They have also been a driver of mass fatalities of some arboreal mammals by breaching their temperature tolerance (Ratnayake et al. 2019), leading for example to the death of more than 70.000 Australian flying foxes in the 2019–2020 summer, triggered first by drought-caused starvation and pup abandonment, second by heat stress related die-offs, and third by wildfires (Mo et al. 2021).

In summary, compound heat and drought extremes events can enhance each other, and can trigger degradation of vegetation, agricultural, animal and human loss, and severe wildfires. These, in turn, can cause cascading impacts with vegetation changes potentially feeding back on climate (see hereafter and Fig. 5). Several factors make the occurrence of local or regional tipping points more likely: the increase in compound hot-dry events across the world; associated changes in disturbance regimes (fires, insect outbreaks, drought-induced mortality); and the fact that carbon losses are fast, but recovery is slow and potentially hampered by climate change. When aggregated spatially and in time, the cumulative effects of such impacts, even beyond the Amazon and boreal regions, are likely to reduce the efficiency of the global land carbon sink beyond what is currently projected (Bustamante et al. 2023; see Humphrey et al. 2018, Green et al. 2019 for examples), thereby further amplifying warming. There is currently medium confidence that such changes can result in positive carbon-climate feedback from climate extreme driven disturbances (Canadell et al. 2021).

We also note that tipping point or threshold behavior may arise in the response by society, such as out-migration (for an example see the dust bowl case study above), which may also occur in response to present day extreme events threatening livelihoods.

2.3 Concurrent/Teleconnected Extremes in Breadbasket Regions

Hot and dry extreme events also have the potential to risk severe consequences on society through their impact on crops. Heat and drought may be especially damaging when extreme

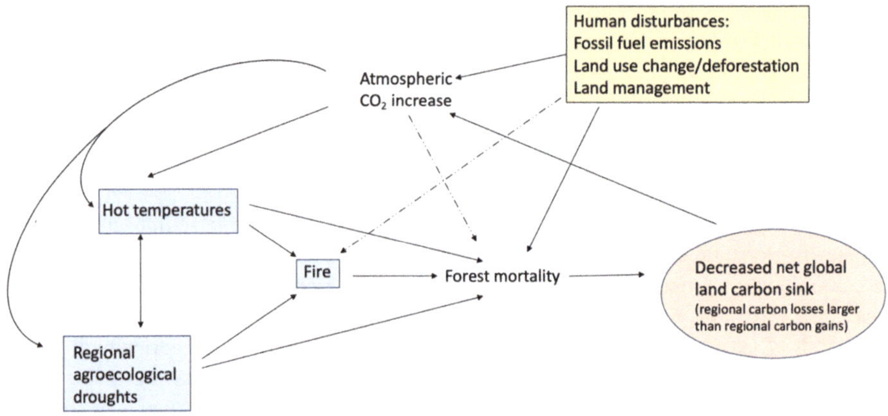

Fig. 5 Links between extreme events, human disturbances, and a potential carbon sink-related tipping point related to spatially and temporally compounding effects of unprecedented impacts of heatwaves, droughts and associated fire and forest mortality discussed in Sect. 2.2. Extremes are indicated by blue rectangles, the tipping element (decreased global land sink) with the orange oval, and the human disturbances with the yellow rectangle. Solid arrows indicate an amplifying mechanism, while dash-dotted lines indicate relationships that can either amplify or dampen the response. Regarding the global land carbon sink tipping point, it corresponds to a globally aggregated impact, and does not exclude the presence of regional carbon gains for some events (e.g., in Scandinavia during the 2018 heat-drought event). On average, and despite these possible regional exceptions, heat-drought events are typically associated with regional carbon losses that can take several years to recover and might be irreversible if extreme events recur faster than recovery times. Note that the figure highlights CO_2, but warming due to other greenhouse gases contributes as well to climate change

conditions occur during the much more vulnerable sprouting and flowering periods in the spring (Senapati et al. 2021; Luo et al. 2023). Where heat and drought events occur in agriculturally important 'breadbasket' regions, they can reduce harvests and threaten global food security (Kornhuber et al. 2020). Recent studies showed that with advancing global warming, the area affected by extreme events increases substantially (Vogel et al. 2019; Batibeniz et al. 2023), including in breadbasket regions (Gaupp et al. 2019; Seneviratne et al. 2021; Biess et al. 2024), and several breadbasket regions may be affected by extremes simultaneously more frequently (Raymond et al. 2022). Recent research also indicates that extreme heat in one region could be dynamically linked to heat in other regions, some of these breadbasket regions (Meehl et al. 2022), highlighting the risk of spatially compound heat events affecting more than one breadbasket region.

Concurrent failure in breadbasket regions can result in increased food prices and limited access to resources, affecting both the population and humanitarian organizations trying to procure and distribute aid (IPCC 2019; Mbow et al. 2019; Kornhuber et al. 2022). Food insecurity associated with pre-existing vulnerabilities can affect human migration patterns. Also, coordinating and delivering humanitarian aid in the aftermath of spatially compounded events within the same country can pose significant logistical challenges. An example of this complexity is evident in the case of Tropical Cyclone Freddy, which impacted Mozambique during two distinct periods: from 6–24 February and 2–14 March 2023 (OCHA 2023).

To our knowledge, literature assessing potential regional or global societal tipping elements associated with concurrent extremes, such as breadbasket failures, is not yet available, and it is an area that should be more investigated in near future. This is in particular the case given the identified underestimation of associate impacts in current climate models (Kornhuber et al. 2022).

2.4 Dangerous Heat and Humidity

Thresholds also exist for the human body. Under concurrent high temperatures and high ambient humidity, evaporative cooling through sweat loses efficiency and we become unable to maintain a stable body temperature, leading to potentially lethal damages (Vecellio et al. 2022; Sherwood and Ramsay 2023). While the relationship between extreme heat and human health is complex (e.g., Lo et al. 2022), recent empirical physiological research in humans finds lower humid heat thresholds than previously thought for which heat stress becomes uncompensated in young, healthy adults performing normal daily activities (Vecellio et al. 2022; Vanos et al. 2023).

Multiple indices exist for evaluating the health risk from heat stress, but the broad picture is similar between them: as global mean temperature increases beyond 1.5, 2 and 3 degrees of warming, an increasing fraction of the population will be exposed to dangerous humid heat, particularly in East Asia (Freychet et al. 2022; Suarez-Gutierrez et al. 2020). Such dangerous heat is not only linked to increased heat-related mortality and morbidity, but has also been connected to large economic losses due to decreased labor productivity (Dunne et al. 2013; García-León et al. 2021). Furthermore, assessments of extreme humid heat stress conditions may underestimate part of the risks due to the incorrect representation of relevant atmospheric and geographical effects in coarse resolution simulations, such as used in Freychet et al. (2022), underestimating humid heat in microclimates near coasts, and in cities. Lastly, as discussed above in the case of Australian flying foxes, severe heat can also cause severe impact on wildlife and with it impact on biodiversity.

The case of humid heat and impacts of extremes on food system illustrate that weather and climate extremes can lead to slow-to-reverse change or tipping in impacts, even where the extremes themselves show no such threshold behavior.

3 Marine Extreme Events from Tipping Point Perspective

3.1 Extremes, Compound Extremes and Tipping Elements in Marine Systems

The rise of atmospheric CO_2 concentrations is a major contributor to ocean warming, ocean deoxygenation and is directly responsible for acidification (Gruber 2011). At the same time, a strengthened hydrological cycle due to warming is expected to increase the intensity and frequency of extreme weather events (see Seneviratne et al. 2021), modulating variations in evaporation and precipitation over both the ocean and the land, hence altering freshwater inputs to the ocean, which impact salinity, stratification, and consequently, circulation and oxygen supply (Fennel and Testa 2019). Modified ocean–atmosphere heat exchanges are also expected to affect wind patterns over the open ocean (Fox-Kemper, et al. 2021) while intrinsic climate variability (e.g., the El Nino-Southern Oscillation (ENSO)) which may be changing regionally (Beobide-Arsuaga et al. 2021) will also play a role in enhancing these forced trends, albeit differently in different ocean basins. All these changes carry high potential impacts for marine ecosystems and ocean biogeochemistry (e.g.: Bopp et al. 2013; Landolfi et al.

2017). As the background state changes, the likelihood of extreme conditions, that lead to global or regional tipping points with severe impacts on marine ecosystems and human services, is expected to increase as well.

Global climate tipping elements in the ocean considered in this paper (Fig. 2) are the Atlantic Meridional Overturning Circulation (AMOC), the North Atlantic subpolar gyre (SPG) mixing, and the ocean oxygen content, while regional tipping elements include low latitude coral reefs (McKay et al., 2022; Wood et al., 2024).

Extending the definitions of extremes discussed in Sect. 1.2 into the ocean realm, marine extreme events (Fig. 2) are defined when ocean properties significantly exceed the typical seasonal variations and persist for at least five consecutive days (Hobday et al. 2016) and extend for several kilometers (Frölicher et al. 2018). Under continuing global warming, marine heatwaves (MHWs—anomalous periods of elevated temperatures), low oxygen (LOX; Gruber et al. 2021), high acidity (OAX; Burger et al. 2020; 2022) and low chlorophyll (LChlX; Le Grix et al. 2021) extremes have become more frequent and some are expected to intensify in frequency, duration and areal coverage under sustained anthropogenic emissions (Gruber et al. 2021; Burger et al. 2022; Frolicher et al. 2018).

Furthermore, multiple extremes may co-occur as "compounds" in space and time, at the ocean surface (Le Grix et al. 2021; Gruber et al. 2021; Burger et al. 2022; Zscheischler et al. 2018) and/or at depth (Gruber et al. 2021; Wang et al. 2023a, b), severely threatening marine organisms and compressing habitability. For example, thermal stress during prolonged MHWs disrupts the relationship between corals and their symbiotic microalgae, causing coral 'bleaching', often resulting in widespread coral death (e.g., Klein et al. 2022 and references therein). Higher acidic conditions are further expected to reduce coral tolerance to heat stress (Anthony et al. 2008), leading to reduced survival of corals under intermediate and high emission scenarios (Klein et al. 2022). According to how compound events unfold in space and time, diverse typologies have been identified: (i) *preconditioned,* where a weather or climate event preconditions the impact/occurrence of the extreme; (ii) *multivariate,* when co-occurrence of multiple extremes leads to an impact; (iii) *temporally compounding*, when temporal succession of multiple extreme leads to an impact; (iv) *spatially compounding,* when extremes are occurring in multiple locations and lead to synchronized/ aggregated impact (Zscheischler et al. 2020). In the following we will use the term "compound" events to refer to multivariate, spatially and temporally compounding extremes. Preconditioned and compound extremes will be further discussed in Sect. 3.3.

In addition to MHWs-OAX-LOX-LChl-primary productivity combinations of extremes, multivariate compound extremes can also coincide with extreme changes in ocean heat uptake, sea ice distribution and mass/heat/salt transports, precipitation, wind stress over the ocean surface, hurricanes, dust storms and other aerosol deposition processes, and sea level rise. Such combinations in conjunction to a changing background can lead to tipping points. For example, extremes in sea ice distribution and sea-ice transport into the subpolar gyre were shown to potentially contribute to the tipping of the AMOC under an extended SSP2-4.5 scenario in the GISS climate model (Romanou et al. 2023) through a feedback loop shown in Fig. 6A (see also discussion in Wood et al. 2024). There, increased GHGs and warming, lead to AMOC declines that triggered intense variations in sea ice and sea ice transport across the Denmark Straits and into the subpolar gyre where they melt, freshen the ocean surface, and shut down mixing and hence deep-water formation (see also Romanou et al 2023).

Combinations of extreme conditions in the ocean can have impacts that exceed the impacts of the singular events (Smith et al 2019; Lucey et al 2022; Liu et al. 2023a, b) and therefore may lead to regional tipping points. For example, acting in unison, MHW and

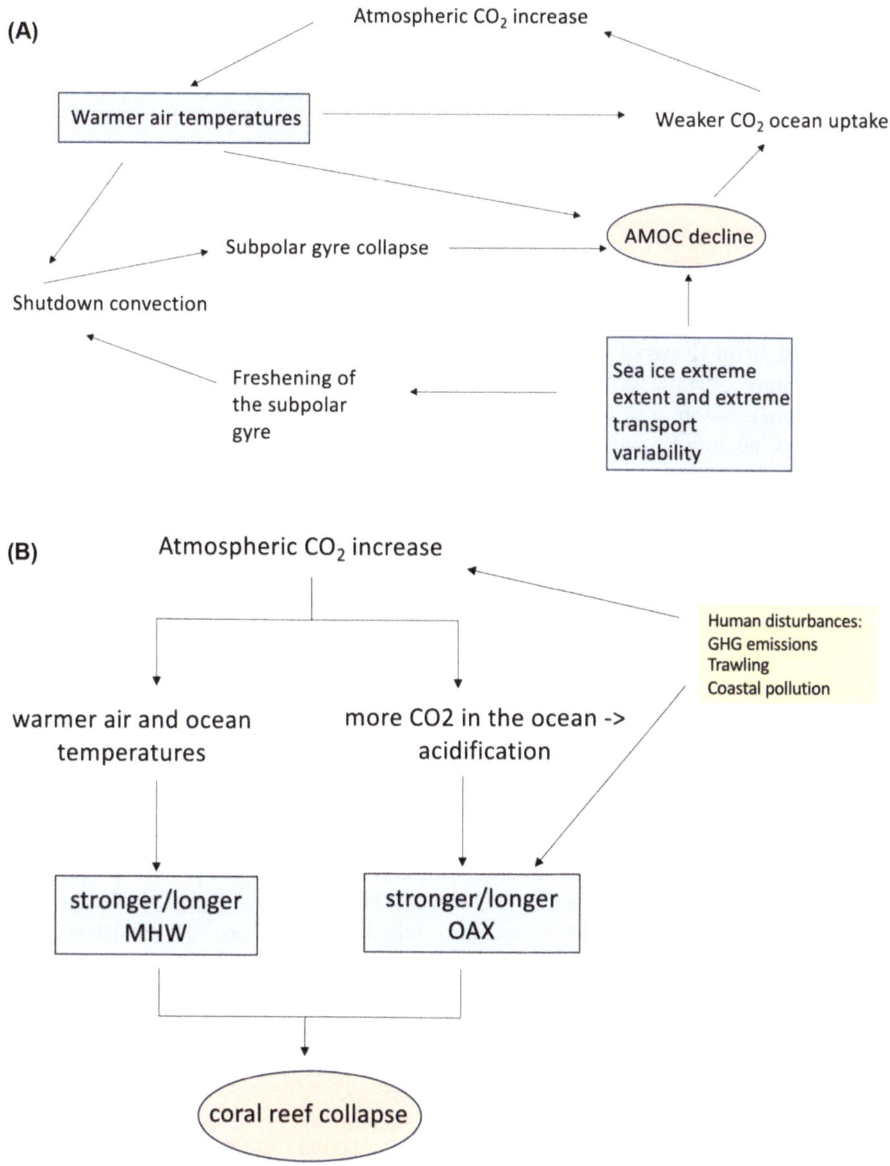

Fig. 6 Examples of compound extremes that can lead to tipping points in the ocean. **A** Compound extreme variability of sea ice extent and sea ice transport out of the Arctic Ocean and into the subpolar gyre, leads to freshening in the SPG and causes AMOC decline (Romanou et al. 2023). **B** Compound extremes such MHWs and OAX, leading to collapse of low latitude coral reefs. Extremes are indicated by blue rectangles, the tipping elements (AMOC decline; coral reef collapse) by orange ovals and human disturbances that exacerbate marine extremes are shown in the yellow box

OAX may cause tipping in coral reefs at 2 °C warming levels, particularly in the tropical oceans (Armstrong McKay et al. 2022; Klein et al. 2022; Hoegh-Guldberg et al. 2018; Heinze et al 2021; Pearce-Kelly et al 2024; Fig. 6B).

3.2 Role of GHG Forcing and Internal Climate Variability in Linking Extremes and Tipping Points in the Ocean

Anthropogenic forcing, i.e. increasing greenhouse gas (GHG) emissions and the ensuing global warming are the dominant drivers of the long-term changes in many extreme events, such as MHW (Frölicher et al. 2018; Oliver et al. 2018; Holbrook et al. 2019) and open ocean low oxygen extremes (Gruber et al. 2021), while CO_2 increase drives high acidity extremes (Gruber et al. 2021; Burger et al 2020; 2022). Yet large-scale natural modes of climate variability, e.g., ENSO, NAO (North Atlantic Oscillation), PDO (Pacific Decadal Oscillation) etc. and their associated atmospheric and oceanic teleconnections, can favor or suppress the occurrence of extremes by modulating local conditions (Köhn et al. 2022; Oliver et al. 2018; Holbrook et al. 2019, 2020; Rodrigues et al. 2019; Le Grix et al. 2021). Therefore anthropogenic forcing trends as well as intrinsic climate variability might trigger and/or amplify extremes or compound extremes (Special Report on the Ocean and Cryosphere in a Changing Climate (SROCC) 2019) and thus lead to tipping points.

While the forced impacts on extremes are better understood, the role of natural climate variability is less clear. Internal climate modes regulate the likelihood of extremes due to a complex interplay of direct and indirect effects on physical, chemical, and biological processes, which are difficult to predict on short time scales. For example, changes in mixed-layer depth and upwelling strength, which are strongly modulated by ENSO, have been shown to affect the frequency of compounding MHW and LChlX events (Le Grix et al. 2021; Le Grix et al. 2022). On the contrary, the likelihood of compound MHW-OAX events is lower in the Equatorial Pacific, where increased temperatures during El Nino directly give rise to lowering pH (due to changes in the carbonate chemistry), but at the same time the reduction in CO_2 solubility, together with the reduction in the upwelling of CO_2-rich waters results in the net CO_2 outgassing and pH increase (Burger et al. 2022). Overall, the combination of underlying trends in ocean warming, acidification, and deoxygenation (Gruber et al. 2021) and large-scale climate modes can make compound extremes more likely, or prolonged or more intense with devastating effects on marine ecosystems. The *Blob*, a major MHW observed in the northeast Pacific during 2013–2015, stands as an example, where regional circulation and ENSO-driven warming led to a persistent MHW (Di Lorenzo nd Mantua 2016), which occurred along with anomalously low oxygen and high $[H^+]$ concentrations generating a triple compound extreme (Gruber et al. 2021).

All these extremes, their combinations and linkages to forced and internal climate variability, are likely to continue to pose a risk to the ocean systems in the future decades even under the more optimistic GHG mitigation scenarios, because their drivers are expected to persist and amplify (Gruber et al. 2021; Holbrook et al. 2019).

Coastal regions are particularly vulnerable to compound extremes because, in addition to anthropogenic warming, other natural or anthropogenic pressures may affect them (e.g., dust and other aerosol deposition, pollution from agricultural practices and from sewage treatments, etc.). Trawling can impose huge pressures to coastal and benthic ocean pH (Atwood et al. 2023). Coastal extremes can also lead to algal blooms that die and decay, consuming oxygen leading to deoxygenation spikes. Globally, climate warming both exacerbates the problems from eutrophication and reduces the introduction of oxygen to the interior of the ocean.

However, unlike extremes on land, our knowledge about compound extremes in the ocean is relatively limited, especially how they manifest themselves, propagate, and interact at depth, and this is due to lack of relevant observations. It is therefore far less understood how marine extremes might lead to tipping points, although the potential for such interactions might be just as large as over land.

3.3 Combination of Extremes in the Ocean and Links to Tipping Elements

Different combinations of extremes, whether preconditioned or multivariate, spatially or temporally compounding, have distinct characteristics, such as temporal and spatial scales of occurrence, regional and/or seasonal development and thus may interact differently with tipping elements. Below we outline some of the most common compound extremes and their characteristics while their specific impacts are addressed in Sect. 3.4.

3.3.1 Marine Heatwaves and High Ocean Acidity Extremes (MHW-OAX)

Compound events of marine heatwaves and high ocean acidity extremes (MHW-OAX) are just starting to receive more attention (Burger et al. 2022). Based on remotely-sensed observations, surface MHW-OAX have been found to occur most likely in the subtropics and less likely in the equatorial Pacific and high latitudes, largely driven by long-term warming and acidification trends (Burger et al. 2022). The Arctic Ocean specifically represents an area where extreme warming due to Arctic amplification and CO_2 uptake lead to faster acidification changes that greatly impact ecosystem diversity and production (Jewett and Romanou 2017). However, MHW-OAX events are hard to fully characterize observationally, in terms of duration, intensity, areal coverage, and frequency, because there are no synoptic, consistent, continuous, daily and adequately long records of ocean acidification as there are for SST. Particularly challenging is the detection of extremes below the surface due to the lack of subsurface observations.

On the other hand, climate models allow detection and analysis of MHW-OAX events at the surface and at depth, from high temporal frequency output, an example shown in Fig. 7

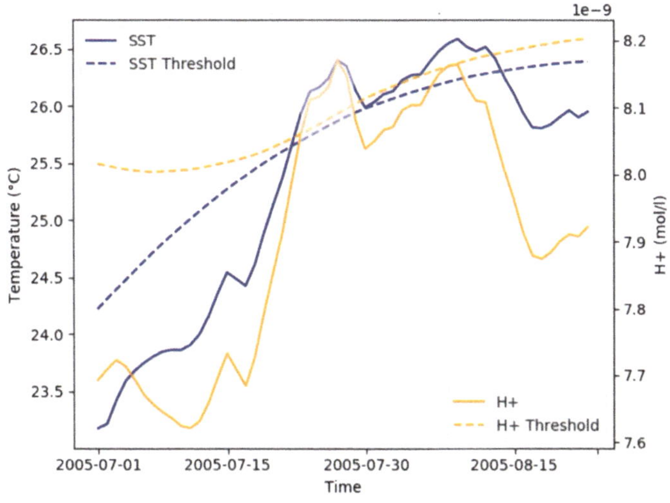

Fig. 7 NASA-GISS modelE historical simulations: an example of a compound MHW (left axis) and OAX (right axis) event in the North Atlantic (highlighted in gray). Thresholds were computed using the 90-th percentile above a present-day baseline

using the NASA-GISS climate model. Burger et al. (2022) in a similar model study showed that under 2 °C warming and associated acidification trends, surface MHW-OAX compound extremes, when computed relative to a fixed preindustrial baseline will last about 265 days per year, in contrast to just 12 days in preindustrial times.

While most studies focus on open ocean MHWs, nearshore extremes are more important for changes in the biogeochemistry and ecology of coastal waters which sustain not just commercial and recreational activities for nearby communities but also fisheries and aquacultures and therefore the food security and economy for large parts of the population. Nevertheless, it is understood that marine heat extremes near the coast, in enclosed bays and harbors, follow more closely the atmospheric variability than their open ocean counterparts (e.g., Cook et al 2022a, b and references therein). Similarly, coastal acidification has much larger variability than in the open ocean and therefore potentially stronger extremes; and is relevant for coastal ecosystems, including their tipping. Freshwater inputs there change seawater chemistry and make it more susceptible to acidification, by changing salinity, dissolved inorganic carbon, total alkalinity, dissolved and particulate organic carbon, and nutrients from riverine, estuarine and glacial melt sources. While these processes have persisted historically, climate-induced increases in glacial melt and high-intensity precipitation events can yield larger freshwater inputs than have occurred in the past. Nutrient runoff can increase coastal acidification by promoting phytoplankton blooms, which, when they die, fuel the bacterial respiration of CO_2. These bacterial blooms can result in extreme acidity events whose intensity depends on local hydrographic conditions, including water column stratification and residence time. Long-term changes in nutrient loading, precipitation, and/or ice melt may also impart long-term changes in the magnitude of coastal acidification. Some high-latitude surface and near-surface waters already experience such corrosive conditions (Sutton et al. 2016; Bates and Mathis 2009; Qi et al. 2017).

This type of compound extreme happens near the coast and near the surface and thus may affect shallow corals and other ecosystems that cannot easily adapt to temperature and acidity changes, and at the same time affect aquacultures and fisheries. Because acidification rises quickly in coastal areas it might be the leading compound extreme for most of the ocean but more nearshore observations will be needed to understand the relationship between MWH-OAX and ecosystem tipping.

3.3.2 Marine Heatwaves and Ocean Deoxygenation Extremes (MHW-LOX)

Warming decreases the oceanic uptake of oxygen from the atmosphere, reduces ventilation of the ocean interior, and also increases the metabolic rate of marine organisms, and hence, their oxygen demand (Deutsch et al. 2015). Quantifying the combined indices of warming and deoxygenation, is done for example via the "metabolic index", which represents the relationship between oxygen supply and demand at a standard metabolic rate (Deutsch et al. 2020) and the "aerobic growth index", which uses growth theory, metabolic theory and biogeography to calculate a theoretical oxygen supply to demand ratio (Clarke et al. 2021).The long-term expansion of low-oxygen waters (Stramma et al. 2008; 2012) and increased metabolic oxygen demand have been found to result in habitat loss of heterotrophic organisms (Breitburg et al. 2018) and reduction of biodiversity, which may culminate in a mass extinction similar to those in Earth's past (Penn and Deutsch 2022).

As the background temperatures and O_2 levels change, O_2-dependent biological processes become extremely susceptible to short-term MHW-LOX compound extremes (Köhn et al. 2022), and may lead to ecosystem tipping when the exposure to extreme conditions go beyond the realized niche limits (Trisos et al. 2020; Harris et al. 2018). Extinction risk is greatest where decadal climate anomalies are strong, or species are living close to their ecophysiological thresholds. Currently, the detection of open ocean LOX is hindered by the lack of O_2 observational data with sufficient spatial and temporal coverage (Grégoire et al. 2021), and is generally carried out in model-data based studies (Gruber et al. 2021; Köhn et al. 2022). The combined effect of warming and deoxygenation on the metabolism and survival of tropical sea urchins and their exposure to compound extreme conditions occurring in their coral reef habitats has been shown to severely affect their survival (Lucey et al. 2022). While these species are able to withstand the extremes of heat or low oxygen separately, their combined effect leads to severe depletion of the population (Lucey et al. 2022). Also, the loss of coral reefs (Altieri et al. 2021) or fauna over seamounts (Ross et al., 2020) has also been attributed to deoxygenation as a cofactor.

3.3.3 Marine Heatwaves and Low Chlorophyll or Low Primary Production Extremes (MHW-LChlX, MHW-NPPX)

MHWs have been shown to lead to reductions in chlorophyll concentrations in the tropics and mid-latitudes and increases in high latitudes (Noh et al. 2022). However, coincident MHW and extremely low chlorophyll (LChlX, Le Grix et al. 2021) hotspots were found in the equatorial Pacific (Hughes et al. 2018), along the boundaries of the subtropical gyres, and in the Arabian Sea and the western Pacific Ocean (Chen et al. 2023).

The Blob was also associated with large negative anomalies in net primary production (NPP; Whitney et al., 2015; Mogen et al. 2022). Le Grix et al. 2023 assessed the characteristics of MHW-NPPX extremes in models and observations and showed that there are hotspots of frequent and long-lasting (> 10 days/yr) events in the equatorial Pacific and in the subtropical Indian Ocean, mostly associated with enhanced nutrient limitation on phytoplankton growth. In the northern high latitudes, they found much shorter events (< 3 days/year) due to enhanced light limitation, whereas large uncertainty between observational NPP products complicated the characterization of events in the Southern Ocean. Such extremes in low ocean productivity can be detrimental to upper trophic chains, e.g., pelagic fish species (Le Grix et al. 2022).

3.3.4 Other Marine Compound Extremes

Low frequency climate variability and air-sea interactions can spawn other combinations of extremes. MHWs in the central tropical Pacific have been observed to co-occur with extreme minima in sea surface salinity during intense precipitation events associated with El Nino (Liu et al. 2023a, b). However, low salinities have been associated with better adaptation to higher levels of acidity, especially in coastal regions (Xu et al. 2022; Li et al. 2022; Röthig et al. 2023) which might alleviate pressures from acidification or acidity extremes.

The co-occurrence of hypoxia and low pH, OAX-LOX, is also likely to occur particularly in shallow habitats, as microbial respiration consumes oxygen and produces CO_2, with detrimental effects for coral survival and contributing, together with temperature anomalies, to seagrass mass mortality events (Altieri et al. 2021). Triple compound events, e.g., MHW-OAX-LOX, are also known to occur. In fact, the MHW known as the *Blob* has been associated with anomalously low oxygen, low pH, (Gruber et al. 2021), resulting in large negative anomalies in phytoplankton NPP (LeGrix et al. 2021; Morgen et al. 2022), as well as shifts in species distribution (Cavole et al. 2016; Cheung and Frölicher 2020) and severe consequences for marine life (Jones et al. 2018; Piatt et al. 2020).

These combinations of extremes are threatening for corals and other ecosystems as well as societal tipping points (collapse of fisheries etc.). For more details see Sect. 3.4.

Furthermore, SROCC (2019) has identified multiple cyclones as a kind of compound extreme, even though they happen sequentially, with potentially severe impacts on land and society. In that sense, sequences of MHW and other extreme events may also be identified as compound extremes. Such prolonged (or sequential) MHWs have been linked to the dynamics of the Pacific Decadal Oscillation (PDO) and the North Pacific Gyre Oscillation (NPGO) which are the two dominant modes of winter sea surface temperature variability in the North Pacific (Joh and DiLorenzo 2017).

3.3.5 Cross-Boundary, Atmosphere–Ocean and Land–Ocean Compound Extremes

Other compound extremes that affect the open-ocean and especially the coastal marine environment might occur over land or in the atmosphere. Examples include series of tropical cyclones in the Atlantic or the Pacific (Yaddanapudi et al. 2022; Swiss Re 2020) or storm surge-precipitation (Wahl et al. 2015; Wu et al. 2018), sea-level rise—storm surge—fluvial flooding (Ganguli and Merz, 2019a, 2019b; Moftakhari et al. 2017), and wind-precipitation events at a regional scale (Huang et al. 2021; Owen et al. 2021).

The *Blob* in the northeast Pacific was also concurrent with a record-breaking drought in California (Griffin and Anchukaitis 2014) that devastated the ecosystem there (see Sect. 3.4 below). Shi et al. (2021) identified similar events in the historical record that took place once in 1917–1918 and once in 1959–1960 but were not as extensive as the 2013–2015 event. Using CMIP6 models they showed that climate change will exacerbate such extremes by 2100.

Cross-boundary compound extremes include also compound flooding events due to extreme sea level rise and storm surge (Hoeke et al. 2013) or compound extreme river discharge and storm surge (Heinrich et al. 2022; Khanal et al. 2019; Couasnon et al. 2020). MHW coupled with extreme sea levels were found to be caused by greenhouse gas induced warming as well as wind decadal climate variability, associated with negative Indian Ocean Dipole (IOD) and a co-occurring La Nina that weakened the coastal upwelling (Han et al. 2022). Compound MHW with increased wind stress that drove enhanced southward transport in the Eastern Australia current may have led to the 2015–2016 Tasman Sea strong MHW that lasted for 251 days (Oliver et al. 2017). Cyclones, fronts and thunderstorms are associated with compound extremes in precipitation and winds, or gustiness and extreme ocean waves (Messmer and Simmonds 2021; Catto and Dowdy 2021). These are mostly found in North America, Japan, the Mediterranean, Australia and regions with high tropical cyclone occurrence. Cyclones associated with compound extremes were found to last longer and wind extremes were exacerbated in the presence of precipitation extremes. Compound low Arctic sea ice and anomalous SSTs can lead to anomalous jet stream meanderings that may have been responsible for the extreme cold spells over North America in 2015 (Bellprat et al., 2016).

In summary, the distinction between land and ocean extremes, single or compound, misses some interlinked extremes, whose consequences can ripple through the Earth system and contribute to long-term change.

3.4 Tipping Points Potentially Triggered by Marine Extremes

3.4.1 Tipping in the Physical System

Compound marine or cross-boundary extremes can have impacts on the regional climate. For example, the Tasman Sea MHW of 2015–2016 led to drought followed by intense rainfall which caused severe bushfires and flooding in northeast Tasmania. Similarly in 2017 in coastal Peru, strong shallow ocean warming that reached 10 °C caused heavy rainfall and flooding (Garreaud 2018). MHWs have also been linked to decreases in sea ice (Laufkötter et al. 2020 and references therein) and could trigger local weather effects that modulate internal variability and therefore may affect local or even regional climate. Sea ice extremes may lead to tipping of the AMOC with impacts to global climate (Fig. 5A, Romanou et al 2023). MHWs have also been linked to changes in the Indian Summer monsoon (Saranya et al. 2022).

While there are few examples of extremes leading to changes in the physical climate, such impacts might be more ubiquitous than presently known and more possible occurrences should be explored in improved and expanded (in areal coverage and spatiotemporal resolution) observational records and in models.

3.4.2 Tipping of Ecosystems, Ecosystem Services and the Blue Economy

Some impacts of isolated extremes on ecosystems and ecosystem services have been well documented (e.g., Smith et al. 2023 for MHWs) and show that extremes can have more significant and more immediate consequences compared to the impacts of the long-term trends due to climate change. Additionally, the ensuing environmental degradation can lead to decline in the quality of food, water and other resources, such as wood, oil and medicinal plants, cultural and commercial importance of water resources (Techera and Winter 2019) and eventually the blue economy, as defined by World Bank (https://www.worldbank.org/en/programs/problue/overview): "the sustainable use of marine resources for economic growth, improved livelihoods, and jobs while preserving the health of ocean ecosystem". Therefore, extremes may be able to lead to tipping of global or regional elements before climate change does.

MHWs cause biomass decrease and shifts in biogeography of fish stocks that are at least four times faster and bigger in magnitude than the effects of decadal scale mean changes throughout the twenty-first century (Cheung and Frölicher 2020). They may also bring on increased mortality of sea birds, salmon, marine mammals, low ocean primary productivity, harmful algal blooms, coral bleaching, defoliation of sea grass and decrease in kelp biomass, changes in biodiversity patterns of sessile invertebrates and demersal fish (although this is debated, Fredston et al. 2023), shifts in community structure and poleward shifts in tropical fish communities, closure of fisheries, locally large carbon outgassing events (Arias-Ortiz 2018). MHWs affect physiological adaptation and cause population and ecosystem changes (Smale et al. 2019; Almeida et al 2024) that are often dependent on the life stage (Dahlke et al. 2020), the organisms' physiology, on population and ecological dynamics and on the co-occurrence with other stressors. On short-term scales, the impact of extremes depends on the plasticity of the organisms (i.e., the physiological mechanisms that permit individual organisms to acclimatize to cope with environmental stress). Over longer timescales, there may be irreversible changes in ecosystem structure and composition (Smale et al. 2019; Straub et al. 2019), as for example, in communities of seagrass (Thomson et al. 2015), coral reefs (Hughes et al. 2019, 2020; Darling et al. 2013); micronekton (Brodeur et al. 2019) and kelp forests (Wernberg et al. 2021; Filbee-Dexter 2020; Michaud et al. 2022; Bell et al. 2023; Tait et al. 2021; Arafeh-Dalmau et al. 2019). Several studies have pointed out that it is often more detrimental for organisms, e.g., corals, when multiple MHWs act simultaneously (Smith et al. 2019; Anthony et al. 2008).

On the other hand, both long-term ocean acidification and extremes of pH (especially in estuaries and coastal waters) reduce the rate of calcification, resulting in decreased ability of marine organisms, such as corals, plankton, and shellfish, to build their shells and skeletal structures, as well as decreased reproduction, disease and death. This can result in ecosystem shifts and loss of the structural complexity and biodiversity, exacerbate existing physiological stresses and can affect the growth and survival, respiration, and reproduction of many non-calcifying species, potentially impacting photosynthesis and primary production from benthic species and corals up to fish (Gao et al. 2019; Tassone et al 2022).

Ocean acidity and warming compound extremes can have synergistic or antagonistic effects although some adaptation might take place. While warming might increase some behavioral responses (such as swimming speed), ocean acidification tends to restrict them (Baag and Mandal 2022). Little is known about impacts on the reproductive ability of many organisms, like copepods, reef fishes, squid, garfish, clams, sea urchins, oysters, the Atlantic cod, the Pacific herring (Baag and Mandal 2022 and references therein). Shell, skeletal and muscle deformities as well as other physiological changes are known to affect swimming behavior and feeding rate. Pathogen growth, disease spreading and parasite transmission is changing under climate change and thermal and acidification stress, as shown for lobsters, gastropods, Atlantic Halibut, and other species (Doney et al 2020 and references therein).

Such impacts will affect both fisheries but also aquaculture with enormous consequences and socio-economic implications on commercially important fisheries and threaten food security, livelihood, and employment to local but also regional communities.

Corals are particularly vulnerable species to MHW-OAX compound extremes. Over the past decades large-scale coral bleaching events in the tropics have become common (Baker et al. 2008; Spalding and Brown 2015; IPCC, Bindoff et al. 2019; Fordyce et al. 2019; Klein et al. 2022) and are linked to the occurrence of MHW and compound events, particularly. The decline in coral production affects the global ecosystem, with biodiversity reduction and collapse of ecosystem services such as food production and coastal protection and socioeconomic impacts (Brander et al. 2012; Robinson et al 2019).

Combined MHW and LOX events will lead to decrease of habitat viability globally (Kim et al. 2023), but will affect different species differently, and may even lead to extinction (Penn and Deutsch 2022). For example, even short duration (~ 1 day) events can be catastrophic for sea urchins (Lucey et al. 2022). Under deoxygenation stresses, most species will lose about 5% of their habitat at 2 °C warming levels (above preindustrial temperatures), some species will lose double or triple that. Deoxygenation is particularly important for mesopelagic and demersal habitats. However large uncertainties exist in the vulnerability and species-specific biophysical thresholds response (Moree et al. 2023).

Extreme changes in ocean stratification and sea ice extent (as seen in Blanchard-Wrigglesworth et al. 2021; Holland et al. 2008) in the subpolar gyre (SPG) can lead to SPG mixing and AMOC collapse which in turn will cause reduction or cessation of deep-water formation and eventually primary production collapse (Schmittner 2005) and deoxygenation (Schmittner et al. 2007; Schmittner and Galbraith 2008).

Prominent compound extreme events such as the *Blob* have had severe impacts on marine life (Cavole et al., 2016), including extreme mortality and reproductive failure of sea birds across the north-west coast of the US (Jones et al. 2018; Piatt et al. 2020), persistent harmful algal blooms, mass stranding of whales in the western Gulf of Alaska and of sea lions in California, and replacements with warm-water species (Cavole et al. 2016; Cheung and Frölicher 2020). Additionally, there were devastating economic and ecological consequences in the terrestrial realm as well (e.g., Lund et al. 2018; Peterson et al. 2017; Santora et al. 2020). Drought led to groundwater overdraft, land subsidence, loss of domestic water supply in the Central Valley (Bee 2015), and multi-billion-dollar agricultural losses from reduced farmland productivity and additional groundwater pumping costs (Howitt et al. 2014, 2015; Medellín-Azuara et al. 2016). This severe and extended drought killed more than 100 million trees, exacerbated the devastating 2015 fire season, and remains a fuel hazard that continues to threaten public safety presently. The economic loss was enormous. For example, about $170 million losses were due to the closure of the Dungeness crab fishery alone (McCabe et al. 2016) while ~$120 million was allocated for drought emergency ecosystem support from state and federal sources (Hanak et al. 2015).

Gruber et al. (2021) point out that deep extensions of compound extremes below the ocean surface will lead to additional habitat compression, and we need to better understand subsurface signatures of the extremes to improve preparedness and reduce disaster risk, including that of tipping points linked to compound ocean extremes discussed above.

In conclusion, marine extremes could have a strong role in short-term but also long-term shifts in the Earth System, and perhaps approaching or exceeding tipping points with potentially serious consequences for marine ecosystem services.

4 Palaeoclimatic Evidence of Link between Tipping Points and Extreme Events

Risk of tipping points is supported by palaeoclimatic evidence, and mechanistic links suggest a role for extremes, although particularly weather extremes have generally not been directly evidenced in palaeoclimatic literature. In contrast, climate events (as defined above) have seasonal signatures and thus can be resolved in some archives.

Drought atlases over the common era (Cook et al. 2022a, b) can begin to shed light on the largest possible seasonal events, and their links with tipping points such as vegetation shifts, and impacts on societies. Megadroughts over the Common Era, with multi-decadal to longer and greater spatial extent than the twentieth century, are documented in records, such as historical documents, tree rings, archaeological remains, lake sediment, and geomorphic data, for the central U.S. (Woodhouse and Overpeck 1998) and southwestern North America (Williams et al. 2020). Fire data are also available across interglacial and recent periods and may help to identify possible links between long-term change and fire. Paleoecological records provide a link of climate, vegetation, and past fires. Frequent fires in boreal Siberia during the late Holocene led to a shift in vegetation to fire-adapted forests (Feurdean et al. 2020). There are also reconstructions of severe storm activity resulting in flooding. A multi-proxy record from central Texas indicated an abrupt and rapid increase in mesoscale convective system activity and the establishment of overall wetter conditions at ~ 14,500 years ago, albeit associated with orbital forcing and deglacial conditions (Sun et al. 2021). Variability in regional intense hurricane activity over the Common era has been detected in lacustrine sediments (Brandon et al. 2013; Rodysil et al. 2020).

Over ocean, corals can provide annual to seasonal resolution and thus resolve climate timescale of extreme events, while other proxies provide less temporal resolution. On the other hand, studies such as Hillyer et al. (2017) and Roach et al. (2021) employ high-resolution sampling of coral skeletons to detect metabolomic changes associated with thermal stress and bleaching, which may help capture signatures of short-term events like MHWs in historical coral records; but also importantly directly document impacts of extremes. This is also the case for tree ring data which record stress on vegetation, something that could be exploited for characterizing impact of extreme events. Advancements in high-resolution sampling, multiproxy approaches, and climate modeling may provide some pathways to infer the occurrence and impacts of these events, although short-term MHWs may remain elusive. For the discussion following below it is an open question if extremes played a role in rapid transitions and one that would be helpful to explore, either from high resolution data or model simulations in order to better enable early warning.

Paleoclimate records also suggest a strongly nonlinear, complex climate system, and reveal instances where there have been abrupt changes in the state or modes of variability of components of the climate system (Rahmstorf 2001; Lockwood 2001; Alley et al. 2003; Rial et al. 2004). These previous shifts point to potential tipping elements of the climate system, are not able to provide, and climate modeling of these transitions as well as analysis of high resolution archives may elucidate the drivers behind these shifts. There are a few candidate periods for doing so:

During the African humid period (AHP) approximately 14,700 to 6,000 years ago, northern Africa experienced a wetter environment in which the hyperarid Sahara was transformed into a mesic landscape, with a northward shift of tropical vegetation by approximately 6–9° latitude compared to the present-day location, large permanent lakes, and extensive river drainage networks (Drake and Bristow 2006; Lézine et al. 2011; Hély et al. 2014; Tierney et al. 2017, Pausata et al. 2020). This favorable climate was influenced by a slightly increased tilt of the Earth's spin axis and perihelion in July, resulting in stronger insolation in the Northern Hemisphere during the summer, which in turn strengthened the West African summer Monsoon. Moreover, increasing greenhouse gas concentrations at the start of the AHP (~ 14,700 years ago) could have further enhanced precipitation in northern Africa (Otto-Bliesner et al., Science 2014). Climate records into and out of the AHP suggest that the transitions were relatively rapid in some regions (deMenocal et al. 2000; Adkins et al. 2006; McGee et al. 2013; Tierney et al. 2017), while they were more synchronized with orbital forcing in other regions (Niedermeyer et al. 2010). New transient climate simulations show spatially heterogeneous vegetation dynamics in northern Africa, with the fastest increase in simulated desert fraction—up to 7 times faster than orbital forcing—occurring around 20°N in the western part and around 14°N in the central and eastern part (Dallmeyer et al. 2020), while the transition was more gradual in other regions in synch with orbital forcing (Shanahan et al. 2015; see also Kutzbach and Guetter 1986). In conclusion, the AHP transition of the Sahara was regionally rapid, suggesting the crossing of a tipping point with respect to changes in the driving forcing. However, when viewed within the context of human lifetimes and local ecosystems, this transition unfolded gradually, although it may have come in the shape of a changing extremes regime, similar to how we experience climate change today. The AHP transition, whether rapid or gradual, brought about enduring and significant transformations. These changes acted as a pivotal moment in human civilizations, giving rise to new cultures along the Nile.

The paleoclimatic record can also inform on the impact of strong high latitude warming on vegetation and the polar ice sheets. A strong northward shift of the boreal vegetation over Canada and Eurasia, combined with forest loss in Southern regions, has been reconstructed in past warm periods, such as the Eemian, although with significant uncertainty on the extent of this retreat, and its spatial homogeneity (e.g., Otto-Bliesner et al. 2020; Thomas et al. 2020 and references within). Presently, changes in the boreal forest toward deciduous vegetation are small (Massey et al. 2023) with difficulty in assessing its resilience using EO data (Smith and Boers 2023). Genetic evidence from a circum-Antarctic octopus indicates that the West Antarctic Ice Sheet (WAIS) WAIS collapsed completely during the Last Interglacial period, when global average temperature was like that of today (Lau et al. 2023). Identifying the precise tipping point of the WAIS during the Last Interglacial, the triggers, and its connection to extreme events are still outstanding questions.

5 Observing, Understanding and Predicting Risk of Extreme Event Related Tipping

5.1 Earth Observation Monitoring Capabilities

When we approach a tipping point, we expect that the climate variability will increase (Scheffer et al. 2009), thus also making some extreme events more likely. A careful monitoring of extreme events could thus help identify situations in which we are close to a critical threshold (see companion paper by Bathiany et al. 2024). When applying this method, it is important to fully consider uncertainty in observations, where changes in the observing system or data availability can also lead to changes in variance; and to fully consider climate variability and decadal drivers of it. Further evidence on tipping risk includes monitoring the ecological or physical health or degradation of a system at risk of tipping.

Earth observations can capture *extreme weather and climate events*. Land surface temperature is recorded by MODIS (Moderate Resolution Imaging Spectroradiometer) on NASA's Terra and Aqua satellites, and by SLSTR (Sea and Land Surface Temperature Radiometer) on Copernicus Sentinel-3 (Nickeson 2023; Yu et al. 2022a, b.; and Good et al. 2022). These can be combined with reanalysis products that date back to the nineteenth century (Slivinski et al. 2019). Reanalyses also capture past extreme heat reasonably well (e.g., Cowan et al. 2020a), and together with satellite data enable identifying extremes in remote or under-recorded regions (e.g. Engdaw et al. 2023). There are offsets between data sources, and retrieval biases, yet identified extreme events show good agreement between reanalysis and satellite data (e.g., Hegedűs et al. 2024). However, Observational products like OISSTv2 and other higher level satellite products have coarse resolution near the coast and thus fail to reproduce observed MHWs in the estuaries and the land–ocean continuum (Cook et al. 2022a, b). This highlights the need for long-term in situ measurements as well as methods to improve the algorithms that produce the satellite effective estimates near the coast.

For precipitation, EO data are useful particularly for remote regions and over the ocean. Blended satellite/in situ records (GPCP; Adler et al. 2018) record global precipitation, although high latitude precipitation retrieval is affected by ice and snow. Precipitation deficit is one possible definition of drought, although accounting for evaporation in P-E or the Palmer drought severity index captures stress on vegetation much better. The latter also shows substantial increase (IPCC 2021; Seneviratne et al. 2021) partly due to added evaporation by global warming.

Multiple satellite-based datasets are available for *monitoring vegetation and fire* (e.g., Jiao et al. 2021), over multiple decades, as AVHRR, Landsat and MODIS, or more recently

Sentinel-2 and Sentinel-3 (Lastovicka et al. 2020; Li et al. 2024). Satellite-based information about vegetation condition can be derived from vegetation indices such as the normalized difference vegetation index (NDVI Tucker 1979), a proxy for leaf area and productivity; normalized difference water index (NDWI Gao 1996), a proxy for drought stress, and passive microwave vegetation optical depth (VOD, Liu et al., 2011), which responds to combined biomass and vegetation water content. NDVI may be problematic over dense vegetation due to saturation effects (Smith et al. 2022). Solar-induced chlorophyll fluorescence is a more recently available satellite-based proxy for photosynthesis (Sun et al. 2023), although it's use to monitor extreme event impacts is still under debate (Martini et al. 2022).

There are also remotely sensed data for active fires and fire radiative power (Xu and Wooster 2023) from Sentinel-3, for normalized burn area (Chuvieco et al. 2018), burn area and fire intensity (Giglio et al. 2016). Passive optical satellite retrievals are affected by cloud cover, which introduces a bias compared to surface air temperatures (e.g. Nickeson 2023). All of these are able, with careful analysis, to evaluate to what extent vegetation recovers and with what delay, and how severe changes are. This applies to all extreme events affecting vegetation, from heat, drought and fire to severe storms, and other vegetation disturbance. For an in-depth discussion of remote-sensing based assessments of vegetation resilience, see Bathiany et al. 2024.

MHW are identified primarily using SST from satellite thermal infrared measurements (e.g. Advanced Very High Resolution Radiometer (AVHRR) on NOAA Polar-orbiting Operational Environmental Satellites (POES), Along-Track Scanning Radiometer (ATSR) aboard the European Remote Sensing Satellite (ERS-2), the Geostationary Operational Environmental Satellite (GOES) Imager, and Moderate Resolution Imaging Spectroradiometer (MODIS) aboard NASA Earth Observing System (EOS) Terra and Aqua satellites, VIIRS on Suomi NPP) or passive microwave measurements (e.g. the Scanning Multichannel Microwave Radiometer (SMMR) carried on Nimbus-7 and Seasat satellites, the Tropical Rainfall Measuring Mission (TRMM) Microwave Imager (TMI), and the Advanced Microwave Scanning Radiometer (AMSR) instrument on the NASA EOS Aqua satellite and on the Japanese Advanced Earth Observing Satellite (ADEOS II), Sea and Land Surface Temperature Radiometer (SLSTR) on Sentinel-3). Infrared radiation of the ocean comes from the top 10 microns of the surface while microwave radiation results from the topmost 1-mm layer. Infrared satellite sensors have better spatial resolution but are more susceptible to cloud contamination than microwave. Today in addition to satellite and shipboard measurements there are thousands of floats in the oceans measuring temperature and salinity and more recently key biogeochemical variables (e.g. the ARGO floats). These are used to validate satellite instruments in addition to sampling throughout the water column. The surface drifters from the NOAA Global Drifter Program (GDP) regularly provide about 60,000 night time SST measurements per month at a shallower depth of 0.2 m, while the Group for High Resolution Sea Surface Temperature (GHRSST) project provides all SST data sets in a common format that allows for easy accessibility across different computer platforms and operating systems and to make it useful to the modeling community also provides a full characterization of the errors associated with each pixel.

The use of satellite-based observations can provide global gap-free SST datasets at high spatial and temporal resolution, typically covering the period from 1979 to present (Yang et al. 2021). Hobday et al. (2016) was the first to define a set of quantitative globally valid metrics to detect and classify MHW events. Since that work, studies on MHWs have been increasing. There is a better understanding (Holbrook et al. 2019; Oliver et al. 2021), although still not complete, of the sensitivity of the MHW detection methods, and the physical processes that drive temperature changes in the mixed layer, leading to the build-up, persistence, and decay of MHWs, and how local internal variability (e.g., eddy instabilities) and large-scale climate variability possibly modulate the severity of these events locally (Collins et al. 2019; Frölicher and Laufkötter 2018; Jacox 2019). Yet, mechanisms leading to subsurface MHW are still poorly constrained. Furthermore, studies investigating the impacts of MHWs on atmospheric circulation and the rest of the physical system are very limited.

Aside from SST, satellites measure sea surface salinity (SSS) and sea level rise (SLR) through the recent advances with ESA's SMOS and NASA's Aquarius and SMAP missions—see Vinogradova et al. (2019) for salinity and the Gravity Recovery and Climate Experiment (GRACE) and Jason – 1 and – 3 missions for SLR.

Satellite-based estimates of chlorophyll and primary productivity come from the Moderate Resolution Imaging Spectroradiometer (MODIS) on NASA's Aqua satellite and ESA's Ocean and Land Colour Instrument (OLCI) on Sentinel-3. However, we lack similar quality, synoptic information for oxygen and H+ concentrations in the ocean, as well as consistent and continuous monitoring of the subsurface (and surface) ocean.

For large scale and regional tipping elements we need consistent and continuous monitoring of the higher spatial and temporal scales to be able to better assess extent, frequency, duration and intensity as well as deep signatures of marine extremes (see Wood et al. 2024).

Emphasis should be placed on global (including high latitudes), high frequency measurements, from satellite missions, e.g. through the anticipated launch of the Plankton, Aerosol, Cloud and Ecosystem (PACE) mission, and the new mission to address submesoscale e.g. the Surface Water and Ocean Topography (SWOT), as well as suborbital platforms, such as planes, ships, moorings, transects, more bioArgo, but also instruments that can capture the development of extreme events e.g. gliders, AUVs through the Lagrangian rather than the Eulerian view of the flow (Gruber et al. 2021). Since the subsurface ocean but also the overlying state of the atmosphere are of critical importance to our understanding of how physical and biogeochemical extremes grow and interact and how they might lead to tipping points, Ocean Super Sites (Clayson et al. 2021) will provide high temporal monitoring to possibly capture early warning signals for tipping points (as described in Scheffer et al. 2009).

An important point that needs to be made abundantly clear (as shown and discussed in Burger et al 2022) is that the choice of baseline is critical in impact studies. Simply a historical baseline in the determination of an extreme event might be sufficient for corals (i.e. immovable species, even though it still does not take into account adaptation capabilities) but still it is not appropriate for assessing impacts on fish (i.e. species that can migrate and change habitat). Even in this case, however, loss of habitat and biodiversity are phenomena that can be illustrated using a constant baseline.

5.2 Climate Modeling and Machine Learning Perspective

Understanding mechanistic connections between extreme events and long-term impacts on the Earth System can also draw on Earth System Models (ESMs), which are scoped to include not only the climate system but also its interaction with the global carbon cycle, terrestrial vegetation, and marine biogeochemical cycles. However, not all modeling groups perform ESM scenario simulations with all interactions switched on. For example, while several models used in AR6 chapter 5 included fire (Table 5.4, Canadell et al.

2021), only 2 models allowed for interactive vegetation and hence a response of vegetation to fire. The potential consequences of fire and drought are therefore likely underestimated, or not captured, in past generations of ESMs. It is also important to evaluate and constrain model responses with observations, as uncertainties are substantial (for example, due to inadequate model complexity; or model ability to capture diverse vegetation/climate systems). Comparisons, for example, between fire models and data indicate underestimated fire responses (e.g., Sanderson and Fisher 2020 for Australia). Major uncertainties exist around ignition and vegetation response (Teckentrup et al. 2019), and the few most complete fire models appear to underestimate the size of fires, which could lead to underestimation of the effect of large extremes and fire (Hantson et al. 2020). EO will provide a vital tool to evaluate ESMs with coupling switched on and determine if they realistically estimate risk from large extreme events and associated fire. Similar issues also arise for other tipping points, such as those associated with ice sheets. Ice sheet models still largely run offline rather than fully coupled and hence will not 'see' extreme warm events or warm ocean incursions, while also being unable to influence ocean dynamics with melt water (Schmidt et al. 2023; Roach et al. 2023). On the ocean side, not all CMIP6 models include fully prognostic and interactive dust or river runoff while subdaily output simulations typically do not cover the breadth of each model's internal variability.

The move toward high resolution has the potential to improve the link between systems due to better resolving the scales required for coupling (e.g., Lee and Hohenegger 2024), although that is not always the case (Ridder et al. 2021). The high computation requirements also affect the potential to sample extreme events in shorter runs, and to fully account for irreversibility.

Ocean modeling, particularly Lagrangian back-trajectory analysis (Holbrook et al. 2020) and data-assimilation of the sparse in situ data, but also forward modeling (climate or process modeling) can provide invaluable insights on the onset, persistence, decay, spatial extent, depth, and intensity as well as future projections of the extremes, their drivers (Burger and Frölicher 2023), and their interactions with the tipping elements.

Very importantly, climate model skill on representing compound extremes needs to be analyzed using multivariate bias-assessment framework, and not singular and sequential individual parameter evaluation, in order to reliably depict the models' capability in assessing risk due to compound extremes (Villalobos-Herrera et al 2021).

In that context, machine learning (ML) enables bridging scales between processes on very fine resolutions, such as ice sheet melting, and larger scale models and also in detecting links between extreme events and tipping. As changes in vegetation or biosphere are generally not mono-causal, AI causality methods can also be used to determine driving factors and responses and identify the most important mechanisms (e.g., Runge et al. 2019).

Furthermore, ML helps in feature detection (Gruber et al. 2021), including space–time coherent extreme events (e.g. Hegedus et al., 2024), identification of precursors, improved prediction (Holbrook et al. 2020), downscaling and bias correction that leads to better study of extremes and their impacts. However, ML limitations include the non-stationarity of the climate solutions, potential for unphysical outcomes, difficulties in uncertainty quantification, and importantly the limited sampling of extremes in observational data due to sparsity of data and short records. In other words, ML can only be as good as the best observational and model data we have, and needs to be accompanied by understanding of mechanisms. To improve predictability of extremes, as well as efficient detection and attribution of the compound extremes, we will need to overcome observational limitations which can be done via emulation of climate predictions at ultra-high resolution (1 km or less) (Molina et al. 2023).

Overall, the large uncertainty in the carbon cycle response to increased warming affects our understanding of the remaining carbon budget. As extremes and tipping points are likely important sources of this uncertainty, better monitoring and simulating of the cascading effect of extremes and their possible impact on tipping points is required to address this uncertainty, and the reversibility of changes associated with extreme events.

6 Conclusions and Outlook

6.1 Summary and Discussion

The importance of extremes, and especially compound extremes, in inducing tipping points is evident from the discussion of both marine and land systems. Additionally, if extremes are defined relative to preindustrial or even late historical conditions, they are expected to become near permanent in large parts of the world even under high mitigation scenarios of greenhouse gases (Batibeniz et al 2023). Efforts should be placed to address cross-system (land–ocean-ice-atmosphere) compounding extremes, e.g. extreme precipitation over land and/or river discharge or wind spells coupled to MHWs, etc. Furthermore, governance and economic markets may create links between global extremes and complicate (exacerbate or alleviate) physical and biogeochemical extremes (Raymond et al 2020). In this context, Earth observations could be of high value to monitoring extremes as well as many of these interactions and tipping points and are therefore a vital tool for early detection. Detection should make use of early warning systems, which could be based on remotely sensed extreme land and ocean surface temperatures, observations of vegetation or crop stress or dieback indicators etc. However, new machine learning and AI techniques may be more successful in identifying early warning systems. Climate prediction systems can provide early warning of atmospheric and ocean variability, which can enhance the probability of individual and compound extremes.

While we might have to reconsider the climate risk assessment framework and target plausibility and not only probability of extremes (Sillmann et al 2021), a potential roadmap for research at the interface between extreme events and tipping points would include the following elements:

- Foster exchanges and the development of a community working at the interface between extreme events and tipping points, to develop interdisciplinary research on the potential of triggering tipping points by extreme events. This should include both land and ocean communities, draw on improved process understanding, and integrate information from models and observations using statistical and ML techniques.
- Ensure detailed discussion and process oriented treatment of high risk events and tipping points in upcoming assessments, e.g. the next report by the Intergovernmental Panel on Climate Change.
- Encourage new, high frequency, spatially and temporally, in-situ observations near the coasts, estuaries and inlets, improve and expand satellite retrieval algorithms to obtain better satellite-based estimates of extremes near the coast.
- Ensure continuous and consistent measurements of key observables to maintain long enough records for climate trends, and also monitoring of high frequency variability to identify potential critical conditions and early warning signs of tipping points.
- Work towards fully interactive coupling of relevant processes in climate models and all their subcomponents, such as prognostic ice sheets/glaciers and their feedbacks with

climate, links between ocean biogeochemistry and dynamics (e.g. effect of mesoscales and submesoscales), two-way feedbacks between land vegetation and climate, coupling between ecological and hydrometerological processes (e.g. Mahecha et al 2024), resolving processes in the land–ocean and the air-sea interfaces. EO data are essential to evaluate the model components individually and in their interactions.

- Explore the potential to use AI to bridge scales and, target in particular high spatial and temporal resolution processes.
- Employ a new methodology for attribution to anthropogenic climate change for compound vs single extreme events (e.g. Zscheischler and Lehner 2022).
- Integrate human factors beyond climate forcing (e.g. land management, deforestation, coastal pollution) to better capture the interaction of climate and other man-made pressures.
- Improve monitoring of extreme events and the recovery from them in space and time, in order to better understand current model limitations and biases, including in determining long-term (multi-decadal) change.

6.2 Open Research Questions

In the following, we highlight some open high priority research questions that would be useful to address in the emerging research area of extremes-induced tipping points; and to better enable early warning and adaptation.

1) What types of extreme events are the most dangerous for triggering long-term change or tipping; what are their early warning signs and is there a possibility to take mitigative action that avoids tipping and also reduces the impact of extreme events on the carbon cycle, climate and society? On the other hand, do some extremes such as cold events (e.g. due to natural variability or volcanic eruptions) increase the resilience to tipping?
2) How realistic are ESMs in capturing the most extreme events and their full impact, and its recovery time, resilience, and reversibility?
3) What are the carbon implications of the most severe extreme events when all cascading effects and recovery timescales are fully considered?
4) What is the potential for socioeconomic tipping in response to severe extreme events, and what are the most important triggering mechanisms?
5) How can the links between extreme events and tipping points be better captured in tipping point and nonlinear response theory?
6) Lastly, in this review, we treated the land- and ocean-based extremes-driven tipping points in isolation, but there are also potential links between the ocean and land tipping points / extreme events, and those links need to be investigated: Examples are marine heatwaves leading to heavy rainfall over land, wildfire smoke plumes causing phytoplankton blooms in the ocean, snow/ice melting on land impacting ocean extremes/ tipping points.

Acknowledgements This paper is an outcome of the Workshop "Tipping Points and Understanding EO data needs for a Tipping Element Model Intercomparison Project " held at the International Space Science Institute (ISSI) in Bern, Switzerland (10–14 October 2022) . We also acknowledge support and funding from the World Climate Research Programme LIghthouse activity 'Safe Landing Climates'. We acknowledge support from several funding agencies in preparing the article. A.B. acknowledges funding by the European Union (ERC StG, ForExD, grant agreement No. 101039567). S.I.S. acknowledges support from the European Union's Horizon 2020 research and innovation program through the XAIDA project under Grant Agreement

101003469. A.L. acknowledges support from European Space Agency projects CareHEAT (ESA Contract No. 4000137121/21/I-DT) and MiTHO (ESA Contract No. 4000142100/23/I-DT). A.R., P.L., J.M. acknowledge support by the NASA-Modeling, Analysis and Prediction program (MAP) (grant number NNX16AC93 G, under Principal Investigator Anastasia Romanou). G.H. acknowledges support by NERC under the GloSAT project (NE/S015698/1). B.L.O-B. acknowledges the support by the National Center for Atmospheric Research (NCAR), which is a major facility sponsored by the National Science Foundation (NSF) under Cooperative Agreement No. 1852977. H.K. acknowledges the support by the National Research Foundation of Korea (NRF) grant Funded by the Korea Government (MSIT)(2021H1D3A2A03097768). L.S.-G. has also received funding from the European Union's Horizon Europe Framework Programme under the Marie Skłodowska-Curie grant agreement No 101064940.

Author Contributions Hegerl, Romanou, Seneviratne: conceptualization of main idea of the paper, writing, reviewing and editing. Abis, Bastos, Conversi, Landolfi, Kim, Lerner, Mekus, Otto-Bliesner, Pausata, Pinto, Suarez-Guiterrez: writing-contributing in different sections.

Declarations

Conflict of interests The authors have no relevant financial or non-financial interests to disclose.

References

Abis B, Brovkin V (2017) Environmental conditions for alternative tree-cover states in high latitudes. Biogeosciences 14(3):511–527. https://doi.org/10.5194/bg-14-511-2017

Abis B, Brovkin V (2019) Alternative tree-cover states of the boreal ecosystem: a conceptual model. Glob Ecol Biogeogr 28(5):612–627. https://doi.org/10.1111/geb.12880

Adkins J, DeMenocal P, Eshel G (2006) The "African Humid Period" and the record of marine upwelling from excess 230Th in Ocean Drilling Program Hole 658C. Paleoceanography 21:PA4203. https://doi.org/10.1029/2005PA001200

Adler R et al (2018) The global precipitation climatology project (GPCP) monthly analysis (New Version 2.3) and a review of 2017 global precipitation. Atmosphere 9.4:138

AghaKouchak A et al (2020) Climate extremes and compound hazards in a warming world. Ann Rev Earth and Plan Sci 48:519–548. https://doi.org/10.1146/annurev-earth-071719-055228

Allen CD, Breshears DD, McDowell NG (2015) On underestimation of global vulnerability to tree mortality and forest die-off from hotter drought in the Anthropocene. Ecosphere 6(8):129. https://doi.org/10.1890/ES15-00203.1

Alley RB, Marotzke J, Nordhaus WD, Overpeck JT, Peteet DM, Pielke RA Jr, Wallace JM (2003) Abrupt climate change. Science 299(5615):2005–2010. https://doi.org/10.1126/science.1081056

Almeida LZ, Laurel BJ, Thalmann HL, Miller JA (2024) Warmer, earlier, faster: cumulative effects of Gulf of Alaska heatwaves on the early life history of Pacific cod. Elementa Sci Anthropocene. https://doi.org/10.1525/elementa.2023.00050

Altieri AH, Johnson MD, Swaminathan SD, Nelson HR, Gedan KB (2021) Resilience of tropical ecosystems to ocean deoxygenation. Trends Ecol Evol 36:227–238. https://doi.org/10.1016/j.tree.2020.11.003

Anoszko E, Frelich LE, Rich RL, Reich PB (2022) Wind and fire: rapid shifts in tree community composition following multiple disturbances in the southern boreal forest. Ecosphere 13(3):e3952. https://doi.org/10.1002/ecs2.3952

Anthony KRN, Kline DI, Diaz-Pulido G, Dove S, Hoegh-Guldberg O (2008) Ocean acidification causes bleaching and productivity loss in coral reef builders. Proc Natl Acad Sci USA 105:17442–17446. https://doi.org/10.1073/pnas.0804478105

Arafeh-Dalmau N, Montaño-Moctezuma G, Martínez JA, Beas-Luna R, Schoeman DS, Torres-Moye G (2019) Extreme marine heatwaves alter kelp forest community near its equatorward distribution limit. Front Mar Sci. https://doi.org/10.3389/fmars.2019.00499

Arias-Ortiz A, Serrano O, Masqué P, Lavery PS, Mueller U, Kendrick GA, Rozaimi M, Esteban A, Fourqurean JW, Marbà NJNCC, Mateo MA (2018) A marine heatwave drives massive losses from the world's largest seagrass carbon stocks. Nat Clim Chang 8(4):338–344. https://doi.org/10.1038/s41558-018-0096-y

Armstrong McKay DI, Staal A, Abrams JF, Winkelmann R, Sakschewski B, Loriani S, Fetzer I, Cornell SE, Rockström J, Lenton TM (2022) Exceeding 1.5 °C global warming could trigger multiple climate tipping points. Sci (NY) 377(6611):eabn7950. https://doi.org/10.1126/science.abn7950

Ashwin P, Wieczorek S, Vitolo R, Cox P (2012) (2012): Tipping points in open systems: bifurcation, noise-induced and rate-dependent examples in the climate system. Trans R Soc A 370:1166–1184. https://doi.org/10.1098/rsta.2011.0306

Atwood TB, Romanou A, DeVries T, Lerner PE, Mayorga JS, Bradley D, Cabral RB, Schmidt GA, Sala E (2023) Atmospheric CO_2 emissions and ocean acidification from bottom-trawling. Front Mar Sci 10:1125137. https://doi.org/10.3389/fmars.2023.1125137

Baag S, Mandal S (2022) Combined effects of ocean warming and acidification on marine fish and shellfish: a molecule to ecosystem perspective. Sci Total Environ. https://doi.org/10.1016/j.scitotenv.2021.149807

Baker AC, Glynn PW, Riegl B (2008) Climate change and coral reef bleaching: an ecological assessment of long-term impacts, recovery trends and future outlook. Estuar Coast Shelf Sci 80(4):435–471. https://doi.org/10.1016/j.ecss.2008.09.003

Bastos A, Orth R, Reichstein M, Ciais P, Viovy N, Zaehle S, Anthoni P, Arneth A, Gentine P, Joetzjer E, Lienert S (2021a) Vulnerability of European ecosystems to two compound dry and hot summers in 2018 and 2019. Earth Syst Dyn 12(4):1015–1035. https://doi.org/10.5194/esd-12-1015-2021

Bastos A, Orth R, Reichstein M, Ciais P, Viovy N, Zaehle S, Anthoni P, Arneth A, Gentine P, Joetzjer E, Lienert S, Loughran T, McGuire PC, PongratzSitch JS (2021b) Vulnerability of European ecosystems to two compound dry and hot summers in 2018 and 2019. Earth Syst Dyn 12(4):1015–1035. https://doi.org/10.1038/s43017-023-00410-3

Bastos A, Sippel S, Frank D, Mahecha MD, Zaehle S, Zscheischler J, Reichstein M (2023) A joint framework for studying compound ecoclimatic events. Nat Rev Earth Environ. https://doi.org/10.1038/s43017-023-00410-3

Bates NR, Mathis JT (2009) The Arctic Ocean marine carbon cycle: evaluation of air-sea CO_2 exchanges, ocean acidification impacts and potential feedbacks. Biogeosciences 6:2433–2459. https://doi.org/10.5194/bg-6-2433-2009

Bathiany S, Bastiaansen R, Bastos A, Blaschke L, Lever J, Loriani S, De Keersmaecker W, Dorigo W, Milenković M, Senf C, Smith T, Verbesselt J, Boers N (2024) Ecosystem resilience monitoring and early warning using EO data: challenges and outlook. Surv Geophys. https://doi.org/10.1007/s10712-024-09833-z

Batibeniz F, Hauser M, Seneviratne SI (2023) Countries most exposed to individual and concurrent extremes and near-permanent extreme conditions at different global warming levels. Earth Syst Dynam 14:485–505. https://doi.org/10.5194/esd-14-485-2023

Beck PSA, Juday GP, Alix C, Barber VA, Winslow SE, Sousa EE et al (2011) Changes in forest productivity across Alaska consistent with biome shift. Ecol Lett 14(4):373–379. https://doi.org/10.1111/j.1461-0248.2011.01598.x

Bee S (2015b) Imperiled fish add to California's drought stress. https://www.sacbee.com/news/california/water-and-drought/article23294532.html

Bell TW, Cavanaugh KC, Saccomanno VR, Cavanaugh KC, Houskeeper HF et al (2023) Kelpwatch: a new visualization and analysis tool to explore kelp canopy dynamics reveals variable response to and recovery from marine heatwaves. PLoS ONE 18(3):e0271477. https://doi.org/10.1371/journal.pone.0271477

Bellprat O, Doblas-Reyes F (2016) Attribution of extreme weather and climate events overestimated by unreliable climate simulations. Geophys Res Lett 43:2158–2164. https://doi.org/10.1002/2015GL067189

Beobide-Arsuaga G, Bayr T, Reintges A, Latif M (2021) Uncertainty of ENSO-amplitude projections in CMIP5 and CMIP6 models. Clim Dyn 56(11–12):3875–3888. https://doi.org/10.1007/s00382-021-05673-4

Biess B, Gudmundsson L, Windisch MG, Senevirante SI (2024) Future changes in spatially compounding hot, wet or dry events and their implications for the world's breadbasket regions. Environ Res Lett 19:064011. https://doi.org/10.1088/1748-9326/ad4619

Bindoff NL, Cheung WWL, Kairo JG, Arístegui J, Guinder VA, Hallberg R, Hilmi N, Jiao N, Karim MS, Levin L, O'Donoghue S, Purca Cuicapusa SR, Rinkevich B, Suga T, Tagliabue A, Williamson P (2019) Changing ocean, marine ecosystems, and dependent communities. SROCC 2019:447–587. https://doi.org/10.1017/9781009157964.007

Binte JT, Hasan S (2023) Understanding the loss in community resilience due to hurricanes using Facebook Data. Int J Disaster Risk Reduct 97:104036. https://doi.org/10.1016/j.ijdrr.2023.104036

Blanchard-Wrigglesworth E, Donohoe A, Roach LA, DuVivier A, Bitz CM (2021) High-frequency sea ice variability in observations and models. Geophys Res Lett 48(14):1–11. https://doi.org/10.1029/2020GL092356

Bochow N, Boers N (2023) The South American monsoon approaches a critical transition in response to deforestation. Sci Adv 9:eadd9973. https://doi.org/10.1126/sciadv.add9973

Boers N, Ghil M, Stocker TF (2022) Theoretical and paleoclimatic evidence for abrupt transitions in the Earth system. Environ Res Lett. https://doi.org/10.1088/1748-9326/ac8944

Bopp L, Resplandy L, Orr JC, Doney SC, Dunne JP, Gehlen M, Halloran P, Heinze C, Ilyina T, Séférian R, Tjiputra J, Vichi M (2013) Multiple stressors of ocean ecosystems in the 21st century: projections with CMIP5 models. Biogeosciences 10:6225–6245. https://doi.org/10.5194/bg-10-6225-2013

Boulton CA, Lenton TM, Boers N (2022) Pronounced loss of Amazon rainforest resilience since the early 2000s. Nat Clim Chang 12:271–278. https://doi.org/10.1038/s41558-022-01287-8

Bowman DMJS et al (2009) Fire in the earth system. Science 324(5926):481–484. https://doi.org/10.1126/science.1163886

Box JE, Fettweis X, Stroeve JC, Tedesco M, Hall DK, Steffen K (2012) Greenland ice sheet albedo feedback: thermodynamics and atmospheric drivers. Cryosphere 6:821–839. https://doi.org/10.5194/tc-6-821-2012

Brander LM, Rehdanz K, Tol RSJ et al (2012) The economic impact of ocean acidification on coral reefs. Clim Chang Econ 3(1):1250002. https://doi.org/10.1142/S2010007812500029

Brandon CM, Woodruff JD, Lane DP, Donnelly JP (2013) Tropical cyclone wind speed constraints from resultant storm surge deposition: a 2500 year reconstruction of hurricane activity from St. Marks, FL. Geochem Geophys Geosyst 14:2993–3008

Breitburg D, Levin LA, Oschlies A, Grégoire M, Chavez FP, Conley DJ, Garçon V, Gilbert D, Gutiérrez D, Isensee K, Jacinto GS, Limburg KE, Montes I, Naqvi SWA, Pitcher GC, Rabalais NN, Roman MR, Rose KA, Seibel BA, Telszewski M, Yasuhara M, Zhang J (2018) Declining oxygen in the global ocean and coastal waters. Science. https://doi.org/10.1126/science.aam7240

Brodeur RD, Auth TD, Phillips AJ (2019) Major shifts in pelagic micronekton and macrozooplankton community structure in an upwelling ecosystem related to an unprecedented marine heatwave. Front Mar Sci 6:212. https://doi.org/10.3389/fmars.2019.00212

Brovkin V, Bartsch A, Hugelius G, Calamita E, Lever JJ, Goo E, Kim H, Stacke T, de Vrese P (2024) Permafrost and freshwater systems in the Arctic as tipping elements of the climate system. Surv. Geophys., submitted

Burger FA, Frölicher TL (2023) Drivers of surface ocean acidity extremes in an Earth system model. Global Biogeochem Cycles 37(9):e2023GB007785. https://doi.org/10.1029/2023GB007785

Burger FA, John JG, Frölicher TL (2020) Increase in ocean acidity variability and extremes under increasing atmospheric CO_2. Biogeosciences 17:4633–4662. https://doi.org/10.5194/bg-17-4633-2020

Burger FA, Terhaar J, Frölicher TL (2022) Compound marine heatwaves and ocean acidity extremes. Nat Commun. https://doi.org/10.1038/s41467-022-32120-7

Bustamante M, Roy J, Ospina D et al (2023) Ten new insights in climate science 2023/2024. Glob Sustain. https://doi.org/10.1017/sus.2023.25

Byrne B, Liu J, Lee M, Yin Y, Bowman KW, Miyazaki K et al (2021) The carbon cycle of southeast Australia during 2019–2020: Drought, fires, and subsequent recovery. AGU Adv 2:e2021AV000469. https://doi.org/10.1029/2021AV000469

Canadell JP et al (2021) Global carbon and other biogeochemical cycles and feedbacks. In: Masson-Delmotte V et al (eds) Climate change 2021: the physical science basis. Contribution of working group I to the sixth assessment report of the intergovernmental panel on climate change. Cambridge University Press. https://hal.science/hal-03336145.

Carpino OA, Berg AA, Quinton WL, Adams JR (2018) Climate change and permafrost thaw-induced boreal forest loss in northwestern Canada. Environ Res Lett 13(8):084018. https://doi.org/10.1088/1748-9326/aad74e

Castellana D, Dijkstra HA (2020) Noise-induced transitions of the atlantic meridional overturning circulation in CMIP5 models. Sci Rep 10(1):1–9. https://doi.org/10.1038/s41598-020-76930-5

Catto JL, Dowdy A (2021) Understanding compound hazards from a weather system perspective. Weather Clim Extrem 32:100313. https://doi.org/10.1016/j.wace.2021.100313

Cavole LM et al (2016) Biological impacts of the 2013–2015 warm-water anomaly in the northeast Pacific: winners, losers, and the future. Oceanography 29:273–285

Chen Q et al (2023) Understanding the compound marine heatwave and low-chlorophyll extremes in the western pacific ocean. Front Marine Sci. https://doi.org/10.3389/fmars.2023.1303663

Chen D, Rojas M, et al (2021) Framing, context, and methods. In: Masson-Delmotte et al (eds) Climate change 2021: the physical science basis. Contribution of WGI to the sixth assessment report of the intergovernmental panel on climate change. Cambridge University Press, Cambridge. pp 147–286. https://doi.org/10.1017/9781009157896.003

Cheung WWL, Frölicher TL (2020) Marine heatwaves exacerbate climate change impacts for fisheries in the northeast Pacific. Sci Rep 10(1):1–10. https://doi.org/10.1038/s41598-020-63650-z

Chuvieco E, Lizundia-Loiola J, Pettinari ML, Ramo R, Padilla M, Tansey K, Mouillot F, Laurent P, Storm T, Heil A, Plummer S (2018) Generation and analysis of a new global burned area product based on MODIS 250 m reflectance bands and thermal anomalies. Earth Syst Sci Data 10(4):2015–2031. https://doi.org/10.5194/essd-10-2015-2018

Ciavarella A, Cotterill D, Stott P, Kew S, Philip S, van Oldenborgh GJ, Skålevåg A, Lorenz P, Robin Y, Otto F, Hauser M (2021) Prolonged Siberian heat of 2020 almost impossible without human influence. Clim Change 166:1–18. https://doi.org/10.1007/s10584-021-03052-w

Clarke TM, Wabnitz CC, Striegel S, Frölicher TL, Reygondeau G, Cheung WL (2021) Aerobic growth index (AGI): an index to understand the impacts of ocean warming and deoxygenation on global marine fisheries resources. Prog Oceanogr 195:102588. https://doi.org/10.1016/j.pocean.2021.102588

Clayson CA, Centurioni L, Cronin MF, Edson J, Gille S, Muller-Karger F, Parfitt R, Riihimaki LD, Smith SR, Swart S, Vandemark D, Bôas ABV, Zappa CJ, Zhang D (2021) Super sites for advancing understanding of the oceanic and atmospheric boundary layers. Mar Technol Soc J 55(3):144–145. https://doi.org/10.4031/MTSJ.55.3.11

Collins M, Sutherland M, Bouwer L, Cheong SM, Frölicher T, Jacot Des Combes H, Koll Roxy M, Losada I, McInnes K, Ratter B, Rivera-Arriaga E, Susanto RD, Swingedouw D, Tibig L (2019) Extremes, abrupt changes and managing risk. In: SROCC 2019

Cook BI, Miller RL, Seager R (2008) Dust and sea surface temperature forcing of the 1930s "Dust Bowl" drought. Geophys Res Lett 35:8. https://doi.org/10.1029/2008GL033486

Cook F, Smith RO, Roughan M, Cullen NJ, Shears N, Bowen M (2022a) Marine heatwaves in shallow coastal ecosystems are coupled with the atmosphere: insights from half a century of daily in situ temperature records. Front Clim. https://doi.org/10.3389/fclim.2022.1012022

Cook BI et al (2022b) Megadroughts in the common era and the anthropocene. Nat Rev Earth Environ 3(11):741–757. https://doi.org/10.1038/s43017-022-00329-1

Couasnon A, Eilander D, Muis S, Veldkamp TI, Haigh ID, Wahl T, Winsemius HC, Ward PJ (2020) Measuring compound flood potential from river discharge and storm surge extremes at the global scale. Nat Hazard 20(2):489–504. https://doi.org/10.5194/nhess-20-489-2020

Cowan T, Hegerl GC, Colfescu I, Bollasina M, Purich A, Boschat G (2017) Factors contributing to record-breaking heat waves over the great plains during the 1930s Dust Bowl. J Clim 30(7):2437–2461. https://doi.org/10.1175/JCLI-D-16-0436.1

Cowan T, Undorf S, Hegerl GC, Harrington LJ, Otto FEL (2020a) Present-day greenhouse gases could cause more frequent and longer Dust Bowl heatwaves. Nat Clim Chang 10(6):505–510. https://doi.org/10.1038/s41558-020-0771-7

Cowan T et al (2020b) Ocean and land forcing of the record-breaking Dust Bowl heatwaves across central United States. Nat Commun 11:1. https://doi.org/10.1038/s41467-020-16676-w

Dahlke FT, Wohlrab S, Butzin M, Pörtner H-O (2020) Thermal bottlenecks in the life cycle define climate vulnerability of fish. Science 369:65–70. https://doi.org/10.1126/science.aaz3658

Dallmeyer A, Claussen M, Lorenz SJ, Shanahan T (2020) The end of the African humid period as seen by a transient comprehensive Earth system model simulation of the last 8000 years. Clim past 16:117–140. https://doi.org/10.5194/cp-16-117-2020

Darling ES, McClanahan TR, Cote IM (2013) Life histories predict coral community disassembly under multiple stressors. Glob Chang Biol 19:1930–1940. https://doi.org/10.1111/gcb.12191

deMenocal P, Ortiz J, Guilderson T, Adkins J, Sarnthein M, Baker L, Yarusinsky M (2000) Abrupt onset and termination of the African Humid Period. Quat Sci Rev 19:347–361. https://doi.org/10.1016/S0277-3791(99)00081-5

Deutsch C, Ferrel A, Seibel B, Pörtner HO, Huey RB (2015) Climate change tightens a metabolic constraint on marine habitats. Science 34(6239):1132–1135. https://doi.org/10.1126/science.aaa1605

Deutsch C, Penn JL, Seibel B (2020) Metabolic trait diversity shapes marine biogeography. Nature 585:557–562. https://doi.org/10.1038/s41586-020-2721-y

Di Lorenzo E, Mantua N (2016) Multi-year persistence of the 2014/15 North Pacific marine heatwave. Nat Clim Chang. https://doi.org/10.1038/nclimate3082
Dial RJ, Maher CT, Hewitt RE et al (2022) Sufficient conditions for rapid range expansion of a boreal conifer. Nature 608:546–551. https://doi.org/10.1038/s41586-022-05093-2
Doney SC, Busch DS, Cooley SR, Kroeker KJ (2020) The impacts of ocean acidification on marine ecosystems and reliant human communities. Annu Rev Environ Resour 45:83–112. https://doi.org/10.1146/annurev-environ-012320-083019
Drake N, Bristow C (2006) Shorelines in the Sahara: geomorphological evidence for an enhanced monsoon from palaeolake Megachad. The Holocene 16:901–911. https://doi.org/10.1191/0959683606hol981rr
Dunne J, Stouffer R, John J (2013) Reductions in labour capacity from heat stress under climate warming. Nature Clim Change 3:563–566. https://doi.org/10.1038/nclimate1827
Engdaw MM, Steiner AK, Hegerl GC, Ballinger AP (2023) Attribution of observed changes in extreme temperatures to anthropogenic forcing using CMIP6 models. Weather Clim Extremes 39:100548. https://doi.org/10.1016/j.wace.2023.100548
EUCRA (2024) European climate risk assessment (EUCRA), https://doi.org/10.2800/204249. https://climate-adapt.eea.europa.eu/en/eu-adaptation-policy/key-eu-actions/european-climate-risk-assessment
Fan L, Wigneron JP, Ciais P et al (2023) Siberian carbon sink reduced by forest disturbances. Nat Geosci 16:56–62. https://doi.org/10.1038/s41561-022-01087-x
Fennel K, Testa JM (2019) Biogeochemical controls on coastal hypoxia. Ann Rev Mar Sci 11:105–130. https://doi.org/10.1146/annurev-marine-010318-095138
Feurdean A et al (2020) Recent fire regime in the southern boreal forests of western Siberia is unprecedented in the last five millennia. Quat Sci Rev. https://doi.org/10.1016/j.quascirev.2020.106495
Filbee-Dexter K, Wernberg T, Grace SP et al (2020) Marine heatwaves and the collapse of marginal North Atlantic kelp forests. Sci Rep 10:13388. https://doi.org/10.1038/s41598-020-70273-x
Fisher RA, Koven CD, Anderegg WRL et al (2018) Vegetation demographics in Earth System Models: a review of progress and priorities. Glob Change Biol 2018(24):35–54. https://doi.org/10.1111/gcb.13910
Fordyce AJ, Ainsworth TD, Heron SF, Leggat W (2019) Marine heatwave hotspots in coral reef environments: physical drivers, ecophysiological outcomes, and impact upon structural complexity. Front Mar Sci 6:498. https://doi.org/10.3389/fmars.2019.00498
Fox-Kemper B et al (2021) Ocean, cryosphere and sea level change. In: Masson-Delmotte et al climate change 2021: the physical science basis. contribution of working group I to the sixth assessment report of the intergovernmental panel on climate change. Cambridge University Press, Cambridge. pp 1211–1362. https://doi.org/10.1017/9781009157896.011.
Frederikse T, Landerer F, Caron L et al (2020) The causes of sea-level rise since 1900. Nature 584:393–397. https://doi.org/10.1038/s41586-020-2591-3
Fredston AL, Cheung WWL, Frölicher TL et al (2023) Marine heatwaves are not a dominant driver of change in demersal fishes. Nature 621:324–329. https://doi.org/10.1038/s41586-023-06449-y
Freychet N, Hegerl GC, Lord N, Lo E, Mitchell D, Collins M (2022) Robust increase in population exposure to heat stress with increasing global warming. Environ Res Lett 17:064049. https://doi.org/10.1088/1748-9326/ac71b9
Frölicher TL, Laufkötter C (2018) Emerging risks from marine heat waves. Nat Commun 9(1):2015–2018. https://doi.org/10.1038/s41467-018-03163-6
Frölicher TL, Fischer EM, Gruber N (2018) Marine heatwaves under global warming. Nature 560:360–364. https://doi.org/10.1038/s41586-018-0383-9
Fu Z, Ciais P, Prentice IC et al (2022) Atmospheric dryness reduces photosynthesis along a large range of soil water deficits. Nat Commun 13:989. https://doi.org/10.1038/s41467-022-28652-7
Ganguli P, Merz B (2019) Extreme coastal water levels exacerbate fluvial flood hazards in Northwestern Europe. Sci Rep 9:1–14. https://doi.org/10.1038/s41598-019-49822-6
Gao K, Beardall J, Häder D-P, Hall-Spencer JM, Gao G, Hutchins DA (2019) Effects of ocean acidification on marine photosynthetic organisms under the concurrent influences of warming, UV radiation, and deoxygenation. Front Mar Sci 6:322. https://doi.org/10.3389/fmars.2019.00322
García-León D, Casanueva A, Standardi G et al (2021) Current and projected regional economic impacts of heatwaves in Europe. Nat Commun 12:5807. https://doi.org/10.1038/s41467-021-26050-z
Garreaud RD (2018) A plausible atmospheric trigger for the 2017 coastal El Niño. Int J Climatol 38:e1296–e1302. https://doi.org/10.1002/joc.5426
Gaupp F, Hall J, Mitchell D, Dadson S (2019) Increasing risks of multiple breadbasket failure under 1.5 and 2°C global warming. Agric Syst 175:34–45. https://doi.org/10.1016/j.agsy.2019.05.010
Giglio L, Schroeder W, Justice CO (2016) The collection 6 MODIS active fire detection algorithm and fire products. Remote Sens Environ 178:31–41. https://doi.org/10.1016/j.rse.2016.02.054

Gladwell M (2000) The tipping point: how little things can make a big difference. Little Brown, New York
Good EJ, Aldred FM, Ghent DJ, Veal KL, Jimenez C (2022) An analysis of the stability and trends in the LST cci land surface temperature datasets over Europe. Earth Space Sci 9:9
Green JK, Seneviratne SI, Berg AM et al (2019) Large influence of soil moisture on long-term terrestrial carbon uptake. Nature 565:476–479. https://doi.org/10.1038/s41586-018-0848-x
Grégoire M, Garçon V, Garcia H, Breitburg D, Isensee K, Oschlies A et al (2021) A global ocean 1884 oxygen database and atlas for assessing and predicting deoxygenation and ocean health in the 1885 open and coastal ocean. Front Mar Sci 8:724913. https://doi.org/10.3389/fmars.2021.724913
Griffin D, Anchukaitis KJ (2014) How unusual is the 2012–2014 California drought? Geophys Res Lett 41:9017–9023. https://doi.org/10.1002/2014GL062433
Gruber N (2011) Warming up turning sour losing breath: ocean biogeochemistry under global change. Philos Trans R Soc A 369:1980–1996. https://doi.org/10.1098/rsta.2011.0003
Gruber N, Boyd PW, Frölicher TL, Vogt M (2021) Biogeochemical extremes and compound events in the ocean. Nature 600(7889):395–407. https://doi.org/10.1038/s41586-021-03981-7
Guiot J, Cramer W (2016) Climate change: the 2015 Paris agreement thresholds and mediterranean basin ecosystems. Science 354:465–468. https://doi.org/10.1126/science.aah5015
Han W, Zhang L, Meehl GA, Kido S, Tozuka T, Li Y, McPhaden MJ, Hu A, Cazenave A, Rosenbloom N, Strand G (2022) Sea level extremes and compounding marine heatwaves in coastal Indonesia. Nat Commun 13(1):6410. https://doi.org/10.1038/s41467-022-34003-3
Hanak E, Mount J, Chappelle C, Lund J, Medellín-Azuara J, Moyle P, Seavy N (2015) What if California's drought continues? Public Policy Institute of California—Water Policy Center. https://www.ppic.org/content/pubs/report/R_815EHR.pdf
Hanna E, Pattyn F, Navarro F, Favier V, Goelzer H, van den Broeke MR, Vizcaino M, Whitehouse PL, Ritz C, Bulthuis K, Smith B (2020) Mass balance of the ice sheets and glaciers—progress since AR5 and challenges. Earth-Sci Rev 201:102976. https://doi.org/10.1016/j.earscirev.2019.102976
Hantson S, Kelley DI, Arneth A, Harrison SP, Archibald S, Bachelet D, Forrest M, Hickler T, Lasslop G, Li F, Mangeon S, Melton JR, Nieradzik L, Rabin SS, Prentice IC, Sheehan T, Sitch S, Teckentrup L, Voulgarakis A, Yue C (2020) Quantitative assessment of fire and vegetation properties in simulations with fire-enabled vegetation models from the Fire Model Intercomparison Project. Geosci Model Dev 13:3299–3318. https://doi.org/10.5194/gmd-13-3299-2020
Harris RMB, Beaumont LJ, Vance TR, Tozer CR, Remenyi TA, Perkins-Kirkpatrick SE, Mitchell PJ, Nicotra AB, McGregor S, Andrew NR (2018) Biological responses to the press and pulse of climate trends and extreme events. Nature Clim Change 8(7):579–587. https://doi.org/10.1038/s41558-018-0187-9
Hartmann H, Bastos A, Das AJ, Esquivel-Muelbert A, Hammond WM, Martínez-Vilalta J, McDowell NG, Powers JS, Pugh TA, Ruthrof KX, Allen CD (2022) Climate change risks to global forest health: emergence of unexpected events of elevated tree mortality worldwide. Annu Rev Plant Biol 73:673–702. https://doi.org/10.1146/annurev-arplant-102820-012804
Hegedűs D, Ballinger AP, Hegerl GC (2024) Observed links between heatwaves and wildfires across northern high latitudes. Environ Res Lett 19(3):034041. https://doi.org/10.1088/1748-9326/ad2b29
Hegerl GC, Brönnimann S, Schurer AP, Cowan T (2018) The early 20th century warming: anomalies, causes, and consequences. Wires Clim Change 9:4. https://doi.org/10.1002/wcc.522
Heinrich P, Hagemann S, Weisse R, Schrum C, Daewel U, Gaslikova L (2022) Compound flood events: analysing the joint occurrence of extreme river discharge events and storm surges in northern and central Europe. Nat Hazards Earth Syst Sci Discuss 2022:1–27. https://doi.org/10.5194/nhess-23-1967-2023
Heinze C, Blenckner T, Martins H (2021) The quiet crossing of ocean tipping points. Proc Natil Acad Sci. https://doi.org/10.1073/pnas.2008478118
Hély C, Lézine A-M (2014) Holocene changes in African vegetation: tradeoff between climate and water availability. Clim past 10:681–686. https://doi.org/10.5194/cp-10-681-2014
Hillyer KE, Dias DA, Lutz A et al (2017) Metabolite profiling of symbiont and host during thermal stress and bleaching in the coral *Acropora aspera*. Coral Reefs 36:105–118. https://doi.org/10.1007/s00338-016-1508-y
Hlásny T, Zimová S, Merganičová K, Štěpánek P, Modlinger R, Turčáni M (2021) Devastating outbreak of bark beetles in the Czech Republic: drivers, impacts, and management implications. For Ecol Manag 490:119075. https://doi.org/10.1016/j.foreco.2021.119075
Hobday AJ, Alexander LV, Perkins SE, Smale DA, Straub SC, Oliver EC et al (2016) A hierarchical approach to defining marine heatwaves. Prog Oceanogr 141:227–238. https://doi.org/10.1016/j.pocean.2015.12.014
Hoegh-Guldberg O et al (2018) Impacts of 1.5 °C global warming on natural and human systems. In: Masson-Delmotte V et al (eds) Global warming of 1.5 °C. An IPCC special report on the impacts

of global warming of 1.5 °C above pre-industrial levels and related global greenhouse gas emission pathways, in the context of strengthening the global response to the threat of climate change, sustainable development, and efforts to eradicate poverty. pp 175–311
Hoeke RK et al (2013) Widespread inundation of Pacific islands triggered by distant-source wind-waves. Glob Planet Change 108:128–138. https://doi.org/10.1016/j.gloplacha.2013.06.006
Holbrook NJ, Scannell HA, Gupta AS, Benthuysen JA, Feng M, Oliver ECJ, Alexander LV, Burrows MT, Donat MG, Hobday AJ, Moore PJ, Perkins-Kirkpatrick SE, Smale DA, Thomas SCS, Thomas W (2019) A global as- sessment of marine heatwaves and their drivers. Nat Commun 10:2624. https://doi.org/10.1038/s41467-019-10206-z
Holbrook NJ, Sen Gupta A, Oliver ECJ, Hobday AJ, Benthuysen JA, Scannell HA, Smale DA, Wernberg T (2020) Keeping pace with marine heatwaves. Nat Rev Earth Environ 1(9):482–493. https://doi.org/10.1038/s43017-020-0068-4
Holland MM, Bitz CM, Tremblay LB, Bailey DA (2008) The role of natural versus forced change in future rapid summer arctic ice loss. Geophys Monogr Ser 180:133–150. https://doi.org/10.1029/180GM10
Howitt R, Medellín-Azuara J, Macewan D (2014) Economic analysis of the 2014 drought for California agriculture. Center for Water- shed Sciences—University of California. http://watershed.ucdavis.edu
Huang WK, Monahan AH, Zwiers FW (2021) Estimating concurrent climate extremes: a conditional approach. Weather Clim Extrem 33:100332. https://doi.org/10.1016/j.wace.2021.100332
Hubau W, Lewis SL, Phillips OL et al (2020) Asynchronous carbon sink saturation in African and Amazonian tropical forests. Nature 579:80–87. https://doi.org/10.1038/s41586-020-2035-0
Hughes TP et al (2018) Spatial and temporal patterns of mass bleaching of corals in the Anthropocene. Science. https://doi.org/10.1126/science.aan8048
Hughes TP, Kerry JT, Connolly SR, Baird AH, Eakin CM, Heron SF et al (2019) Ecological memory modifies the cumulative impact of recurrent climate extremes. Nat Clim Chang 9:40–43. https://doi.org/10.1038/s41558-018-0351-2
Hughes DJ, Alderdice R, Cooney C, Kühl M, Pernice M, Voolstra CR, Suggett DJ (2020) Coral reef survival under accelerating ocean deoxygenation. Nat Clim Chang 10(4):296–307. https://doi.org/10.1038/s41558-020-0737-9
Humphrey V, Zscheischler J, Ciais P, Gudmundsson L, Sitch S, Seneviratne SI (2018) Sensitivity of atmospheric CO_2 growth rate to observed changes in terrestrial water storage. Nature 560:628–631. https://doi.org/10.1038/s41586-018-0424-4
IPCC (2012) Managing the risks of extreme events and disasters to advance climate change adaptation. In: Field CB, Barros V, Stocker TF, Qin D, Dokken DJ, Ebi KL, Mastrandrea MD, Mach KJ, Plattner GK, Allen SK, Tignor M, Midgley PM (eds) A special report of working groups I and II of the intergovernmental panel on climate change. Cambridge University Press, Cambridge. pp 582
IPCC (2019) Summary for policymakers. In: Shukla PR, Skea J, Calvo Buendia E, Masson-Delmotte V, Pörtner HO, Roberts DC, Zhai P, Slade R, Connors S, van Diemen R, Ferrat M, Haughey E, Luz S, Neogi S, Pathak M, Petzold J, Portugal Pereira J, Vyas P, Huntley E, Kissick K, Belkacemi M, Malley J (eds) Climate change and land: an IPCC special report on climate change, desertification, land degradation, sustainable land management, food security, and greenhouse gas fluxes in terrestrial ecosystems. https://doi.org/10.1017/9781009157988.001
IPCC (2021) Climate change 2021: the physical science basis. In: Masson-Delmotte V, Zhai P, Pirani A, Connors SL, Péan C, Berger S, Caud N, Chen Y, Goldfarb L, Gomis MI, Huang M, Leitzell K, Lonnoy E, Matthews JBR, Maycock TK, Waterfield T, Yelekçi O, Yu R, Zhou B (eds) Contribution of working group I to the sixth assessment report of the intergovernmental panel on climate change. Cambridge University Press, Cambridge. p 2391. https://doi.org/10.1017/9781009157896.
Jacox MG (2019) Marine heatwaves in a changing climate. Nature 571(7766):485–487. https://doi.org/10.1038/d41586-019-02196-1
Jewett L, Romanou A (2017) Ocean acidification and other ocean changes. In: Wuebbles DJ, (eds). Climate science special report: fourth national climate assessment. Vol 1. U.S. Global Change Research Program, Washington, pp 364–392
Jiao W, Wang L, McCabe MF (2021) Multi-sensor remote sensing for drought characterization: current status, opportunities and a roadmap for the future. Remote Sens Environ 256:112313. https://doi.org/10.1016/j.rse.2021.112313
Joh Y, Di Lorenzo E (2017) Increasing coupling between NPGO and PDO leads to prolonged marine heatwaves in the northeast pacific. Geophys Res Lett. https://doi.org/10.1002/2017GL075930
Jones T, Parrish JK, Peterson WT, Bjorkstedt EP, Bond NA, Ballance LT, Bowes V, Hipfner JM, Burgess HK, Dolliver JE, Lindquist K, Lindsey J, Nevins HM, Robertson RR, Roletto J, Wilson L,

Joyce T, Harvey J (2018) Massive mortality of a planktivorous seabird in response to a marine heatwave. Geophys Res Lett 45:3193–3202. https://doi.org/10.1002/2017GL076164
Junior S et al (2020) Persistent collapse of biomass in Amazonian forest edges following deforestation leads to unaccounted carbon losses. Sci Adv 6:eaaz8360. https://doi.org/10.1126/sciadv.aaz8360
Kannenberg SA, Novick KA, Alexander HR, Maxwell JT, Moore DJP, Phillips RP, Anderegg WRL (2019) Linking drought legacy effects across scales: from leaves to tree rings to ecosystems. Glob Change Biol 25:2978–2922. https://doi.org/10.1111/gcb.14710
Khanal S, Ridder N, De Vries H, Terink W, Van den Hurk B (2019) Storm surge and extreme river discharge: a compound event analysis using ensemble impact modeling. Front Earth Sci 7:224. https://doi.org/10.3389/feart.2019.00224
Khanna J, Medvigy D, Fueglistaler S et al (2017) Regional dry-season climate changes due to three decades of Amazonian deforestation. Nat Clim Change 7:200–204. https://doi.org/10.1038/nclimate3226
Kim H, Franco AC, Sumaila UR (2023) A selected review of impacts of ocean deoxygenation on fish and fisheries. Fishes 8(6):316. https://doi.org/10.3390/fishes8060316
Kimutai J, Barnes C, Zachariah M, Philip S, Kew S, Pinto I, Wolski P, Koren G, Vecchi G, Yang W, Li S, Vahlberg M, Singh R, Heinrich D, Pereira CM, Arrighi J, Thalheimer L, Kane C, Otto F (2023) Human-induced climate change increased drought severity in Horn of Africa [Report]. https://doi.org/10.25561/103482
Klein SG, Geraldi NR, Anton A, Schmidt-Roach S, Ziegler M, Cziesielski MJ, Martin C, Rädecker N, Frölicher TL, Mumby PJ, Pandolfi JM, Suggett DJ, Voolstra CR, Aranda M, Duarte CM (2022) Projecting coral responses to intensifying marine heatwaves under ocean acidification. Glob Change Biol 28(5):1753–1765. https://doi.org/10.1111/gcb.15818
Kohn EE, Munnich M, Vogt M, Desmet F, Gruber N (2022) Strong habitat compression by extreme shoaling events of hypoxic waters in the eastern pacific. J Geophys Research-Oceans 127:6. https://doi.org/10.1029/2022JC018429
Kornhuber K, Coumou D, Vogel E et al (2020) Amplified Rossby waves enhance risk of concurrent heatwaves in major breadbasket regions. Nat Clim Chang 10:48–53. https://doi.org/10.1038/s41558-019-0637-z
Kornhuber K et al (2022) Risks of synchronized low yields are underestimated in climate and crop model projections. Nat Commun. https://doi.org/10.1038/s41467-023-38906-7
Kutzbach JE, Guetter PJ (1986) The influence of changing orbital parameters and surface boundary conditions on climate simulations for the past 18,000 years. J Atmos Sci 43:1726–1759. https://doi.org/10.1175/1520-0469(1986)043%3c1726:TIOCOP%3e2.0.CO;2
Kwon MJ, Ballantyne A, Ciais P, Bastos A, Chevallier F, Liu Z, Green JK, Qiu C, Kimball JS (2021) Siberian 2020 heatwave increased spring CO_2 uptake but not annual CO_2 uptake. Environ Res Lett 16(12):24030. https://doi.org/10.1088/1748-9326/ac358b
Landolfi A, Somes CJ, Koeve W, Zamora LM, Oschlies A (2017) Oceanic nitrogen cycling and N_2O flux perturbations in the anthropocene, global biogeochem. Cycles 31:1236–1255. https://doi.org/10.1002/2017GB005633
Lapola D et al (2023) The drivers and impacts of Amazon forest degradation. Science 379:8622. https://doi.org/10.1126/science.abp8622
Lastovicka J, Svec P, Paluba D, Kobliuk N, Svoboda J, Hladky R, Stych P (2020) Sentinel-2 data in an evaluation of the impact of the disturbances on forest vegetation. Remote Sens 12(12):1914. https://doi.org/10.3390/rs12121914
Lau SC, Wilson NG, Golledge NR, Naish TR, Watts PC, Silva CN, Cooke IR, Allcock AL, Mark FC, Linse K, Strugnell JM (2023) Genomic evidence for West Antarctic Ice sheet collapse during the last interglacial period. Science 382:1384–1389. https://doi.org/10.1101/2023.01.29.525778
Laufkötter C, Zscheischler J, Frölicher TL (2020) High-impact marine heatwaves attributable to human-induced global warming. Science 369(6511):1621–1625. https://doi.org/10.1126/science.aba0690
Lawrence D, Coe M, Walker W, Verchot L, Vandecar K (2022) The unseen effects of deforestation: biophysical effects on climate. Front for Glob Change. https://doi.org/10.3389/ffgc.2022.756115
Le Grix N, Zscheischler J, Laufkötter C, Rousseaux CS, Frölicher TL (2021) Compound high-temperature and low-chlorophyll extremes in the ocean over the satellite period. Biogeosciences 18(6):2119–2137. https://doi.org/10.5194/bg-18-2119-2021
Le Grix N et al (2022) Hotspots and drivers of compound marine heatwaves and low net primary production extremes. Biogeosciences. https://doi.org/10.5194/bg-19-5807-2022
Le Grix N, Cheung WL, Reygondeau G, Zscheischler J, Frölicher TL (2023) Extreme and compound ocean events are key drivers of projected low pelagic fish biomass. Glob Change Biol 29:6478–6492. https://doi.org/10.1111/gcb.16968

Lee J, Hohenegger C (2024) Weaker land-atmosphere coupling in global storm-resolving simulation. PNAS. https://doi.org/10.1073/pnas.2314265121
Lenton TM, Held H, Kriegler E, Hall JW, Lucht W, Rahmstorf S, Schellnhuber HJ (2008) Tipping elements in the Earth's climate system. Proc Natl Acad Sci 105(6):1786–1793. https://doi.org/10.1073/pnas.0705414105
Lenton TM, Armstrong McKay DI, Loriani S, Abrams JF, Lade SJ, Donges JF, Milkoreit M, Powell T, Smith SR, Zimm C, Buxton JE, Bailey E, Laybourn L, Ghadiali A, Dyke JG (2023) The global tipping points report 2023. University of Exeter, Exeter. https://global-tipping-points.org
Lézine A-M, Hély C, Grenier C, Braconnot P, Krinner G (2011) Sahara and Sahel vulnerability to climate changes, lessons from Holocene hydrological data. Quat Sci Rev 30:3001–3012. https://doi.org/10.1016/j.quascirev.2011.07.006
Li X, Shen Y, Bao Y, Wu Z, Yang B, Jiao L, Zhang C, Tocher DR, Zhou Q, Jin M (2022) Physiological responses and adaptive strategies to acute low-salinity environmental stress of the euryhaline marine fish black seabream (*Acanthopagrus schlegelii*). Aquaculture 554:738117. https://doi.org/10.1016/j.aquaculture.2022.738117
Li Z, Ding L, Shen B, Chen J, Xu D, Wang X, Fang W, Pulatov A, Kussainova M, Amarjargal A, Isaev E (2024) Quantifying key vegetation parameters from Sentinel-3 and MODIS over the eastern Eurasian steppe with a Bayesian geostatistical model. Sci Total Environ 909:168594. https://doi.org/10.1016/j.scitotenv.2023.168594
Liu B, Liang Y, He HS, Liu Z, Ma T, Wu MM (2022) Wildfire affects boreal forest resilience through post-fire recruitment in Northeastern China. Ecol Ind 145:109705. https://doi.org/10.1016/j.ecolind.2022.109705
Liu H, Nie X, Cui C, Wei Z (2023a) Compound marine heatwaves and low sea surface salinity extremes over the tropical Pacific Ocean. Environ Res Lett. https://doi.org/10.1088/1748-9326/acd0c4
Liu L, Ciais P, Wu M, Padron RS, Friedlingstein P, Schwaab J, Gudmundsson L, Seneviratne SI (2023b) Increasingly negative tropical water-interannual CO_2 growth rate coupling. Nature. https://doi.org/10.1038/s41586-023-06056-x
Lo YT, Mitchell DM, Thompson R, O'Connel E, Gasparrini A (2022) Estimating heat-related mortality in near real time for national heatwave plans. Environ Res Lett 17:024017. https://doi.org/10.1088/1748-9326/ac4cf4
Lockwood JG (2001) Abrupt and sudden climatic transitions and fluctuations: a review. Int J Climatol J R Meteorol Soc 21(9):1153–1179. https://doi.org/10.1002/joc.630
Lucey N, Aube C, Herwig A, Collin R (2022) Compound extreme events induce rapid mortality in a tropical sea urchin. Biol Bull 243(2):239–254. https://doi.org/10.1086/722283
Lund J, Medellin-Azuara J, Durand J, Stone K (2018) Lessons from California's 2012–2016 drought. J Water Resour Plan Manag 144:04018067. https://doi.org/10.1061/(asce)wr.1943-5452.0000984
Luo N, Mueller N, Zhang Y, Feng P, Huang S, Liu DL, Yu Y, Wang X, Wang P, Meng Q (2023) Short-term extreme heat at flowering amplifies the impacts of climate change on maize production. Environ Res Lett 18(8):084021. https://doi.org/10.1088/1748-9326/ace7e3
Mack M et al (2021) Carbon loss from boreal forest wildfires offset by increased dominance of deciduous trees. Science 372:280–283. https://doi.org/10.1126/science.abf3903
Mahecha MD, Bastos A, Bohn FJ, Eisenhauer N, Feilhauer H, Hickler T et al (2024) Biodiversity and climate extremes: Known interactions and research gaps. Earth's Future 12:e2023EF003963. https://doi.org/10.1029/2023EF003963
Martini D, Sakowska K, Wohlfahrt G, Pacheco-Labrador J, van der Tol C, Porcar-Castell A, Magney TS, Carrara A, Colombo R, El-Madany TS, Gonzalez-Cascon R, Martín MP, Julitta T, Moreno G, Rascher U, Reichstein M, Rossini M, Migliavacca M (2022) Heatwave breaks down the linearity between sun-induced fluorescence and gross primary production. New Phytol 233(6):2415–2428. https://doi.org/10.1111/nph.17920
Massey R, Rogers BM, Berner LT et al (2023) Forest composition change and biophysical climate feedbacks across boreal North America. Nat Clim Chang 13:1368–1375. https://doi.org/10.1038/s41558-023-01851-w
Mbow C, Rosenzweig C, Barioni LG, Benton TG, Herrero M, Krishnapillai M, Liwenga E, Pradhan P, Rivera-Ferre MG, Sapkota T, Tubiello FN, Xu Y (2019) Food security. In: Shukla PR, Skea J, Calvo Buendia E, Masson-Delmotte V, Portner HO, Roberts DC, Zhai P, Slade R, Connors S, van Diemen R, Ferrat M, Haughey E, Luz S, Neogi S, Pathak M, Petzold J, Portugal Pereira J, Vyas P, Huntley E, Kissick K, Belkacemi M, Malley J, (eds) Climate change and land: an IPCC special report on climate change, desertification, land degradation, sustainable land management, food security, and greenhouse gas fluxes in terrestrial ecosystems. https://doi.org/10.1017/9781009157988.007

McCabe RM, Hickey BM, Kudela RM, Lefebvre KA, Adams NG, Bill BD, Gulland FM, Thomson RE, Cochlan WP, Trainer VL (2016) An unprecedented coastwide toxic algal bloom linked to anomalous ocean conditions. Geophys Res Lett 43(19):10–366. https://doi.org/10.1002/2016GL070023

McGee D, deMenocal PB, Winckler G, Stuut JBW, Bradtmiller LI (2013) The magnitude, timing and abruptness of changes in North African dust deposition over the last 20,000 yr. Earth Planet Sci Lett 371–372:163–176. https://doi.org/10.1016/j.epsl.2013.03.054

Medellín-Azuara J, Macewan D, Howitt RE, Sumner DA, Lund JR, Scheer J et al (2016) Economic analysis of the 2016 California drought on agriculture. Center for Watershed Sciences University of California

Meehl GA, Teng H, Rosenbloom N, Hu A, Tebaldi C, Walton G (2022) How the Great Plains Dust Bowl drought spread heat extremes around the Northern Hemisphere. Sci Rep 12:1. https://doi.org/10.1038/s41598-022-22262-5

Messmer M, Simmonds I (2021) Global analysis of cyclone-induced compound precipitation and wind extreme events. Weather Clim Extremes 32:100324. https://doi.org/10.1016/j.wace.2021.100324

Michaud KM, Reed DC, Miller RJ (2022) The Blob marine heatwave transforms California kelp forest ecosystems. Commun Biol 5:1143. https://doi.org/10.1038/s42003-022-04107-z

Mo M, Roache M, Davies J, Hopper J, Pitty H, Foster N, Guy S, Parry-Jones K, Francis G, Koosmen A, Colefax L (2021) Estimating flying-fox mortality associated with abandonments of pups and extreme heat events during the austral summer of 2019–20. Pac Conserv Biol 28(2):124–139. https://doi.org/10.1071/PC21003

Moftakhari HR, Salvadori G, AghaKouchak A, Sanders BF, Matthew RA (2017) Compound effects of sea level rise and fluvial flooding. Proc Natl Acad Sci USA 114:9785–9790. https://doi.org/10.1073/pnas.1620325114

Mogen SC, Lovenduski NS, Dallmann AR, Gregor L, Sutton AJ, Bograd SJ, Quiros NC, Di Lorenzo E, Hazen EL, Jacox MG, Buil MP, Yeager S (2022) Ocean biogeochemical signatures of the north pacific blob. Geophys Res Lett 49:e2021GL096938. https://doi.org/10.1029/2021GL096938

Molina MJ, O'Brien TA, Anderson G, Ashfaq M, Bennett KE, Collins WD, Dagon K, Restrepo JM, Ullrich PA (2023) A review of recent and emerging machine learning applications for climate variability and weather phenomena. Artif Intell Earth Syst 2(4):1–17. https://doi.org/10.1175/aies-d-22-0086.1

Moree A et al (2023) Impact of deoxygenation and warming on global marine species in the 21st century. Biogeosciences. https://doi.org/10.5194/bg-20-2425-2023

Myers-Smith IH, Elmendorf SC, Beck PS, Wilmking M, Hallinger M, Blok D, Tape KD, Rayback SA, Macias-Fauria M, Forbes BC, Speed JD (2015) Climate sensitivity of shrub growth across the tundra biome. Nat Clim Chang 5(9):887–891. https://doi.org/10.1038/nclimate2697

Nickeson J (2023) MODIS land science team (homepage). https://modis-land.gsfc.nasa.gov/. Accessed 13 Apr 2023

Niedermeyer EM, Schefuß E, Sessions AL, Mulitza S, Mollenhauer G, Schulz M, Wefer G (2010) Orbital- and millennial-scale changes in the hydrologic cycle and vegetation in the western African Sahel: insights from individual plant wax dD and d13C. Quat Sci Rev 29:2996–3005. https://doi.org/10.1016/j.quascirev.2010.06.039

Nobre CA, Sampaio G, Borma LS, Castilla-Rubio JC, Silva JS, Cardoso M (2016) Land-use and climate change risks in the Amazon and the need of a novel sustainable development paradigm. Proc Natl Acad Sci USA 113(39):10759–10768. https://doi.org/10.1073/pnas.1605516113

Noh KM, Lim HG, Kug JS (2022) Global chlorophyll responses to marine heatwaves in satellite ocean color. Environ Res Lett 17(6):064034. https://doi.org/10.1088/1748-9326/ac70ec

OCHA (2023) Southern Africa: tropical cyclone Freddy—flash Update No. 7. Reliefweb. https://reliefweb.int/report/mozambique/southern-africa-tropical-cyclone-freddy-flash-update-no-7-14-march-2023. Retrieved 14 Jan 2024

Oliver ECJ, Benthuysen JA, Bindoff NL, Hobday AJ, Holbrook NJ, Mundy CN, Perkins-Kirkpatrick SE (2017) The unprecedented 2015/16 Tasman Sea marine heatwave. Nat Commun. https://doi.org/10.1038/ncomms16101

Oliver ECJ, Donat MG, Burrows MT, Moore PJ, Smale DA, Alexander LV, Benthuysen JA, Feng M, Sen Gupta A, Hobday AJ, Holbrook NJ, Perkins-Kirkpatrick SE, Scannell HA, Straub SC, Wernberg T (2018) Longer and more frequent marine heatwaves over the past century. Nat Commun 9(1):1–12. https://doi.org/10.1038/s41467-018-03732-9

Oliver ECJ, Benthuysen JA, Darmaraki S, Donat MG, Hobday AJ, Holbrook NJ, Schlegel RW, Sen Gupta A (2021) Marine Heatwaves. Ann Rev Mar Sci 13:313–342. https://doi.org/10.1146/annurev-marine-032720-095144

Otto-Bliesner BL, Russell JM, Clark PU, Liu Z, Overpeck JT, Konecky B, deMenocal P, Nicholson SE, He F, Lu Z (2014) Coherent changes of southeastern equatorial and northern African rainfall during the last deglaciation. Science 346:1223–1227. https://doi.org/10.1126/science.1259531
Otto-Bliesner BL, Brady EC, Tomas RA, Albani S, Bartlein PJ, Mahowald NM, Shafer SL, Kluzek E, Lawrence PJ, Leguy G, Rothstein M, Sommers AN (2020) A comparison of the CMIP6 midHolocene and lig127k simulations in CESM2. Paleoceanogr Paleoclimatol 35:10. https://doi.org/10.1029/2020PA003957
Owen LE, Catto JL, Stephenson DB, Dunstone NJ (2021) Compound precipitation and wind extremes over Europe and their relationship to extratropical cyclones. Weather Clim Extrem 33:100342. https://doi.org/10.1016/j.wace.2021.100342
Pausata FSR, Gaetani M, Messori G, Berg A, de Souza DM, Sage RF, DeMenocal PB (2020) The greening of the Sahara: past changes and future implications. One Earth 2(3):235–250. https://doi.org/10.1016/j.oneear.2020.03.002
Pearce-Kelly P, Altier AH, Bruno JF, Cornwall CE, McField M, Muñiz-Castillo AI, Rocha J, Setter RO, Sheppard C, Roman-Cuesta RM, Yesson C (2024) Assessment of warm-water coral reef tipping point thresholds. Earth Syst Dyn Discuss. https://doi.org/10.5194/esd-2023-35
Penn JL, Deutsch C (2022) Avoid ocean mass extinction from climate warming. Science 376(6592):524–526. https://doi.org/10.1126/science.abe9039
Peterson WT et al (2017) The pelagic ecosystem in the Northern California Current off Oregon during the 2014–2016 warm anomalies within the context of the past 20 years. J Geophys Res Oceans 122(9):7267–7290. https://doi.org/10.1002/2017jc012952
Piatt JF, Parrish JK, Renner HM, Schoen SK, Jones TT, Arimitsu ML et al (2020) Extreme mortality and reproductive failure of common murres resulting from the northeast Pacific marine heatwave of 2014–2016. PLoS ONE 15:e0226087. https://doi.org/10.1371/journal.pone.0226087
Qi D, Chen L, Chen B, Gao Z, Zhong W, Feely RA, Anderson LG, Sun H, Chen J, Chen M, Zhan L, Zhang Y, Cai W-J (2017) Increase in acidifying water in the western Arctic Ocean. Nat Clim Chang 7:195–199. https://doi.org/10.1038/nclimate3228
Qin Y, Xiao X, Wigneron JP, Ciais P, Canadell JG, Brandt M, Li X, Fan L, Wu X, Tang H, Dubayah R (2022) Large loss and rapid recovery of vegetation cover and aboveground biomass over forest areas in Australia during 2019–2020. Remote Sens Environ 278:113087. https://doi.org/10.1016/j.rse.2022.113087
Rahmstorf S (2001) Abrupt climate change. In: Steele J, Thorpe S, Turekian K (eds). Encyclopedia of ocean sciences. Academic, London, pp 1–6
Ratnayake HU, Kearney MR, Govekar P, Karoly D, Welbergen JA (2019) Forecasting wildlife die-offs from extreme heat events. Anim Conserv 22(4):386–395. https://doi.org/10.1111/acv.12476
Raymond C, Horton RM, Zscheischler J et al (2020) Understanding and managing connected extreme events. Nat Clim Chang 10:611–621. https://doi.org/10.1038/s41558-020-0790-4
Raymond C, Suarez-Gutierrez L, Kornhuber K, Pascolini-Campbell M, Sillmann J, Waliser DE (2022) Increasing spatiotemporal proximity of heat and precipitation extremes in a warming world quantified by a large model ensemble. Environ Res Lett 17:035005
Swiss Re (2020) Natural catastrophes in times of economic accumulation and climate change (Sigma report No. No 2/2020). Swiss Re Institute. https://www.swissre.com/dam/jcr:85598d6e-b5b5-4d4b-971e-5fc9eee143fb/sigma-2-2020-en.pdf
Rial JA, Pielke RA, Beniston M, Claussen M, Canadell J, Cox P, Held H, de Noblet-Ducoudré N, Prinn R, Reynolds JF, Salas JD (2004) Nonlinearities, feedbacks and critical thresholds within the Earth's climate system. Clim Change 65:11–38. https://doi.org/10.1023/B:CLIM.0000037493.89489.3f
Ridder NN, Pitman AJ, Ukkola AM (2021) Do CMIP6 climate models simulate global or regional compound events skillfully? Geophys Res Lett 48(2):1–11. https://doi.org/10.1029/2020GL091152
Ridder NN, Ukkola AM, Pitman AJ et al (2022) Increased occurrence of high impact compound events under climate change. NPJ Clim Atmos Sci 5:3. https://doi.org/10.1038/s41612-021-00224-4
Roach TNF, Dilworth J et al (2021) Metabolomic signatures of coral bleaching history. Nat Ecol Evol 5:495–503. https://doi.org/10.1038/s41559-020-01388-7
Roach LA, Mankoff KD, Romanou A, Blanchard-Wrigglesworth E, Haine TW, Schmidt GA (2023) Winds and meltwater together lead to Southern Ocean surface cooling and sea ice expansion. Geophys Res Lett 50(24):e2023GL105948. https://doi.org/10.1029/2023GL105948
Robinson JPW, Wilson SK, Jennings S, Graham NAJ (2019) Thermal stress induces persistently altered coral reef fish assemblages. Glob Change Biol. https://doi.org/10.1111/gcb.14704
Rodrigues RR, Subramanian A, Zanna L, Berner J (2019) ENSO bimodality and extremes. Geophys Res Lett 46(9):4883–4893. https://doi.org/10.1029/2019GL082270

Romanou A, Rind D, Jonas J, Miller R, Kelley M, Russell G, Orbe C, Nazarenko L, Latto R, Schmidt GA (2023) Stochastic bifurcation of the north atlantic circulation under a midrange future climate scenario with the NASA-GISS ModelE. J Clim 36(18):6141–6161. https://doi.org/10.1175/jcli-d-22-0536.1
Rosan TM, Sitch S, Mercado LM, Heinrich V, Friedlingstein P, Aragão LEOC (2022) Fragmentation-driven divergent trends in burned area in Amazonia and Cerrado. Front for Glob Change 5:801408. https://doi.org/10.3389/ffgc.2022.801408
Ross T (2020) Rapid deep ocean deoxygenation and acidification threaten life on Northeast Pacific seamounts. Global Change Biol 26(11):6424–6444. https://doi.org/10.1111/gcb.15307
Rotbarth R, Van Nes EH, Scheffer M et al (2023) Northern expansion is not compensating for southern declines in North American boreal forests. Nat Commun 14:3373. https://doi.org/10.1038/s41467-023-39092-2
Röthig T, Trevathan-Tackett SM, Voolstra CR, Ross C, Chaffron S, Durack PJ, Warmuth LM, Sweet M (2023) Human-induced salinity changes impact marine organisms and ecosystems. Glob Change Biol 29(17):4731–4749. https://doi.org/10.1111/gcb.16859
Ruehr NK, Grote R, Mayr S, Arneth A (2019) Beyond the extreme: recovery of carbon and water relations in woody plants following heat and drought stress. Tree Physiol 39:1285–1299. https://doi.org/10.1093/treephys/tpz032
Runge J, Nowack P, Kretschmer M, Flaxman S, Sejdinovic D (2019) Detecting and quantifying causal associations in large nonlinear time series datasets. Sci Adv 5(11):eaau4996. https://doi.org/10.1126/sciadv.aau4996
Saatchi S, Longo M, Xu L, Yang Y, Abe H, Andre M, Aukema JE, Carvalhais N, Cadillo-Quiroz H, Cerbu GA, Chernela JM (2021) Detecting vulnerability of humid tropical forests to multiple stressors. One Earth 4(7):988–1003. https://doi.org/10.1016/j.oneear.2021.06.002
Sakschewski B, von Bloh W, Boit A et al (2016) Resilience of Amazon forests emerges from plant trait diversity. Nat Clim Change 6:1032–1036. https://doi.org/10.1038/nclimate3109
Sanderson B, Fisher R (2020) A fiery wake-up call for climate science. Nat Clim Change. https://doi.org/10.1038/s41558-020-0707-2
Santora JA, Mantua NJ, Schroeder ID et al (2020) Habitat compression and ecosystem shifts as potential links between marine heatwave and record whale entanglements. Nat Commun 11:536. https://doi.org/10.1038/s41467-019-14215-w
Saranya JS, Roxy MK, Dasgupta P, Anand A (2022) Genesis and trends in marine heatwaves over the tropical indian ocean and their interaction with the indian summer monsoon. JGR Oceans 127:e2021JC017427. https://doi.org/10.1029/2021JC017427
Schär C, Vidale PL, Lüthi D, Frei C, Häberli C, Liniger MA, Appenzeller C (2004) The role of increasing temperature variability in European summer heatwaves. Nature 427(6972):332–336. https://doi.org/10.1038/nature02300
Scheffer M, Bascompte J, Brock WA, Brovkin V, Carpenter SR, Dakos V, Held H, van Nes EH, Rietkerk M, Sugihara G (2009) Early-warning signals for critical transitions. Nature 461:53–59. https://doi.org/10.1038/nature08227
Scheffer M, Hirota M, Holmgren M, Van Nes EH, Chapin FS III (2012) Thresholds for boreal biome transitions. Proc Natl Acad Sci 109(52):21384–21389. https://doi.org/10.1073/pnas.1219844110
Schmidt GA, Romanou A, Roach L, Mankoff K, Li Q, Rye CD, Kelley M, Marshall JC, Busecke JJM (2023) Anomalous meltwater from ice sheets and ice shelves is a historical forcing. Geophys Res Lett 50(24):e2023GL106530. https://doi.org/10.1029/2023GL106530
Schmittner A (2005) Decline of the marine ecosystem caused by a reduction in the Atlantic overturning circulation. Nature 434:628–633. https://doi.org/10.1038/nature03476
Schmittner A, Galbraith ED (2008) Glacial greenhouse-gas fluctuations controlled by ocean circulation changes. Nature 456:373–376. https://doi.org/10.1038/nature07531
Schmittner A, Galbraith ED, Hostetler SW, Pedersen TF, Zhang R (2007) Large fluctuations of dissolved oxygen in the Indian and Pacific oceans during Dansgaard-Oeschger oscillations caused by variations of North Atlantic Deep Water subduction: OXYGEN DURING D/O OSCILLATIONS. Paleoceanography. https://doi.org/10.1029/2006pa001384
Scholten RC, Coumou D, Luo F, Veraverbeke S (2022) Early snowmelt and polar jet dynamics co-influence recent extreme Siberian fire seasons. Science 378(6623):1005–1009. https://doi.org/10.1126/science.abn4419
Schubert SD, Suarez MJ, Pegion PJ, Koster RD, Bacmeister JT (2004) On the cause of the 1930s dust bowl. Science 303(5665):1855–1859. https://doi.org/10.1126/science.1095048
Seidl R, Turner MG (2022) Post-disturbance reorganization of forest ecosystems in a changing world. PNAS 119(28):e2202190119. https://doi.org/10.1073/pnas.22021901

Senapati N, Halford NG, Semenov MA (2021) Vulnerability of European wheat to extreme heat and drought around flowering under future climate. Environ Res Lett 16(2):024052. https://doi.org/10.1088/1748-9326/abdcf3

Seneviratne SI, Lüthi D, Litschi M, Schär C (2006) Land-atmosphere coupling and climate change in Europe. Nature 443:205–209. https://doi.org/10.1038/nature05095

Seneviratne SI, Corti T, Davin EL, Hirschi M, Jaeger EB, Lehner I, Orlowsky B, Teuling AJ (2010) Investigating soil moisture-climate interactions in a changing climate: a review. Earth Sci Rev 99(3–4):125–161. https://doi.org/10.1016/j.earscirev.2010.02.004

Seneviratne SI, Zhang X, Adnan M, Badi W, Dereczynski C, Di Luca A, Ghosh S, Iskandar I, Kossin J, Lewis S, Otto F, Pinto I, Satoh M, Vicente-Serrano SM, Wehner M, Zhou B (2021) Weather and climate extreme events in a changing climate. In: Masson-Delmotte V, Zhai P, Pirani A, Connors SL, Péan C, Berger S, Caud N, Chen Y, Goldfarb L, Gomis MI, Huang M, Leitzell K, Lonnoy E, Matthews JBR, Maycock TK, Waterfield T, Yelekçi O, Yu R, Zhou B (eds) Climate change 2021: the physical science basis. Contribution of working group I to the sixth assessment report of the intergovernmental panel on climate change. Cambridge University Press, Cambridge. pp 1513–1766. https://doi.org/10.1017/9781009157896.013

Shanahan TM, McKay NP, Hughen KA, Overpeck JT, Otto-Bliesner B, Heil CW, King J, Scholz CA, Peck J (2015) The time- transgressive termination of the African Humid Period. Nat Geosci 8:140–144. https://doi.org/10.1038/ngeo2329

Sherwood SC, Ramsay EE (2023) Closer limits to human tolerance of global heat. Proc Natl Acad Sci 120:43. https://doi.org/10.1073/pnas.2316003120

Shi H, García-Reyes M, Jacox MG, Rykaczewski RR, Black BA, Bograd SJ, Sydeman WJ (2021) Co-occurrence of California drought and northeast pacific Marine heatwaves under climate change. Geophys Res Lett 48:e2021GL092765. https://doi.org/10.1029/2021GL092765

Sillmann J, Shepherd TG, van den Hurk B, Hazeleger W, Martius O, Slingo J, Zscheischler J (2021) Event-based storylines to address climate risk. Earth's Future 9:e2020EF001783. https://doi.org/10.1029/2020EF001783

Silva Junior CHL, Pessôa ACM, Carvalho NS et al (2021) The Brazilian Amazon deforestation rate in 2020 is the greatest of the decade. Nat Ecol Evol 5:144–145. https://doi.org/10.1038/s41559-020-01368-x

Slivinski LC et al (2019) Towards a more reliable historical reanalysis: improvements for version 3 of the twentieth century reanalysis system. Q J R Meteorol Soc 145(724):2876–2908. https://doi.org/10.1002/qj.3598

Smale DA, Wernberg T, Oliver EC, Thomsen M, Harvey BP, Straub SC, Burrows MT, Alexander LV, Benthuysen JA, Donat MG, Feng M (2019) Marine heatwaves threaten global biodiversity and the provision of ecosystem services. Nat Clim Chang 9(4):306–312. https://doi.org/10.1038/s41558-019-0412-1

Smith T, Boers N (2023) Reliability of vegetation resilience estimates depends on biomass density. Nat Ecol Evol 7:11. https://doi.org/10.1038/s41559-023-02194-7

Smith KM, Payton TG, Sims RJ, Stroud CS, Jeanes RC, Hyatt TB, Childress MJ (2019) Impacts of consecutive bleaching events and local algal abundance on transplanted coral colonies in the Florida Keys. Coral Reefs 38(4):851–861. https://doi.org/10.1007/s00338-019-01823-7

Smith SL, O'Neill HB, Isaksen K et al (2022) The changing thermal state of permafrost. Nat Rev Earth Environ 3:10–23. https://doi.org/10.1038/s43017-021-00240-1

Smith KE, Burrows MT, Hobday AJ, King NG, Moore PJ, Sen Gupta A, Thomsen MS, Wernberg T, Smale DA (2023) Biological impacts of marine heatwaves. Ann Rev Mar Sci 15:119–145. https://doi.org/10.1146/annurev-marine-032122-121437

Spalding MD, Brown BE (2015) Warm-water coral reefs and climate change. Science 350(6262):769–771. https://doi.org/10.1126/science.aad0349

SROCC (2019) Extremes, abrupt changes and managing risks. In: Pörtner HO, Roberts DC, Masson-Delmotte V, Zhai P, Tignor M, Poloczanska E, Mintenbeck K, Alegría A, Nicolai M, Okem A, Petzold J, Rama B, Weyer NM (eds) IPCC special report on the ocean and cryosphere in a changing climate. Cambridge University Press, Cambridge, pp 589–655. https://doi.org/10.1017/9781009157964.008

Stramma L, Johnson GC, Sprintall J, Mohrholz V (2008) Expanding oxygen-minimum zones in the tropical oceans. Science 320(5876):655–658. https://doi.org/10.1126/science.1153847

Stramma L, Prince ED, Schmidtko S, Luo J, Hoolihan JP, Visbeck M et al (2012) Expansion of oxygen minimum zones may reduce available habitat for tropical pelagic fishes. Nat Clim Chang 2(1):33–37. https://doi.org/10.1038/nclimate1304

Straub SC, Wernberg T, Thomsen MS, Moore PJ, Burrows MT, Harvey BP, Smale DA (2019) Resistance, extinction, and everything in between–the diverse responses of seaweeds to marine heatwaves. Front Mar Sci. https://doi.org/10.3389/fmars.2019.00763
Suarez-Gutierrez L, Müller WA, Li C et al (2020) Hotspots of extreme heat under global warming. Clim Dyn 55:429–447. https://doi.org/10.1007/s00382-020-05263-w
Suarez-Gutierrez L, Müller WA, Marotzke J (2023) Extreme heat and drought typical of an end-of-century climate could occur over Europe soon and repeatedly. Commun Earth Environ 4:415. https://doi.org/10.1038/s43247-023-01075-y
Sun C, Shanahan TM, DiNezio PN et al (2021) Great Plains storm intensity since the last glacial controlled by spring surface warming. Nat Geosci 14:912–917. https://doi.org/10.1038/s41561-021-00860-8
Sun Y, Gu L, Wen J, van der Tol C, Porcar-Castell A, Joiner J, Chang CY, Magney T, Wang L, Hu L, Rascher U, Zarco-Tejada P, Barrett CB, Lai J, Han J, Luo Z (2023) From remotely sensed solar-induced chlorophyll fluorescence to ecosystem structure, function, and service: Part I—Harnessing theory. Glob Change Biol 29:2926–2952. https://doi.org/10.1111/gcb.16634
Sutton AJ, Sabine CL, Feely RA, Cai WJ, Cronin MF, McPhaden MJ, Morell JM, Newton JA, Noh JH, Ólafsdóttir SR, Salisbury JE, Send U, Vandemark DC, Weller RA (2016) Using present-day observations to detect when anthropogenic change forces surface ocean carbonate chemistry outside preindustrial bounds. Biogeosciences 13:5065–5083. https://doi.org/10.5194/bg-13-5065-2016
Tait LW, Thoral F, Pinkerton MH, Thomsen MS, Schiel DR (2021) Loss of giant Kelp, *Macrocystis pyrifera*, driven by marine heatwaves and exacerbated by poor water clarity in New Zealand. Front Mar Sci. https://doi.org/10.3389/fmars.2021.721087
Tassone SJ, Besterman AF, Buelo CD et al (2022) Co-occurrence of aquatic heatwaves with atmospheric heatwaves, low dissolved oxygen, and low pH events in estuarine ecosystems. Estuar Coasts 45:707–720. https://doi.org/10.1007/s12237-021-01009-x
Techera E, Winter G (2019) Marine extremes: ocean safety, marine health and the blue economy. Routledge, Oxon. https://doi.org/10.4324/9780429491023
Teckentrup L, Harrison SP, Hantson S, Heil A, Melton JR, Forrest M, Li F, Yue C, Arneth A, Hickler T, Sitch S, Lasslop G (2019) Response of simulated burned area to historical changes in environmental and anthropogenic factors: a comparison of seven fire models. Biogeosciences 16:3883–3910. https://doi.org/10.5194/bg-16-3883-2019
Tedesco M, Fettweis X, van den Broeke MR, van de Wal RS, Smeets CJPP, van de Berg WJ, Serreze MC, Box JE (2011) The role of albedo and accumulation in the 2010 melting record in Greenland. Environ Res Lett 6:014005. https://doi.org/10.1088/1748-9326/6/1/014005
Thomas ZA, Jones RT, Turney CSM, Golledge N, Fogwill C, Bradshaw CJA et al (2020) Tipping elements and amplified polar warming during the last interglacial. Quatern Sci Rev 233:106222. https://doi.org/10.1016/j.quascirev.2020.106222
Thomson JA, Burkholder DA, Heithaus MR, Fourqurean JW, Fraser MW, Statton J, Kendrick GA (2015) Extreme temperatures, foundation species, and abrupt ecosystem change: an example from an iconic seagrass ecosystem. Global Change Biol 21:1463–1474. https://doi.org/10.1111/gcb.12694
Tierney JE, Pausata FSR, De Menocal PB (2017) Rainfall regimes of the Green Sahara. Sci Adv 3:e1601503. https://doi.org/10.1126/sciadv.1601503
Trisos CH, Merow C, Pigot AL (2020) The projected timing of abrupt ecological disruption from climate change. Nature 580:7804. https://doi.org/10.1038/s41586-020-2189-9
Tucker CJ (1979) Red and photographic infrared linear combinations for monitoring vegetation. Remote Sens Environ 8(2):127–150. https://doi.org/10.1016/0034-4257(79)90013-0
Turner MG (2010) Disturbance and landscape dynamics in a changing world. Ecology 91:2833–2849. https://doi.org/10.1890/10-0097.1
Turner MG, Seidl R (2023) Novel disturbance regimes and ecological responses. Annu Rev Ecol Evol Syst 54:63–83. https://doi.org/10.1146/annurev-ecolsys-110421-101120
van Oldenborgh GJ et al (2021) Attribution of the Australian bushfire risk to anthropogenic climate change. Nat Hazard. https://doi.org/10.5194/nhess-21-941-2021
Vanos J, Guzman-Echavarria G, Baldwin JW et al (2023) A physiological approach for assessing human survivability and liveability to heat in a changing climate. Nat Commun 14:7653. https://doi.org/10.1038/s41467-023-43121-5
Vecellio DJ, Wolf ST, Cottle RM, Kenney WL (2022) Evaluating the 35Â °C wet-bulb temperature adaptability threshold for young, healthy subjects (PSU HEAT Project). J Appl Physiol 132:2. https://doi.org/10.1152/japplphysiol.00738.2021
Villalobos-Herrera R, Bevacqua E, Ribeiro AFS, Auld G, Crocetti L, Mircheva B, Ha M, Zscheischler J, De Michele C (2021) Towards a compound-event-oriented climate model evaluation: a

decomposition of the underlying biases in multivariate fire and heat stress hazards. Nat Hazards Earth Syst Sci 21:1867–1885. https://doi.org/10.5194/nhess-21-1867-2021
Vogel MM, Zscheischler J, Wartenburger R, Dee D, Seneviratne SI (2019) Concurrent 2018 hot extremes across Northern Hemisphere due to human-induced climate change. Earth's Future 7:692–703. https://doi.org/10.1029/2019EF001189
Wahl H, Galler R, Winiwarter W (2015) Demonstration of the generic software framework for the integrated intelligent computer-assisted language learning (iiCALL) environment. In: Proceedings of the 17th international conference on information integration and web-based applications and services (iiWAS '15). vol 60. Association for Computing Machinery, New York, pp 1–6. https://doi.org/10.1145/2837185.2837217
Wang S, Foster A, Lenz EA, Kessler JD, Stroeve JC, Anderson LO, Turetsky M, Betts R, Zou S, Liu W, Boos WR, Hausfather Z (2023a) Mechanisms and impacts of earth system tipping elements. Rev Geophys 61(1):1–81. https://doi.org/10.1029/2021RG000757
Wang Z, Wang Z, Zou Z et al (2023b) Severe global environmental issues caused by Canada's record-breaking wildfires in 2023. Adv Atmos Sci. https://doi.org/10.1007/s00376-023-3241-0
Wernberg T (2021) Marine heatwave drives collapse of kelp forests in Western Australia. In: Canadell JG, Jackson RB (eds). Ecosystem collapse and climate change. Ecological studies, vol 241. Springer, Cham. https://doi.org/10.1007/978-3-030-71330-0_12
White RH et al (2023) The unprecedented Pacific Northwest heatwave of June 2021. Nat Commun 14:1. https://doi.org/10.1038/s41467-023-36289-3
Whitney FA (2015) Anomalous winter winds decrease 2014 transition zone productivity in the NE Pacific. Geophys Res Lett 42:428–431. https://doi.org/10.1002/2014GL062634
Williams AP et al (2020) Large contribution from anthropogenic warming to a developing North American megadrought. Science 368:314–318. https://doi.org/10.1126/science.aaz960
Wood RA, Baker JA, Beaugrand G, Boutin J, Conversi A, Donner RV, Frenger I, Goberville E, Hayashida H, Koeve W, Kvale K, Landolfi A, Maslowski W, Oschlies A, Romanou A, Somes CJ, Stocker TF, Swingedouw D (2024) Opportunities for earth observation to inform risk management for ocean tipping points. Surv. Geophys. https://doi.org/10.1007/s10712-024-09859-3
Woodhouse CA, Overpeck JT (1998) 2000 years of drought variability in the central United States. Bull Am Meteorol Soc 79:2693–2714
Worster D (1979) Dust bowl: the southern plains in the 1930s. Oxford University Press
Wu W, McInnes K, O'grady J, Hoeke R, Leonard M, Westra S (2018) Mapping dependence between extreme rainfall and storm surge. J Geophys Res Oceans 123(4):2461–2474. https://doi.org/10.1002/2017JC013472
Xu W, Wooster MJ (2023) Sentinel-3 SLSTR active fire (AF) detection and FRP daytime product-Algorithm description and global intercomparison to MODIS, VIIRS and landsat AF data. Sci Remote Sens. https://doi.org/10.1016/j.srs.2023.100087
Xu Y, Kalra TS, Ganju NK, Fagherazzi S (2022) Modeling the dynamics of salt marsh development in coastal land reclamation. Geophys Res Lett 49:e2021GL095559. https://doi.org/10.1029/2021GL095559
Yaddanapudi R, Mishra A, Huang W, Chowdhary H (2022) Compound wind and precipitation extremes in global coastal regions under climate change. Geophys Res Lett. https://doi.org/10.1029/2022GL098974
Yang YT, Lam KM, Dong JY, Sun X, Jian MW (2021) Super-resolution on remote sensing images. In: Nakajima M, Kim JG, Lie WN, Kemao Q (eds) International workshop on advanced imaging technology (IWAIT) 2021, p 11766. https://doi.org/10.1117/12.2590197
Yu P et al (2022a) Global spatiotemporally continuous MODIS land surface temperature dataset. Sci Data 9:1. https://doi.org/10.1038/s41597-022-01214-8
Yu X et al (2022b) Contrasting drought legacy effects on gross primary productivity in a mixed versus pure beech forest. Biogeosciences 19:4315–4329. https://doi.org/10.5194/bg-19-4315-2022
Zemp D, Schleussner CF, Barbosa H et al (2017) Self-amplified Amazon forest loss due to vegetation-atmosphere feedbacks. Nat Commun 8:14681. https://doi.org/10.1038/ncomms14681
Zhang P, Jeong JH, Yoon JH, Kim H, Wang SYS, Linderholm HW, Fang K, Wu X, Chen D (2020) Abrupt shift to hotter and drier climate over inner East Asia beyond the tipping point. Science 370(6520):1095–1099. https://doi.org/10.1126/science.abb3368
Zheng B, Ciais P, Chevallier F, Yang H, Canadell JG, Chen Y, van der Velde IR, Aben I, Chuvieco E, Davis SJ, Deeter M (2023) Record-high CO_2 emissions from boreal fires in 2021. Science 379(6635):912–917. https://doi.org/10.1126/science.ade0805
Zscheischler J, Lehner F (2022) Attributing compound events to anthropogenic climate change. BAMS. https://doi.org/10.1175/BAMS-D-21-0116.1

Zscheischler J, Seneviratne SI (2017) Dependence of drivers affects risks associated with compound events. Sci Adv 2017(3):e1700263. https://doi.org/10.1126/sciadv.1700263
Zscheischler J, Westra S, Van Den Hurk BJ, Seneviratne SI, Ward PJ, Pitman A, AghaKouchak A, Bresch DN, Leonard M, Wahl T, Zhang X (2018) Future climate risk from compound events. Nat Clim Chang 8(6):469–477. https://doi.org/10.1038/s41558-018-0156-3
Zscheischler J, Martius O, Westra S et al (2020) A typology of compound weather and climate events. Nat Rev Earth Environ 1:333–347. https://doi.org/10.1038/s43017-020-0060-z

Authors and Affiliations

A. Romanou[1] · G. C. Hegerl[2] · S. I. Seneviratne[3] · B. Abis[4] · A. Bastos[5,6] · A. Conversi[7] · A. Landolfi[8] · H. Kim[9] · P. E. Lerner[10] · J. Mekus[1,11] · B. L. Otto-Bliesner[12] · F. S. R. Pausata[13] · I. Pinto[14] · L. Suarez-Guiterrez[3,15]

✉ G. C. Hegerl
gabi.hegerl@ed.ac.uk

1 NASA Goddard Institute for Space Studies, and Department of Applied Physics and Applied Mathematics, Columbia University, New York, NY 10025, USA

2 School of GeoSciences, University of Edinburgh, Edinburgh EH8 9XP, UK

3 Institute for Atmospheric and Climate Science, ETH Zurich, 8006 Zurich, Switzerland

4 Starion Group, 00044 Frascati, Italy

5 Institute for Earth System Science and Remote Sensing, Leipzig University, 04013 Leipzig, Germany

6 Max Planck Institute for Biogeochemistry, 07745 Jena, Germany

7 National Research Council of Italy, CNR - ISMAR - Lerici, Forte Santa Teresa, Loc. Pozzuolo, 19032 Lerici, SP, Italy

8 National Research Council of Italy, CNR - ISMAR - Rome, 00133 Rome, Italy

9 Moon Soul Graduate School of Future Strategy, Korea Advanced Institute of Science and Technology, Daejeon 305-701, Republic of Korea

10 Columbia University, New York, NY 10027, USA

11 Autonomic Integra, New York, NY 10025, USA

12 Climate and Global Dynamics Laboratory, NSF National Center for Atmospheric Research, Boulder, CO 80307-3000, USA

13 Department of Earth and Atmospheric Sciences, Centre ESCER (Etude Et Simulation du Climat À L'Echelle Regionale) and GEOTOP (Research Center on the Dynamics of the Earth System), University of Quebec in Montréal, Montréal (Québec) H3C 3P8, Canada

14 Royal Netherlands Meteorological Institute (KNMI), NL-3731 GA De Bilt, The Netherlands

15 Institut Pierre-Simon Laplace, CNRS, 75252 Paris Cedex 05, France

Surveys in Geophysics (2025) 46:421–442
https://doi.org/10.1007/s10712-024-09844-w

Reflecting on the Science of Climate Tipping Points to Inform and Assist Policy Making and Address the Risks they Pose to Society

T. F. Stocker[1,2] · R. G. Jones[3] · M. I. Hegglin[4,5] · T. M. Lenton[6] · G. C. Hegerl[7] · S. I. Seneviratne[8] · N. van der Wel[9] · R. A. Wood[3]

Received: 19 September 2023 / Accepted: 8 May 2024 / Published online: 4 June 2024

Abstract
There is a diverging perception of climate tipping points, abrupt changes and surprises in the scientific community and the public. While such dynamics have been observed in the past, e.g., frequent reductions of the Atlantic meridional overturning circulation during the last ice age, or ice sheet collapses, tipping points might also be a possibility in an anthropogenically perturbed climate. In this context, high impact—low likelihood events, both in the physical realm as well as in ecosystems, will be potentially dangerous. Here we argue that a formalized assessment of the state of science is needed in order to establish a consensus on this issue and to reconcile diverging views. This has been the approach taken by the Intergovernmental Panel on Climate Change (IPCC). Since 1990, the IPCC has consistently generated robust consensus on several complex issues, ranging from the detection and attribution of climate change, the global carbon budget and climate sensitivity, to the projection of extreme events and their impact. Here, we suggest that a scientific assessment on tipping points, conducted collaboratively by the IPCC and the Intergovernmental Science-Policy Platform on Biodiversity and Ecosystem Services, would represent an ambitious yet necessary goal to be accomplished within the next decade.

Keywords Tipping points · HILL events · IPCC assessment · Scientific consensus

✉ T. F. Stocker
stocker@climate.unibe.ch

1 Climate and Environmental Physics, Physics Institute, University of Bern, 3012 Bern, Switzerland

2 Oeschger Centre for Climate Change Research, University of Bern, 3012 Bern, Switzerland

3 MetOffice, Hadley Centre, Exeter EX1 3PB, United Kingdom

4 Institut für Energie- und Klimaforschung, Forschungszentrum Jülich, 52428 Jülich, Germany

5 Meteorology Department, University of Reading, Reading RG6 6ET, United Kingdom

6 Global Systems Institute, University of Exeter, Exeter EX4 4RN, United Kingdom

7 School of Geosciences, University of Edinburgh, Edinburgh EH9 3FE, United Kingdom

8 Institute of Atmospheric and Climate Science, ETH Zurich, 8092 Zürich, Switzerland

9 World Climate Research Programme, WMO, 1211 Geneva 2, Switzerland

Article Highlights

- A scientific consensus regarding tipping points, abrupt changes and surprises in the climate system will address diverging perceptions in the scientific community and the public, and accelerate scientific progress
- The Intergovernmental Panel on Climate Change (IPCC) has a successful history of generating consensus through a formalized process across a wide range of topics, from the physical science basis to mitigation of climate change
- A broader view on tipping points, including high impact—low likelihood events, provided more comprehensive information for policymakers and the public

1 Introduction

With a flurry of high-profile papers in the past few years (Boers et al. 2017; Lenton et al. 2019; Pattyn and Morlighem 2020; Boers 2021; Boers and Rypdal 2021; Caesar et al. 2020; DeConto et al. 2021; Armstrong McKay et al. 2022; Ditlevsen and Ditlevsen 2023) the conversation on tipping points, abrupt changes and surprises (terms defined in (IPCC 2021a)) has been picked up by the media. The Global Tipping Points Report (Lenton et al. 2023), an interdisciplinary status report written by a group of scientists, was published during COP28 in December 2023 and received much media coverage worldwide. This is just the most recent example where media regularly amplify such potentially catastrophic consequences of anthropogenic climate change to "doomsday" scenarios. This contrasts starkly with the more detailed presentations in scientific publications (Weijer et al. 2019; Heinze et al. 2021; Wang et al. 2023) and the more nuanced, perhaps even conservative, assessments of this highly policy-relevant topic presented in successive reports of the Intergovernmental Panel on Climate Change (IPCC) since 1990. Furthermore, some authors criticize the overly narrow tipping point framing and argue in an opinion paper that focusing on them would distract from urgent action (Kopp et al. 2023). Given the potential risks from tipping points, we argue that this topic now requires focused attention and a formalized process to achieve a scientific consensus where possible.

The purpose of this paper is therefore to discuss ways to reduce the widely diverging perception among the public, policymakers, and even scientists, of tipping points, abrupt changes and surprises in the climate system that appears to have grown in recent years. This divergence was also enhanced by social movements engaged in climate action when the urgent need for action was argued with the approach of a threshold beyond which irreversible changes would be unleashed (Hagedorn et al. 2019). The concern regarding the approach of the Earth System toward tipping points or thresholds is real and scientifically fully justified, all the more since recent research has identified nonlinear transitions that may occur earlier than previously suggested (Ditlevsen and Ditlevsen 2023), or that involve interactions not considered in previous work such as vegetation dieback (Bochow and Boers 2023) and its impacts on regional climate (Nepstad et al. 2008). Nevertheless, we advise not to neglect the consensus finding process in the scientific community in this hotly debated topic in order to better quantify the probability of reaching tipping points, and identify early warning signs, increasingly using space-based Earth observations (Boulton et al. 2022) and employing the most advanced climate models (Schär et al. 2020). Consensus processes have been applied successfully to other low probability outcomes, such as extreme events. This has led to a widely shared awareness about the growing risk of extreme weather and climate events in the coming decades.

Scientific consensus finding is the starting point of effective and coherent communication. It is essential for the maintenance of the credibility of science, robust policy advice, and long-term implementation of mitigation. There is ample experience in the scientific community and the policymaker arena on how to successfully pave the way to scientific consensus, and there is strong appreciation for the value of a scientific consensus that emerges from a formalized process. A case in point is the global carbon budget. This concept was first presented in the scientific literature around 2009 (Allen et al. 2009; Matthews et al. 2009). By 2013 the scientific consensus had solidified to the point that the concept could be defended against strong resistance by some policymakers (IPCC 2013b). Without the formalized assessment process carried out by the IPCC, which has been applied since its inception in 1988, the scientific basis of important milestones to protect the climate system, such as the UN Framework Convention on Climate Change (UNFCCC), the Kyoto Protocol, and the historic Paris Agreement, could not have been achieved.

This paper is organized as follows. First, we consider the Earth System as a dynamical system with a range of possible behaviors, in particular abrupt changes. A well-documented and simulated example of a climate tipping point is then discussed. Key gaps in the understanding of climate surprises are identified, and an impact-oriented approach is advocated. Four examples of scientific progress enabled by consensus finding in a formalized assessment process are presented. Finally, it is argued that clear statements and simple language, firmly rooted in the science, render communication effective. In conclusion, a combined, interdisciplinary assessment process involving both the IPCC and the Intergovernmental Science-Policy Platform on Biodiversity and Ecosystem Services (IPBES) is proposed. Inclusion of the latter is important as both climate and biodiversity are tightly linked.

2 The Complex Climate System as a Dynamical System

The climate system can be understood as a dynamical system that is modeled by a set of nonlinear partial differential equations and parameterizations, including state equations. As such, the climate system is expected to exhibit, in principle, at least some features of the wide range of behavior documented by the mathematical analysis of such dynamical systems (Ghil and Lucarini 2020). These involve multiple equilibrium states, both stable and unstable, damped and self-sustained oscillations, attractors, limit cycles and chaotic behavior. This applies for the individual components of the climate system, such as the atmosphere, ocean, cryosphere, and biosphere, as well as for the fully coupled system. Early examples are multiple equilibria in the ocean's deep circulation (Stommel 1961), the complexity of the meridional atmospheric circulation that evolves from thermal instabilities (Lorenz 1963), the planetary energy balance (Sellers 1969) which shows multiple equilibria caused by the nonlinear dependence of planetary albedo on surface temperature, the El Niño-Southern Oscillation variability in the equatorial Pacific caused by ocean-atmosphere instabilities (Zebiak and Cane 1987), or glacial–interglacial cycles (Saltzman and Maasch 1991).

The Lorenz model has revolutionized the science of dynamical systems and provoked a paradigm shift in our understanding of the climate system. Although the model is entirely deterministic, predictability of the internal structure of the circulation is fundamentally limited. This is because the system can abruptly change from one mode of variability to another which is characterized by different amplitude and frequency. For an observer, such a change may come as a surprise, if warning signals are unknown, inconclusive, or ignored. Loss of predictability does not imply the absence of predictive power to estimate some global properties and their changes in the system. For instance, the preference of one mode over another, when some external forcing is applied to the system (Palmer 1999), is an example of predictability in the presence of chaos. Furthermore, depending on the specific parameter values the behavior may change from stable to oscillatory to chaotic, suggesting that external forcing resulting in different climate states may also lead to different dynamical characteristics. The Lorenz model teaches us that such different regimes may lie very close to one another and therefore, slight changes in the background state could, in principle, lead to completely different dynamical behavior.

Abrupt shifts, surprises, and irreversibility are not just characteristics of simplified models (Weijer et al. 2019). Indeed, there is now a wide range of such dynamical behavior that can be found in the currently most comprehensive coupled climate models (Drijfhout et al. 2015). In their survey of the coupled models assessed by the IPCC in 2013, they found many regional climate indicators that exhibited abrupt change in response to future warming, even for global warming levels below 2°C. The model generation at that time, however, did not yield a consistent picture across different models. The next generation of coupled models was analyzed with respect to the stability of the ocean circulation to freshwater perturbations and showed a wide range of response (Baker et al. 2023; Jackson et al. 2023).

Not only the warming, but also small stochastic fluctuations at the ocean's surface could destabilize ocean circulation and result in large changes, as earlier demonstrated in models of reduced complexity (Knutti and Stocker 2002), and most recently in state-of-the-art coupled models (Romanou et al. 2023), with consequent impacts on atmospheric circulation and regional climate (Orbe et al. 2023). The coupled ocean-atmosphere system may thus be less stable than previously assessed. Another possibility of dynamical behavior are self-sustained oscillations under current and past climate states (Kuniyoshi et al. 2022; Vettoretti et al. 2022; Izumi et al. 2023). As such, complex dynamics are found in the entire hierarchy of climate models; thus we argue that such, behavior is an intrinsic part of the functioning of the Earth System (Weijer et al. 2019; Malmierca-Vallet et al. 2023). Ongoing modeling intercomparison projects, e.g., TipMIP (Loriani et al. 2023), will help in elucidating the dynamical behavior of the current climate models much more systematically and provide new insight into which of the many tipping elements in Lenton et al. (2023) may be likely or plausible and hence need to be monitored.

3 The Atlantic Meridional Overturning Circulation: An Iconic Tipping Element

3.1 Combining Evidence from the Paleoclimatic Records and Models

Based on the general mathematical knowledge of nonlinear dynamical systems, it is no surprise that the paleoclimatic record, in particular reconstructions at decadal, annual and sub-annual resolutions, has been interpreted through the lens of dynamical systems. An early example is the interpretation of an abrupt millennial cooling during the transition from the last ice age to the current Holocene, about 12,900 years ago, found both in a Greenland ice core (Dansgaard et al. 1993) and in lake sediments in Switzerland (Oeschger et al. 1984). These authors suggested that an ocean flip-flop system would be the origin of these climate swings and thus revived an elegant, almost forgotten idea involving two stable equilibria of the meridional overturning circulation (Stommel 1961). One circulation state, typical for the modern Atlantic Ocean, shows strong meridional overturning and an associated substantial meridional heat flux. The other stable state has no overturning and, hence, the northward heat flux is absent. The iconic view of the deep ocean circulation as a heat "conveyor belt" as devised by Broecker (1987a), motivated researchers to seek further evidence for an unstable ocean circulation in the past (Broecker et al. 1985), and potential implications for the future (Broecker 1987b).

Strong initial support for ocean-mediated abrupt climate change was given by idealized model simulations that exhibited multiple equilibria (Bryan 1986; Stocker and Wright 1991; Stocker et al. 1992). Since then, it has been recognized that such abrupt coolings and warmings are a common feature of climate variability during ice ages. They were identified in high-resolution records of marine sediments, polar ice cores, lake sediments, speleothems, and tree rings throughout the entire northern hemisphere (Li and Born 2019; Rousseau et al. 2022). The important role of the ocean in these abrupt temperature swings registered in Greenland ice cores was further evidenced in the characteristic century-to-millennial scale, more gradual warmings and coolings found in ice cores from Antarctica for each of the abrupt changes in the north (EPICA Community Members 2006). This has been interpreted as the fingerprint of a thermal bipolar seesaw mediated by the Atlantic meridional overturning circulation (AMOC) (Crowley 1992; Stocker and Johnsen 2003; Pedro et al. 2018). Taken together, the past 40 years of paleoclimate analysis and modeling have demonstrated the important role of ocean circulation in limiting the stability of the global climate system (Clark et al. 2002; Weijer et al. 2019). They also emphasize the value of paleoclimatic evidence for identifying possible tipping points, and the need to interpret these considering the global system (e.g., Brovkin et al. 2021).

What is valid for the past is also relevant for the future, in particular in an anthropogenically forced future. If the climate system has exhibited limited stability to perturbations and responded with abrupt changes in the past, it would be no surprise if such behavior were to develop in the future. In fact, it would be quite remarkable if this same physical system were to behave strictly linearly and reversibly despite the potentially very large and rapid anthropogenic perturbations. Indeed, early model simulations based on idealized scenarios suggested that the AMOC would reduce in response to the warming (Stouffer et al. 1989). In the most recent family of scenarios (Riahi et al. 2017), business-as-usual would result in global mean heating of 3.3–5.7°C relative to pre-industrial levels by the end of this century (IPCC 2021b). This approaches the size of a typical glacial-to-interglacial warming of 5–7°C. The evident difference is the speed of change: The natural global mean warming out of the last ice age took about 10,000 years, whereas the ongoing anthropogenic heating would reach the same magnitude in just 200 years in a business-as-usual scenario, appearing like a shock compared to the gentler evolution of the climate system during the last several million years. But even lower scenarios are problematic since so-called temporary overshoots are now also considered to be compatible with the goals of the Paris Agreement, at least in the long-term (Tokarska et al. 2019). Such overshoots can generate hysteresis (Jeltsch-Thömmes et al. 2020; Wunderling et al. 2023), and hence surprises after time of larger forcing has ended.

By the end of the 1980s the paleoclimatic evidence for large-scale changes in AMOC was well established, and early model simulations demonstrated the relevance of such

Table 1 Occurrence of terms relevant to "tipping points" in the six assessment reports of IPCC Working Group I (WGI) since 1990. The definitions of the terms can be found in the respective IPCC glossaries

Term in WGI	FAR (1990)		SAR (1995)			TAR (2001)			AR4 (2007)			AR5 (2013)			AR6 (2021)		
	SPM	*AR*	*SPM*	*TS*	*AR*	*SPM*	*TS*	*AR*	*SPM*	*TS*	*AR*	*SPM*	*TS*	*AR*	*SPM*	*TS*	*AR*
Multiple equilibria		✓			✓			✓			✓			✓			
Abrupt change[1]		✓		✓				✓	✓	✓	✓	✓	✓	✓	✓	✓	✓
Irreversibility[2]		✓			✓		✓	✓		✓	✓	✓	✓	✓	✓	✓	✓
Surprise[3]		✓		✓			✓	✓			✓			✓		✓	✓
Tipping point[4]											✓		✓	✓	✓	✓	✓
HILL[5]							✓				✓			✓	✓	✓	✓

[1]abrupt change, or abrupt climate change

[2]irreversibility or irreversible change

[3]surprise, or unexpected [event, change]

[4]tipping point, or tipping element

[5]high impact event, low likelihood-high impact [event, outcome], or vice versa

reorganizations for future climate change (Stouffer et al. 1989). Consequently, this has aroused policy makers' interest, as it would have grave consequences for regional climate, both in the atmosphere and in the ocean in the North Atlantic regions with potentially global repercussions. Already the first assessment report of the IPCC reflected this (Table 1), as relevant terms entered the main report (IPCC 1990). Although a reduction or possible shutdown of the AMOC was already a consensus finding in the in the Summary for Policymakers of Working Group I (WG I) in the Third Assessment Report (IPCC 2001b), the terms listed in Table 1 have not yet made it into this top-level, government approved document. The projection chapter of the 4th Assessment Report featured a box that explained the different dynamical behaviors that were loosely referred to as "climate surprises" (Meehl et al. 2007). The box also pointed to additional elements in the climate system that may be prone to exhibit "surprises" in response to anthropogenic forcing. In addition to the AMOC, Arctic sea ice, glaciers and ice caps, the polar ice sheets, vegetation and atmosphere, and atmosphere-ocean regimes were mentioned. The overview has motivated the coinage of the terms "tipping elements" and "tipping points" in the climate system that were featured in the inaugural article by Lenton et al. (2008). These terms have since been widely used in science, by the media and in the public, and finally in 2021, found their way into the Summary for Policymakers (Table 1).

The recent IPCC 6th Assessment Report (IPCC 2021b) concluded that there was *medium confidence* that an AMOC collapse will not occur before 2100. The medium confidence, despite widespread model agreement, results from a number of missing processes in current climate model projections, and some evidence of a bias toward excessive AMOC stability in the models. Beyond 2100, the IPCC Special Report on Oceans, Cryosphere and Climate Change (IPCC 2019) concluded that an AMOC collapse was *about as likely as not* by 2300, for high emissions scenarios. Some model simulations also suggest the possibility of a temporary, multi-decadal to century scale AMOC reduction, followed by a recovery (Stocker and Schmittner 1997; Hu et al. 2013; Koven et al. 2022; Pöppelmeier et al. 2023). These assessments and recent studies emphasize the still incomplete nature of current scientific understanding of the relevant processes.

3.2 Five Reasons for Patchy Knowledge of the Future Evolution of AMOC and its Impact

The current state of knowledge on tipping points and their climatic impact on regional to global scales is patchy for five reasons. These include a lack of comprehensive observational data, the incomplete physical understanding of trigger factors, the limitations of models and scenarios, and the complexity of impacts of tipping points. This is best illustrated by considering the AMOC, which has been studied extensively, but it essentially applies to all systems that may exhibit tipping potential (Wang et al. 2023).

First, patchiness arises from the fact that two monitoring campaigns (Rapid-AMOC since 2004, and Overturning in the Subpolar North Atlantic Program (OSNAP) since 2014) with direct oceanographic measurements only cover a relatively short time period. They show an unexpected high variability but no trend (Li et al. 2021; Worthington et al. 2021; Jackson et al. 2022). This contrasts with long-term, paleoclimatic records that are indicators of surface temperature, deep ocean temperature, water mass distribution, and deep currents (Broecker 1997). However, as these are proxy records, evidence remains circumstantial. A selection of such marine records covering the last millennium suggests that a decline of AMOC is well underway and that today's strength is the lowest (Caesar et al. 2021a). However, there are several open issues. A number of locations in the North Atlantic, where additional records exist, have not been considered in this study and remain to be analyzed and confronted with the suggestion of AMOC weakening (Halimeda et al. 2022). Then, climate signals from, for example, the southern hemisphere could provide independent evidence for an AMOC evolution and eventual reduction during the last millennium, via a teleconnection (Stocker and Johnsen 2003; Pedro et al. 2018), with no indication so far. Finally, some argue that the observed North Atlantic cooling is instead caused by local heat loss due to a changing large-scale atmospheric circulation (Li et al. 2022), or by remote influences not directly reflecting the AMOC (Hu and Fedorov 2020; Keil et al. 2020). Hence the paleoclimatic evidence for an ongoing AMOC slow-down remains debated and a consensus is not yet reached.

Second, theory and simple models suggest that critical AMOC thresholds can arise when temperature and salinity anomalies generate different atmospheric responses that result in surface buoyancy anomalies leading to instabilities. However, it is not established how vertical and horizontal mixing in the ocean interior, the wind-driven circulations of the sub-tropical and sub-polar gyres, and high-latitude deep water formation would modify the overall stability of the circulation system. In fact, some models show a preponderance toward instability in very narrow windows of parameter space, particularly with respect to vertical mixing (Vettoretti et al. 2022; Malmierca-Vallet et al. 2023). This shows that there are gaps in the process understanding and quantification that must be closed before a consensus for the long-term fate of AMOC is reached.

Third, the currently most complete climate models show a very wide spread of the response of the AMOC to anthropogenic warming and to changes in the meridional water transport through the atmosphere by the end of the 21st century (IPCC 2021b). While some decrease in the AMOC is simulated consistently, the amount of reduction is very uncertain (Weijer et al. 2019). Furthermore, the observational record is still too short to constrain these model simulations meaningfully. None of the models shows an abrupt reduction or a collapse during this century which would be one of the characteristics of crossing a tipping point.

Fourth, commonly reported multi-model means are inadequate to investigate the consequences of threshold crossings of the AMOC. Impact models forced with multi-model output, or regional analyses of multi-model output would not be appropriate to inform about risks and regional consequences associated with an AMOC collapse. Therefore, regional surprises may go unnoticed in such products if not specifically and systematically searched for. Here, the analysis of apparent "outlier members" of multi-model ensembles might be worthwhile, a path similar to what has been followed recently in the analysis of heat waves (Fischer et al. 2021).

Fifth, current IPCC scenarios used to assess impacts are incomplete with respect to the possible crossing of thresholds in the climate system. IPCC WGII and WGIII base their assessment on projections which often use multi-model mean outcomes from CMIP efforts. For a comprehensive risk analysis, individual global simulations showing tipping behavior should be used for regional model projections or impact studies. The ongoing comparison project TIPMIP will likely provide such simulations that could generate new input to impact models (Loriani et al. 2023).

4 A Broader View on Climate Tipping: High Impact—Low Likelihood (HILL) Events

To clarify the discussion, it is worthwhile to recall the common definition of "tipping point" according to IPCC's WG I (IPCC 2021a): "A critical threshold beyond which a system reorganizes, often abruptly and/or irreversibly." WG II gives a somewhat more detailed definition: "A level of change in system properties beyond which a system reorganizes, often abruptly, and does not return to the initial state even if the drivers of the change are abated" (IPCC 2022). In terms of impacts, these definitions seem to be too restrictive and do not comprehensively cover the relevant climate system behaviors that could occur and that are not simply a linear response to an anthropogenic perturbation. Recently, a new term has been proposed for such behavior: High impact—low likelihood (HILL) events (Wood et al. 2023). This term shifts the focus and emphasizes the impact rather than the dynamical behavior of abrupt events, irreversible changes, surprises, etc. It is useful to visually clarify the perspective that underlies the notions of "tipping" and "HILL." This is illustrated in Fig. 1 which compares the dynamical or time perspective with the impact perspective. The relationship between forcing and response in the time perspective provides a basis for "tipping," which denotes a change that is more rapid than the forcing causing the change. In the impact perspective, the relationship between forcing and impact determines whether the system responds in a resilient way or whether a high impact—low likelihood (HILL) event results. Obviously, many quantities may also show linear behavior for a large range of forcing speeds and levels, respectively.

Focusing on the impacts of HILL events directs the attention to the regional consequences of large-scale events such as the reduction or collapse of the AMOC, the gradual decrease in mass of the polar ice sheets with potentially irreversible melting, the shift of atmospheric circulation systems, or the possible large-scale decline of the Amazon rainforest. This calls for a new generation of atmosphere and ocean models (Hewitt et al. 2022; Slingo et al. 2022) featuring much higher resolution and more realistic coupling, commensurate with the physical processes that trigger HILL events. It requires realistic inclusion, in Earth System models, of interactive processes such as fire and vegetation disturbance and loss. Some events previously considered to occur with low likelihood have already become more likely, as illustrated by increased vulnerability of some tropical forests (e.g.,

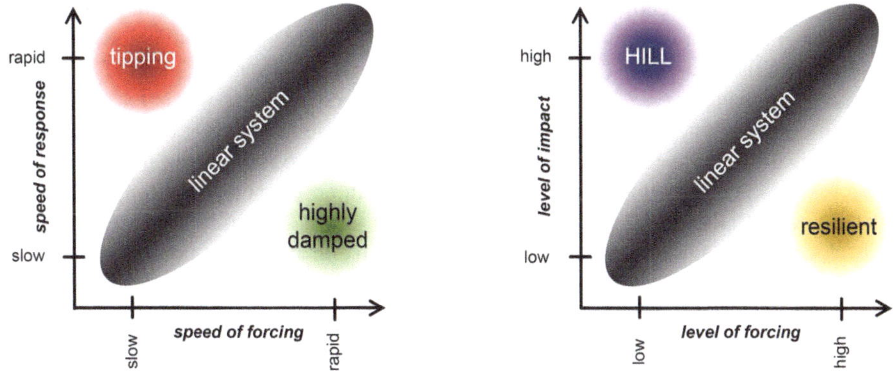

Fig. 1 Two alternative views on nonlinear behavior in the Earth System, the dynamical view (left) and the impact view (right), respectively. HILL denotes high impact—low likelihood event

Saatchi et al. 2021), observations of severe coral bleaching (Hughes et al. 2017), and the occurrence of unprecedented extreme events (e.g., Thompson et al. 2022). With models that can resolve processes and scales relevant for HILL events and tipping points, a fresh look must be taken at the question of the existence of thresholds and their location, the sensitivity of large-scale and regional-scale atmosphere, coupled atmosphere-ocean, and coupled ocean–cryosphere systems to the ongoing global heating.

HILL events go far beyond the physical science basis, they occur in geochemical and biological systems (Bastos et al. 2023), and they are relevant for impacts on ecosystems (Dakos et al. 2019; Willcock et al. 2023). Generally, evaluating HILL events requires dense sampling of extreme events, and coverage of long enough timescales to capture the statistics of rare events and to address slower processes such as ice sheet melting. At the same time, new high-resolution simulations should be carried out to gain a more detailed understanding of the dynamics that can generate HILL events.

Deep uncertainties are generally associated with tipping elements. The process and timing of boreal forest recovery after drought and fire is poorly understood. Land surface and vegetation models still do not capture the full complexity of the coupled Earth System (Fisher and Koven 2020). Amazon drying in response to climate change varies between climate models (Parry et al. 2022), and is affected by uncertainties in vegetation response and societal choices such as deforestation (Nobre et al. 2016). For other potential tipping points, such as Amazon forest loss or boreal forest shifts, observations are still relatively short and uncertain. Finally, current understanding of ice sheet tipping points is largely based on offline modeling (Pattyn et al. 2018), as are the effects of permafrost melting and methane releases in the high latitudes of the northern hemisphere (Kleinen et al. 2021).

In summary, taking a broader view on surprises in the climate system, combining tipping points, HILL events and irreversible change, would accelerate the understanding of the Earth System and the responses, from global to regional and local, to the ongoing anthropogenic perturbation.

5 Consensus Finding Process Promotes Scientific Progress

Anthropogenic climate change has been a topic of scientific attention since the late 1960s (Manabe and Wetherald 1967; Broecker 1975). To the wider public, the topic was known through the landmark report of the Club of Rome (Meadows et al. 1972), and later it was covered by the media worldwide. Public awareness of the greenhouse effect has increased steadily since the mid-1980s (Nisbet and Myers 2007). Many different aspects of anthropogenic climate change have featured in the news media and innovative ways of analyzing the growing corpus of media coverage are being applied (Hase et al. 2021; Schäfer and Hase 2023). Regarding climate tipping points, the media coverage has steeply increased since about 2017, when also digital news media showed an accelerating mention of this and related terms (Bellamy 2023). Often, more dramatic terms, such as “climate emergency” and “climate breakdown” were used when reporting on climate tipping points.

Generally, disasters are prone to distortion in media reports, possible future disasters even more. It is important to emphasize that the distortion can go both ways. Reporting could exaggerate the consequences and invoke doomsday scenarios, but it could equally downplay the gradually increasing preponderance of disasters. Society sometimes ‘discounts’ future risk in their decision-making, particularly in regions and situations when domestic pressures are high. While this is common practice in classical cost-benefit analyses of measures to counter anthropogenic climate change, research shows that this is inappropriate when it comes to climate surprises (Dasgupta 2008; Weitzman 2011). We feel that the same mechanism is at play in the current discussion on tipping points in the climate system and on climate change projections. Some media inflate scientific results in order to generate attention, others regularly ignore such findings despite their potential humanitarian and economic consequences. "Catastrophizing" science can lead to despair and hopelessness which eventually leads to inaction. In the scientific community, when it comes to climate surprises, we observe the development of groups with diverging views, here loosely termed as "tippists" and "linearists."

"Tippists" argue that tipping points are ubiquitous in the nonlinear climate system and that there is unequivocal evidence for an approach to tipping points in a number of climate system components, such as the AMOC, the Greenland ice sheet, and the Amazon rainforest. Some even suggest that one tipping point could trigger another one or unleash an entire cascade (Steffen et al. 2018; Lenton et al. 2019; Brovkin et al. 2021; Wunderling et al. 2021; Wunderling et al. 2024). Currently, these arguments rest on very simple nonlinear mathematical models; thus, it is not yet clear whether a consensus will emerge on this issue. Tipping cascades are often used to invoke a state of emergency of the climate which requires swift action of mitigation, or some even advocate geoengineering to avoid disaster (Heutel et al. 2016; Helwegen et al. 2019). Evidently, this view generates media coverage and high attention. Looking beyond the physical climate system, thresholds seem to be abundant, e.g., physiological limits in vegetation and the biosphere (Scheffer et al. 2001; Higgins and Scheiter 2012; Biggs et al. 2015; IPBES 2018), including for humans (Vecellio et al. 2023), or in snow and ice cover which exhibit nonlinear responses to a change in ambient conditions or in forcing.

"Linearists," on the other hand, emphasize the complexity of the climate system that has many dissipative elements that would prevent large-scale instabilities from unfolding. Some hold the view that tipping points are artifacts of simplified or incomplete climate models, and comprehensive models would not show such behavior (Stouffer and Manabe 2003; Wunsch 2006). The argument of absence of tipping is often connected with the equilibrium response of the system long-term fate of the climate system, e.g., when an initially reduced AMOC has recovered (Stouffer and Manabe 2003; Dijkstra et al. 2004; Stouffer et al. 2006), or the bifurcation occurs for only very large warming (Hu et al. 2013; Hu et al. 2023). Others argue that the concept of tipping is employed too liberally, may not be dynamically accurate, and often confuses the issues of climate system response (Kopp et al. 2023). Clearly, such a view would be readily taken by some others to call off the urgency for mitigation and stress that anthropogenic perturbations would merely cause some manageable impacts.

Diverging views in the scientific community, framed, and amplified by the media as controversies, must be viewed critically. They should not be confused with indispensable scientific discussion and debate without which knowledge gain is impossible. Controversy generally bears the potential for progress and deeper insight. However, any controversy must follow the scientific approach, i.e., the production of reproducible facts, observations, and theoretical foundation. This is precisely the reason why consensus building in a formalized process has a special value. Within the agreed framework this process brings about facts and findings more sharply and allows a transparent exposition of the arguments of both tippists, linearists, and all in between.

One successful model is the IPCC process in which a diverse group of elected authors assesses a topic comprehensively and submits drafts of their report to multiple stages of open review. This raises awareness of questions, gaps, and controversies which need to be addressed in the course of revision. This process has a high chance of convergence because the selection process of the topics happens at the juncture of two different departure points: bottom-up from the scientific community, and top-down as requested by policymakers. Since the first IPCC report in 1990 the scientific community has found robust consensus in many difficult issues of climate science. If this happens, very strong and robust messages emerge. On the other hand, some topics may not yet have reached maturity for consensus. This is then presented as a finding of low confidence, or the absence of consensus is declared, and gaps of knowledge are explicitly enumerated. Often, this has stimulated further research. The community of climate scientists looks back on more than 35 years of experience in formalized IPCC assessments of complex climate science. We just mention four representative examples.

The first example concerns the detection and attribution of climate change. The question to what extent the increase in greenhouse gas concentrations is responsible for the observed changes in the climate system, has found a robust consensus over several successive assessments. After a few seminal studies in the early 1990s, the first cautious statements on global mean temperature increase were formulated for top-level documents, identifying a discernible human influence (IPCC 1996; Santer et al. 1996). With an increasing number of scientific findings and studies using multiple datasets and methods, longer and better observations, and improved climate models, the statements could be quantified, first for global mean temperature increase (IPCC 2001a), and subsequently for a growing number of observed variables–examples are large-scale precipitation, polar sea ice extent, worldwide glacier reduction, heat wave statistics and other extreme events (IPCC 2013a; 2021b).

The second example is the global carbon budget that is now accepted as a scientific fact owing to a deep understanding and detailed observations of the global carbon cycle, specifically the continued monitoring of atmospheric CO_2 increase and the uptake of anthropogenic CO_2 in the ocean. The global carbon budget was presented by scientists of IPCC WGI in the 5th Assessment Report in 2013 (IPCC 2013a). It met fierce opposition by some policymakers but was finally approved in consensus thanks to multiple lines of independent evidence, the bar that science sets for robust findings. As a result, the global carbon budget was included for the first time in AR5 and was accompanied by a compelling figure in the WGI Summary for Policymakers (IPCC 2013b). This result has stood the test of time: Since then, each successive report has included the updated carbon budget with the latest estimates in their highest-level documents (IPCC 2014; 2021b; 2023). The concept has thus become the most policy-relevant scientific finding for the Paris Agreement. It is the basis for estimating the required emissions pathways toward climate stabilization (e.g., Stocker 2013), and for assessing the stocktaking of emissions reductions by the UNFCCC.

The third example revolves around equilibrium climate sensitivity, i.e., the increase in global mean temperature upon a doubling of the atmospheric CO_2 concentration and equilibration of the climate system. It was first assessed in 1979 by a small group of scientists who were tasked by the US National Academy of Science (Charney et al. 1979). Equilibrium climate sensitivity has been at center stage in both the scientific and public debate ever since. The formalized consensus finding process of the IPCC has forced the scientific community to periodically revisit this issue and comprehensively assess the range based on the latest findings. This process has been going on since the beginning of the IPCC. Successive cycles could narrow the likely range of climate sensitivity, but in the 5th Assessment Report, the range had to be widened again, and a best estimate could not be delivered. This was due to the advent of new quantitative estimates from early instrumental records that yielded lower estimates than those from climate modeling, and other lines of evidence, such as paleoclimatic archives. At the same time, the consensus building process also fostered collaboration across usually separate communities such as paleoclimate science, cloud physics and climate modeling, attribution analysis and analysis of feedbacks. It was discovered that the pattern of warming matters, and that feedbacks evolve with the warming, rendering a constant feedback estimate from the observed warming uncertain (Sherwood et al. 2020). Overall, this significantly increased understanding, in particular, of the multiple sources of uncertainties, and of how to combine results across multiple lines of evidence and resulted, again, in a narrower range of climate sensitivity (IPCC 2021b). In the scientific debate, more pertinent metrics for global warming were developed to better inform policymakers such as transient climate response, and transient climate response to cumulative emissions (IPCC 2013a).

As the fourth example we mention the science of extreme events. In 2008 when the IPCC Special Report on Extreme Events was proposed, knowledge and understanding were patchy, and the relationship to anthropogenic warming was poorly explored. It was recognized that the impacts caused by extreme climate events created high losses and damages, and policymakers needed more scientific knowledge about projected changes in intensity, frequency, and regional expression of extreme events. The mere announcement of the forthcoming Special Report, carried out by two IPCC Working Groups jointly, accelerated research significantly, notably through internationally coordinated research initiatives such as the World Climate Research Programme, or Future Earth. It led to new collaborations across disciplines, and the new risk framing, proposed in this report, helped communicate the findings and their uncertainties more effectively (IPCC 2012). Since then, the science has evolved rapidly to the extent that individual extreme events can now be attributed to anthropogenic climate change (Otto 2017), and that the tight links between physical climate science and impacts are recognized for extreme events. The synergy of assessment and production of new science is evident in this example. A similar effect also resulted from IPCC (2018), where the scientific literature rapidly expanded to address the questions posed in response to the government request via the UNFCCC for the IPCC Special Report on 1.5°C warming to which the Paris Agreement aims to limit global warming. In both cases, this rapid advance has led to a step change in the scientific understanding of important and policy-relevant questions. The recent joint IPBES-IPCC workshop is a further important step toward cross-discipline assessments (Pörtner et al. 2021).

In summary, these four examples demonstrate that formal assessments play an important role beyond informing the policymakers and the public. Scientists focusing their attention on a specific topic, that is unresolved, has substantial uncertainties, or generates diverging views, with the clear goal to arrive at a consensus, or map out where there is none, is a healthy process that stimulates further research to address the open issues. We have seen in numerous examples that the consensus finding process also brings together scientists from diverse fields, when it becomes clear that a comprehensive understanding is only possible beyond the individual silos. While cumbersome and intensive, assessments have accelerated the progress of scientific knowledge generation in these issues by bringing on board funding agencies and policymakers.

6 Effective Communication in Support of Understanding and Action

Communication is central to making the results of a scientific assessment available to policymakers and to the broader public. In past IPCC Assessment Reports the Summary for Policymakers generated criticism due to the complexity of the language used (Barkemeyer et al. 2016). Likewise, the scientific diagrams of these documents have been found not to deliver on the promise of assisting and illustrating assessment findings (Harold et al. 2016; 2020). A significant improvement in the accessibility of language was introduced with "headline statements" (Stocker and Plattner 2016). These are short and succinct summaries of complex findings presented in the top-level documents of the IPCC. They use plain language without jargon, and in most cases without numbers. In short, they should be statements that could be quoted directly by newspapers, radio, TV, and in social media. These headline statements form part of the approved text and have therefore been elevated from statements by the scientists to statements of the entire panel, i.e., the 195 governments participating in the IPCC. The improvement in readability was demonstrated by simple metrics of text analysis (Stocker and Plattner 2016). IPCC has recognized the value of "headline statements" in the top-level documents and has adopted this instrument as an integral element of their products (IPCC 2016).

An example of a particularly effective headline statement that was approved by the panel in 2013 was the affirmation that "Human influence on the climate system is clear." We do not go into the details of how the authors deliberated for an hour in a closed preparatory meeting ahead of the approval plenary about the most suitable and appropriate adjective describing human influence. Along with "clear" about ten alternative adjectives were proposed, debated and pondered before "clear" was the preferred choice. This concise statement is the one-sentence summary of the complete detection and attribution science assessed in the WG I contribution to the 5th assessment cycle. An entire chapter of 87 pages was dedicated to detection and attribution, then condensed to 19 pages in the Technical Summary, and afterward summarized in one section of 2 pages and one figure in the Summary for Policymakers. In this one section, three groups of headlines summarized up to 12 bulleted paragraphs each, containing all the complexities, numbers, and uncertainties. The three headline groups were finally distilled into two overarching headline statements of which the opening statement is the quoted one above. The successive hierarchical approval, from full complexity to super-distillation, eventually permitted the consensus approval of a statement that seemed unapprovable at first. Such headline statements would be most useful for HILL events, but this would first require a robust scientific consensus.

Taken together, a well nuanced assessment of nonlinearities, thresholds and tipping points must be carried out across all three IPCC working groups, i.e., from the physical climate science to ecological tipping and social tipping in decarbonization. In order to be comprehensive, such an assessment would include social tipping points, often considered as "positive" (Tàbara et al. 2018; Lenton et al. 2022), but would also critically address aspects of climate justice (Pereira et al. 2024). This could accelerate scientific understanding, converge into clear communication and help avoid the most dangerous aspects of climate change.

7 Conclusions

Based on scientific evidence HILL events, which include many of the 'traditional' climate tipping points but emphasize their impact rather than the dynamics, are a plausible outcome of continued anthropogenic forcing. Therefore, they must be factored into the development of strategies to adapt to or prevent impacts. Currently, their effect is largely missing in analyses of the widely used IPCC scenarios; and hence, the associated risks are poorly quantified. Moreover, there are significant gaps in our knowledge of the impacts of HILL events at the regional level. This is due to the limited resolution of climate models, the incompleteness of small-scale and coupled processes in the atmosphere, ocean, cryosphere and land surface, as well as limitations in observational data records. Proper consideration of these in a new generation of climate models, along with new long-term Earth Observations that are becoming available through a rapidly expanding space sector, will refine, and in some cases revise, the assessment of the stability of the climate system and the possibility of nonlinear behavior and surprises. Further, evaluation of the impacts of climate change has to date focused largely on the most likely ranges of climate drivers, resulting in incomplete knowledge of the consequences of HILL climate outcomes. These gaps in knowledge could be closed by implementing strong and enhanced research programs focusing on tipping points. A step increase in understanding will come from related projects, e.g., within EC's Horizon 2020 and Horizon Europe programs, continued monitoring efforts in the ocean (ARGO, RAPID and OSNAP, and future campaigns), in the atmosphere (GCOS and GAW, and future campaigns), satellite observations of land surface properties and ocean status, and the development of the next generation of global climate models (Schär et al. 2020; Jungclaus et al. 2022). Targeted global programs under the auspices of WCRP and ESA, which plans work on tipping points in its new climate program "Climate-Space," as well as Future Earth, could significantly boost the scientific understanding of this critical feature of the earth system.

However, a broad consensus on tipping elements, tipping points, HILL events, and climate surprises in general, in short "tipping points (*sensu lato*)," is not yet present in the scientific community. This is due to the absence of a comprehensive and formalized assessment of the peer-reviewed literature in this topic. In the past 20 years the relevant scientific literature has grown rapidly but only a few individual studies have dominated the public discussion and perception on tipping points. It seems to us that the time is mature to commission a focused assessment following a procedure that is time-tested by IPCC. Clearly, a cross-working group Special Report on Tipping Points (*sensu lato*), would provide the most effective process for a comprehensive assessment and the building of a robust scientific consensus. However, at its 60th plenary in January 2024 negotiating the work program, IPCC decided not to prepare more than the special report on cities that was already decided in 2016. In a survey carried out by the IPCC in preparation of that plenary (IPCC-LX/Doc.4), tipping points ranked top of the list for additional IPCC products, calling for a response in the forthcoming scoping process. This offers the opportunity for each working group to scope a dedicated chapter on tipping points (*sensu lato*) and combine the findings and consensus in the synthesis report.

As we have argued, HILL events go far beyond the physical science basis. They are most relevant for impacts and even more in the area of ecosystems and biodiversity. We therefore suggest that a joint IPCC/IPBES assessment could be envisaged as an ambitious goal to be achieved within the next decade. Furthermore, rapid changes and nonlinearities can also occur in the area of climate mitigation, from rapid acceptance of new technologies to political instability derailing agreements; and some tipping points such as Amazon or boreal rapid forest change would have implications on the carbon budgets ultimately affecting mitigation pathways. This cross-working group/cross-panel process could be assisted by a series of workshops, similar to that hosted by ISSI, the International Space Science Institute in Bern, in October 2022. These workshops should be convened jointly by the three working groups of the IPCC and IPBES. For the IPCC this could be an innovative new element for its forthcoming 7th assessment cycle.

Acknowledgements This paper is an outcome of the Workshop: "Tipping Points and Understanding EO data needs for a Tipping Element Model Intercomparison Project" held at the International Space Science Institute (ISSI) in Bern, Switzerland (10-14 October 2022). Constructive comments by T. Frölicher and F. Pöppelmeier are gratefully acknowledged. This study was supported by the European Union's Horizon 2020 research and innovation program under grant agreement No 820970 (project TiPES) and the Swiss National Science Foundation (project 200020_200492) awarded to TFS. MIH was supported by the ESA Water Vapour Climate Change Initiative (Contract No. 4000123554). RGJ and RAW were supported by the Hadley Centre Climate Programme funded by UK Government departments BEIS and Defra.

Funding Open access funding provided by University of Bern. The authors have no relevant financial or non-financial interests to disclose.

Declarations

Conflict of Interest The authors have no competing interests to declare that are relevant to the content of this article.

References

Allen MR, Frame DJ, Huntingford C, Jones CD, Lowe JA, Meinshausen M, Meinshausen N (2009) Warming caused by cumulative carbon emissions towards the trillionth tonne. Nature 458:1163–1166

Armstrong McKay DI, Staal A, Abrams JF, Winkelmann R, Sakschewski B, Loriani S, Fetzer I, Cornell SE, Rockström J, Lenton TM (2022) Exceeding 1.5°C global warming could trigger multiple climate tipping points. Science 377:7950

Baker JA, Bell MJ, Jackson LC, Renshaw R, Vallis GK, Watson AJ, Wood RA (2023) Overturning pathways control AMOC weakening in CMIP6 models. Geophys Res Lett 50:e2023GL103381

Barkemeyer R, Dessai S, Monge-Sanz B, Renzi BG, Napolitano G (2016) Linguistic analysis of IPCC summaries for policymakers and associated coverage. Nature Clim Change 6:311–316

Bastos A, Sippel S, Frank D, Mahecha MD, Zaehle S, Zscheischler J, Reichstein M (2023) A joint framework for studying compound ecoclimatic events. Nat Rev Earth Environ 4:333–350

Bellamy R (2023) Public perception of climate tipping points. Public Underst Sci 32:1033–1047

Biggs R, Gordon L, Raudsepp-Hearne C, Schlüter M, Walker B (2015) Principle 3—Manage slow variables and feedbacks. In: Biggs R et al (eds) Principles for building resilience: sustaining ecosystem services in social-ecological systems. Cambridge University Press, Cambridge, pp 105–141

Bochow N, Boers N (2023) The South American monsoon approaches a critical transition in response to deforestation. Science Adv 9:eadd9973

Boers N (2021) Observation-based early-warning signals for a collapse of the Atlantic meridional overturning circulation. Nat Clim Chang 11(680):688

Boers N, Rypdal M (2021) Critical slowing down suggests that the western Greenland ice sheet is close to a tipping point. Proc Natl Acad Sci USA 118:e2024192118
Boers N, Marwan N, Barbosa HMJ, Kurths J (2017) A deforestation-induced tipping point for the South American monsoon system. Sci Rep 7:41489
Boulton CA, Lenton TM, Boers N (2022) Pronounced loss of Amazon rainforest resilience since the early 2000s. Nat Clim Chang 12:271–278
Broecker WS (1975) Climatic change: are we on the brink of a pronounced global warming? Science 189:460–463
Broecker WS (1987) The biggest chill. Nat Hist 96:74–82
Broecker WS (1987) Unpleasant surprises in the greenhouse? Nature 328:123–126
Broecker WS (1997) Thermohaline circulation, the Achilles heel of our climate system: will man-made CO_2 upset the current balance? Science 278:1582–1588
Broecker WS, Peteet DM, Rind D (1985) Does the ocean-atmosphere system have more than one stable mode of operation? Nature 315:21–25
Brovkin V, Brook E, Williams JW, Bathiany S, Lenton TM, Barton M, DeConto RM, Donges JF, Ganopolski A, McManus J, Praetorius S, de Vernal A, Abe-Ouchi A, Cheng H, Claussen M, Crucifix M, Gallopin G, Iglesias V, Kaufman DS, Kleinen T, Lambert F, van der Leeuw S, Liddy H, Loutre MF, McGee D, Rehfeld K, Rhodes R, Seddon AWR, Trauth MH, Vanderveken L, Yu ZC (2021) Past abrupt changes, tipping points and cascading impacts in the earth system. Nat Geosci 14:550–558
Bryan F (1986) High-latitude salinity effects and interhemispheric thermohaline circulations. Nature 323:301–304
Caesar L, McCarthy GD, Thornalley DJR, Cahill N, Rahmstorf S (2021) Current Atlantic meridional overturning circulation weakest in last millennium. Nat Geosci 14:118–120
Caesar L, Rahmstorf S, Feulner G (2020) On the relationship between Atlantic meridional overturning circulation slowdown and global surface warming. Env Res Lett. 15, 024003
Charney JG, Arakawa A, Baker DJ, Bolin B, Dickinson RE, Goody RM, Leith CE, Stommel HM, Wunsch CI (1979) Carbon dioxide and climate: a scientific assessment. National Academy of Science, Washington, DC
Clark PU, Pisias NG, Stocker TF, Weaver AJ (2002) The role of the thermohaline circulation in abrupt climate change. Nature 415:863–869
Crowley TJ (1992) North Atlantic deep water cools the southern hemisphere. Paleoceanography 7:489–497
Dakos V, Matthews B, Hendry AP, Levine J, Loeuille N, Norberg J, Nosil P, Scheffer M, De Meester L (2019) Ecosystem tipping points in an evolving world. Nat Ecol Evol 3:355–362
Dansgaard W, Johnsen SJ, Clausen HB, Dahl-Jensen D, Gundestrup NS, Hammer CU, Hvidberg CS, Steffensen JP, Sveinbjörnsdottir AE, Jouzel J, Bond G (1993) Evidence for general instability of past climate from a 250-kyr ice-core record. Nature 364:218–220
Dasgupta P (2008) Discounting climate change. J Risk Uncertain 37:141–169
DeConto RM, Pollard D, Alley RB, Velicogna I, Gasson E, Gomez N, Sadai S, Condron A, Gilford DM, Ashe EL, Kopp RE, Li DW, Dutton A (2021) The Paris Climate Agreement and future sea-level rise from Antarctica. Nature 593:83–89
Dijkstra HA, Te Raa L, Weijer W (2004) A systematic approach to determine thresholds of the ocean's thermohaline circulation. Tellus 56A:362–370
Ditlevsen P, Ditlevsen S (2023) Warning of a forthcoming collapse of the Atlantic meridional overturning circulation. Nat Comm 14:4254
Drijfhout S, Bathiany S, Beaulieu C, Brovkin V, Claussen M, Huntingford C, Scheffer M, Sgubin G, Swingedouw D (2015) Catalogue of abrupt shifts in Intergovernmental Panel on Climate Change climate models. Proc Natl Acad Sci USA 112:E5777–E5786
EPICA Community Members (2006) One-to-one coupling of glacial climate variability in Greenland and Antarctica. Nature 444:195–198
Fischer EM, Sippel S, Knutti R (2021) Increasing probability of record-shattering climate extremes. Nat Clim Change 11:689–695
Fisher RA, Koven CD (2020) Perspectives on the future of land surface models and the challenges of representing complex terrestrial systems. J Adv Model Earth Syst 12:e2018MS001453
Ghil M, Lucarini V (2020) The physics of climate variability and climate change. Rev Mod Phys 92:035002
Hagedorn G, Loew T, Seneviratne SI, Lucht W, Beck ML, Hesse J, Knutti R, Quaschning V, Schleimer JH, Mattauch L, Breyer C, Hübener H, Kirchengast G, Chodura A, Clausen J, Creutzig F, Darbi M, Daub CH, Ekardt F, Göpel M, Hardt JN, Hertin J, Hickler T, Köhncke A, Köster S, Krohmer J, Kromp-Kolb H, Leinfelder R, Mederake L, Neuhaus M, Rahmstorf S, Schmidt C, Schneider C, Schneider G, Seppelt R, Spindler U, Springmann M, Staab K, Stocker TF, Steininger K, von Hirschhausen E, Winter S,

Wittau M, Zens J (2019) The concerns of the young protesters are justified A statement by Scientists for future concerning the protests for more climate protection. Gaia-Ecol Perspect Sci Soc 28:79–87

Halimeda K, Wanamaker AD, Paola S, Reynolds DJ, Amrhein DE, Butler PG, Gebbie G, Goes M, Jansen MF, Little CM, Mette M, Moreno-Chamarro E, Ortega P, Otto-Bliesner BL, Rossby T, Scourse J, Whitney NM (2022) Atlantic circulation change still uncertain. Nat Geosc 15:165–167

Harold J, Lorenzoni I, Shipley TF, Coventry KR (2016) Cognitive and psychological science insights to improve climate change data visualization. Nat Clim Change 6:1080–1089

Harold J, Lorenzoni I, Shipley TF, Coventry KR (2020) Communication of IPCC visuals: IPCC authors' views and assessments of visual complexity. Clim Change 158:255–270

Hase V, Mahl D, Schäfer MS, Keller TR (2021) Climate change in news media across the globe: an automated analysis of issue attention and themes in climate change coverage in 10 countries (2006–2018). Glob Environ Change-Hum Policy Dimens 70:102353

Heinze C, Blenckner T, Martins H, Rusiecka D, Doscher R, Gehlen M, Gruber N, Holland E, Hov O, Joos F, Matthews JBR, Rodven R, Wilson S (2021) The quiet crossing of ocean tipping points. Proc Natl Acad Sci USA 118:e2008478118

Helwegen KG, Wieners CE, Frank JE, Dijkstra HA (2019) Complementing CO_2 emission reduction by solar radiation management might strongly enhance future welfare. Earth Syst Dyn 10:453–472

Heutel G, Moreno-Cruz J, Shayegh S (2016) Climate tipping points and solar geoengineering. J Econ Behav Org 132:19–45

Hewitt H, Fox-Kemper B, Pearson B, Roberts M, Klocke D (2022) The small scales of the ocean may hold the key to surprises. Nat Clim Change 12:496–499

Higgins SI, Scheiter S (2012) Atmospheric CO_2 forces abrupt vegetation shifts locally, but not globally. Nature 488:209–212

Hu S, Fedorov AV (2020) Indian ocean warming as a driver of the North Atlantic warming hole. Nat Comm 11:4785

Hu AX, Meehl GA, Han WQ, Lu JH, Strand WG (2013) Energy balance in a warm world without the ocean conveyor belt and sea ice. Geophys Res Lett 40:6242–6246

Hu AX, Meehl GA, Abe-Ouchi A, Han WQ, Otto-Bliesner B, He F, Wu TW, Rosenbloom N, Strand WG, Edwards J (2023) Dichotomy between freshwater and heat flux effects on oceanic conveyor belt stability and global climate. Commun Earth Environ 4:246

Hughes TP, Kerry JT, Alvarez-Noriega M, Alvarez-Romero JG, Anderson KD, Baird AH, Babcock RC, Beger M, Bellwood DR, Berkelmans R, Bridge TC, Butler IR, Byrne M, Cantin NE, Comeau S, Connolly SR, Cumming GS, Dalton SJ, Diaz-Pulido G, Eakin CM, Figueira WF, Gilmour JP, Harrison HB, Heron SF, Hoey AS, Hobbs JPA, Hoogenboom MO, Kennedy EV, Kuo CY, Lough JM, Lowe RJ, Liu G, Cculloch MTM, Malcolm HA, McWilliam MJ, Pandolfi JM, Pears RJ, Pratchett MS, Schoepf V, Simpson T, Skirving WJ, Sommer B, Torda G, Wachenfeld DR, Willis BL, Wilson SK (2017) Global warming and recurrent mass bleaching of corals. Nature 543:373–377

IPBES, The IPBES regional assessment report on biodiversity and ecosystem services for Africa, [E Archer, *et al.* (eds.)], pp. 492, 2018.

IPCC (2012) Managing the risks of extreme events and disasters to advance climate change adaptation. In: Field CB et al (eds) A special report of Working Groups I and II of the Intergovernmental Panel on Climate Change. Cambridge University Press, Cambridge

IPCC (1996) Climate Change. The Science of Climate Change. In: Houghton JT et al (eds) Contribution of Working Group I to the Second Assessment Report of the Intergovernmental Panel on Climate Change. Cambridge University Press, Cambridge

IPCC (2013a) Climate Change 2013. The Physical Science Basis. In: Stocker TF et al (eds) Contribution of Working Group I to the Fifth Assessment Report of the Intergovernmental Panel on Climate Change. Cambridge University Press, Cambridge

IPCC (1990) Climate Change. The IPCC Scientific Assessment. In: Houghton JT et al (eds) Intergovernmental Panel on Climate Change. Cambridge University Press, Cambridge

IPCC (2001) The Scientific Basis. In: Houghton JT et al. (eds.), Contribution of Working Group I to the Third Assessment Report of the Intergovernmental Panel on Climate Change. Cambridge University Press, Cambridge

IPCC (2001b) Summary for Policymakers of the Third Assessment Report Working Group I

IPCC (2013b) Summary for Policymakers, In: Stocker TF et al. (eds.), Climate Change 2013: The Physical Science Basis. Contribution of Working Group I to the Fifth Assessment Report of the Intergovernmental Panel on Climate Change, pp. 1–30

IPCC, Climate Change 2014: Synthesis Report, [Pachauri RK et al. (eds.)], 151 pp., Cambridge University Press

IPCC, Meeting Report of the Intergovernmental Panel on Climate Change Expert Meeting on Communication, [Lynn J et al (eds.)], 229 pp., 2016
IPCC, Special Report on Global Warming of 1.5°C, [Masson-Delmotte V et al (eds.)], 562 pp., Intergovernmental Panel on Climate Change, 2018
IPCC, Special Report on the Ocean and Cryosphere in a Changing Climate, [Pörtner HO et al. (eds.)], 755 pp., Intergovernmental Panel on Climate Change, 2019
IPCC, Annex VII: Glossary, in Climate Change 2021: The Physical Science Basis. Contribution of Working Group I to the Sixth Assessment Report of the Intergovernmental Panel on Climate Change, [Matthews JBR et al. (eds.)], pp. 2215–2256, 2021a
IPCC, Climate Change 2021: The Physical Science Basis. Contribution of Working Group I to the Sixth Assessment Report of the Intergovernmental Panel on Climate Change, [Masson-Delmotte V et al. (eds.)], 2058 pp, Cambridge University Press, Cambridge, 2021b
IPCC, Annex II: Glossary, in Climate Change 2022: Impacts, Adaptation and Vulnerability. Contribution of Working Group II to the Sixth Assessment Report of the Intergovernmental Panel on Climate Change, [Agard J and Schipper ELF (eds.)], pp 1757-1776, 2022.
IPCC, Climate Change 2023: Synthesis Report. In: Core Writing Team, et al. (eds.), Contribution of Working Group I, II and III to the Sixth Assessment Report of the Intergovernmental Panel on Climate Change, 81 pp
Izumi K, Armstrong E, Valdes P (2023) Global footprints of Dansgaard-Oeschger oscillations in a GCM. Quat Sci Rev 305:108016
Jackson LC, Biastoch A, Buckley MW, Desbruyeres DG, Frajka-Williams E, Moat B, Robson J (2022) The evolution of the North Atlantic meridional overturning circulation since 1980. Nat Rev Earth Environ 3:241–254
Jackson LC, de Asenjo EA, Bellomo K, Danabasoglu G, Haak H, Hu AX, Jungclaus J, Lee W, Meccia VL, Saenko O, Shao A, Swingedouw D (2023) Understanding AMOC stability: the North Atlantic hosing model intercomparison project. Geosci Mod Devel 16:1975–1995
Jeltsch-Thömmes A, Stocker TF, Joos F (2020) Hysteresis of the earth system under positive and negative CO_2 emissions. Env Res Lett 15:124026
Jungclaus JH, Lorenz SJ, Schmidt H, Brovkin V, Bruggemann N, Chegini F, Cruger T, De-Vrese P, Gayler V, Giorgetta MA, Gutjahr O, Haak H, Hagemann S, Hanke M, Ilyina T, Korn P, Kroger J, Linardakis L, Mehlmann C, Mikolajewicz U, Muller WA, Nabel J, Notz D, Pohlmann H, Putrasahan DA, Raddatz T, Ramme L, Redler R, Reick CH, Riddick T, Sam T, Schneck R, Schnur R, Schupfner M, Storch JS, Wachsmann F, Wieners KH, Ziemen F, Stevens B, Marotzke J, Claussen M (2022) The ICON earth system model version 1.0. J Adv Model Earth Syst 14:e2021MS002813
Keil P, Mauritsen T, Jungclaus J, Hedemann C, Olonscheck D, Ghosh R (2020) Multiple drivers of the North Atlantic warming hole. Nat Clim Change 10:667–671
Kleinen T, Gromov S, Steil B, Brovkin V (2021) Atmospheric methane underestimated in future climate projections. Env Res Lett 16:094006
Knutti R, Stocker TF (2002) Limited predictability of the future thermohaline circulation close to an instability threshold. J Clim 15:179–186
Kopp RE, Gilmore EA, Shwom RL, Adler C, Adams A, Oppenheimer M, Patwardhan A, Schmidt DN, York R (2023) "Tipping Points" confuse and can distract from urgent climate action. ESS Open Archive. https://doi.org/10.22541/essoar.170542965.59092060/v1
Koven CD, Arora VK, Cadule P, Fisher RA, Jones CD, Lawrence DM, Lewis J, Lindsay K, Mathesius S, Meinshausen M, Mills M, Nicholls Z, Sanderson BM, Seferian R, Swart NC, Wieder WR, Zickfeld K (2022) Multi-century dynamics of the climate and carbon cycle under both high and net negative emissions scenarios. Earth Syst Dyn 13:885–909
Kuniyoshi Y, Abe-Ouchi A, Sherriff-Tadano S, Chan WL, Saito F (2022) Effect of climatic precession on dansgaard-oeschger-like oscillations. Geophys Res Lett 49:e2021GL095695
Lenton TM, Held H, Kriegler E, Hall JW, Lucht W, Rahmstorf S, Schellnhuber HJ (2008) Tipping elements in the earth's climate system. Proc Natl Acad Sci USA 105:1786–1793
Lenton TM, Rockstrom J, Gaffney O, Rahmstorf S, Richardson K, Steffen W, Schellnhuber HJ (2019) Climate tipping points—too risky to bet against. Nature 575:592–595
Lenton TM, Benson S, Smith T, Ewer T, Lanel V, Petykowski E, Powell TWR, Abrams JF, Blomsma F, Sharpe S (2022) Operationalising positive tipping points towards global sustainability. Global Sustain 5:e1
Lenton TM, Armstrong McKay DI, Loriani S, Abrams JF, Lade SJ, Donges JF, Milkoreit M, Powell T, Smith SR, Zimm C, Buxton JE, Bailey E, Laybourn L, Ghadiali A, Dyke JG (2023) The Global Tipping Points Report. University of Exeter, Exeter

Li C, Born A (2019) Coupled atmosphere-ice-ocean dynamics in Dansgaard-Oeschger events. Quat Sci Rev 203:1–20
Li F, Lozier MS, Bacon S, Bower AS, Cunningham SA, de Jong MF, deYoung B, Fraser N, Fried N, Han G, Holliday NP, Holte J, Houpert L, Inall ME, Johns WE, Jones S, Johnson C, Karstensen J, Le Bras IA, Lherminier P, Lin X, Mercier H, Oltmanns M, Pacini A, Petit T, Pickart RS, Rayner D, Straneo F, Thierry V, Visbeck M, Yashayaev I, Zhou C (2021) Subpolar North Atlantic western boundary density anomalies and the Meridional overturning circulation. Nature Comm 12:3002
Li LF, Lozier MS, Li FL (2022) Century-long cooling trend in subpolar North Atlantic forced by atmosphere: an alternative explanation. Clim Dyn 58:2249–2267
Lorenz EN (1963) Deterministic non-periodic flow. J Atm Sci 20:130–141
Loriani S, Sakschewski B, Donges J, Winkelmann R (2023) Systematic assessment of climate tipping points, In: *EGU General Assembly*
Malmierca-Vallet I, Slime LC (2023) Dansgaard-Oeschger events in climate models: review and baseline marine isotope stage 3 (MIS3) protocol. Clim Past 19:915–942
Manabe S, Wetherald RT (1967) Thermal equilibrium of the atmosphere with a given distribution of relative humidity. J Atm Sci 50:241–259
Matthews HD, Gillett NP, Stott PA, Zickfeld K (2009) The proportionality of global warming to cumulative carbon emissions. Nature 459:829–833
Meadows, DH, DL Meadows, J Randers, WW Behrens III (1972), *The Limits to Growth*, 205 pp., Potomac Associates Book
Meehl, GA, TF Stocker, WD Collins, P Friedlingstein, AT Gaye, JM Gregory, A Kitoh, R Knutti, JM Murphy, A Noda, SCB Raper, IG Watterson, WA J, and Z-C Zhao (2007) Global Climate Projections, In: Solomon S et al. (eds.) Climate Change 2007: The Physical Science Basis. Contribution of Working Group I to the Fourth Assessment Report of the Intergovernmental Panel on Climate Change, pp. 747–845
Nepstad DC, Stickler CM, Filho BS, Merry F (2008) Interactions among Amazon land use, forests and climate: prospects for a near-term forest tipping point. Phil Trans Roy Soc B B363:1737–1746
Nisbet MC, Myers T (2007) Twenty years of public opinion about global warming. The Public Opin Q 71:444–470
Nobre CA, Sampaio G, Borma LS, Castilla-Rubio JC, Silva JS, Cardoso M (2016) Land-use and climate change risks in the Amazon and the need of a novel sustainable development paradigm. Proc Natl Acad Sci USA 113:10759–10768
Oeschger H, Beer J, Siegenthaler U, Stauffer B, Dansgaard W, Langway CC (1984) Late glacial climate history from ice cores. In: Hansen JE, Takahashi T (eds) Climate Processes and Climate Sensitivity. Wiley, Hoboken, pp 299–306
Orbe C, Rind D, Miller RL, Nazarenko LS, Romanou A, Jonas J, Russell GL, Kelley M, Schmidt GA (2023) Atmospheric response to a collapse of the north atlantic circulation under a mid-range future climate scenario: a regime shift in northern hemisphere dynamics. J Clim 36:6669–6693
Otto FEL (2017) Attribution of weather and climate events. Ann Rev Env Resour 42:627–646
Palmer TN (1999) A nonlinear dynamical perspective on climate prediction. J Clim 12:575–591
Parry IM, Ritchie PDL, Cox PM (2022) Evidence of localised Amazon rainforest dieback in CMIP6 models. Earth Syst Dyn 13:1667–1675
Pattyn F, Morlighem M (2020) The uncertain future of the Antarctic ice sheet. Science 367:1331–1335
Pattyn F, Ritz C, Hanna E, Asay-Davis X, DeConto R, Durand G, Favier L, Fettweis X, Goelzer H, Golledge NR, Munneke PK, Lenaerts JTM, Nowicki S, Payne AJ, Robinson A, Seroussi H, Trusel LD, van den Broeke M (2018) The Greenland and Antarctic ice sheets under 1.5°C global warming. Nature Clim Change 8:1053–1061
Pedro JB, Jochum M, Buizert C, He F, Barker S, Rasmussen SO (2018) Beyond the bipolar seesaw: toward a process understanding of interhemispheric coupling. Quat Sci Rev 192:27–46
Pereira LM, Gianelli I, Achieng T, Amon D, Archibald S, Arif S, Castro A, Chimbadzwa TP, Coetzer K, Field TL, Selomane O, Sitas N, Stevens N, Villasante S, Armani M, Kimuyu DM, Adewumi IJ, Lapola DM, Obura D, Pinho P, Roa-Clavijo F, Rocha J, Sumaila UR (2024) Equity and justice should underpin the discourse on tipping points. Earth Syst Dyn 15:341–366
Pöppelmeier F, Joos F, Stocker TF (2023) The coupled ice sheet-earth system model Bern3D v3.0. J Clim 36:7563–7582
Pörtner HO, Scholes RJ, Agard J, Archer E, Arneth A, Bai X, Barnes D, Burrows M, Chan L, Cheung WL, Diamond S, Donatti C, Duarte C, Eisenhauer N, Foden W, Gasalla MA, Handa C, Hickler T, Hoegh-Guldberg O, Ichii K, Jacob U, Insarov G, Kiessling W, Leadley P, Leemans R, Levin L, Lim M, Maharaj S, Managi S, Marquet PA, McElwee P, Midgley G, Oberdorff T, Obura D, Osman E, Pandit R, Pascual U, Pires APF, Popp A, Reyes-García V, Sankaran M, Settele J, Shin YJ, Sintayehu DW,

Smith P, Steiner N, Strassburg B, Sukumar R, Trisos C, Val AL, Wu J, Aldrian E, Parmesan C, Pichs-Madruga R, Roberts DC, Rogers AD, Díaz S, Fischer M, Hashimoto S, Lavorel S, Wu N, Ngo HT (2021) Scientific outcome of the IPBES-IPCC co-sponsored workshop on biodiversity and climate change. IPBES Secretariat, Bonn Germany. https://doi.org/10.5281/zenodo.4659158

Riahi K, van Vuuren DP, Kriegler E, Edmonds J, O'Neill BC, Fujimori S, Bauer N, Calvin K, Dellink R, Fricko O, Lutz W, Popp A, Cuaresma JC, Samir KC, Leimbach M, Jiang LW, Kram T, Rao S, Emmerling J, Ebi K, Hasegawa T, Havlik P, Humpenöder F, da Silva LA, Smith S, Stehfest E, Bosetti V, Eom J, Gernaat D, Masui T, Rogelj J, Strefler J, Drouet L, Krey V, Luderer G, Harmsen M, Takahashi K, Baumstark L, Doelman JC, Kainuma M, Klimont Z, Marangoni G, Lotze-Campen H, Obersteiner M, Tabeau A, Tavoni M (2017) The Shared socioeconomic pathways and their energy, land use, and greenhouse gas emissions implications: an overview. Glob Environ Change-Human Policy Dimens 42:153–168

Romanou A, Rind D, Jonas J, Miller R, Kelley M, Russell G, Orbe C, Nazarenko L, Latto R, Schmidt GA (2023) Stochastic bifurcation of the north atlantic circulation under a midrange future climate scenario with the NASA-GISS Model E. J Clim 36:6141–6161

Rousseau DD, Bagniewski W, Ghil M (2022) Abrupt climate changes and the astronomical theory: are they related? Clim Past 18:249–271

Saatchi S, Longo M, Xu L, Yang Y, Abe H, Andre M, Aukema JE, Carvalhais N, Cadillo-Quiroz H, Cerbu GA, Chernela JM, Covey K, Sanchez-Clavijo LM, Cubillos IV, Davies SJ, De Sy V, De Vleeschouwer F, Duque A, Durieux AMS, Fernandes KD, Fernandez LE, Gammino V, Garrity DP, Gibbs DA, Gibbon L, Gowae GY, Hansen M, Harris NL, Healey SP, Hilton RG, Johnson CM, Kankeu RS, Laporte-Goetz NT, Lee H, Lovejoy T, Lowman M, Lumbuenamo R, Malhi Y, Martinez J, Nobre C, Pellegrini A, Radachowsky J, Roman F, Russell D, Sheil D, Smith TB, Spencer RGM, Stolle F, Tata HL, Torres DD, Tshimanga RM, Vargas R, Venter M, West J, Widayati A, Wilson SN, Brumby S, Elmore AC (2021) Detecting vulnerability of humid tropical forests to multiple stressors. One Earth 4:988–1003

Saltzman B, Maasch KA (1991) A first-order global model of late Cenozoic climatic change. II. Further analysis based on a simplification of CO_2 dynamics. Clim Dyn 5:201–210

Santer BD, Taylor KE, Wigley TML, Johns TC, Jones PD, Karoly DJ, Mitchell JFB, Oort AH, Penner JE, Ramaswamy V, Schwarzkopf MD, Stouffer RJ, Tett S (1996) A search for human influences on the thermal structure of the atmosphere. Nature 382:39–46

Schäfer MS, Hase V (2023) Computational methods for the analysis of climate change communication: towards an integrative and reflexive approach. WIREs Clim Change 14:e806

Schär C, Fuhrer O, Arteaga A, Ban N, Charpilloz C, Di Girolamo S, Hentgen L, Hoefler T, Lapillonne X, Leutwyler D, Osterried K, Panosetti D, Rüdisühli S, Schlemmer L, Schulthess T, Sprenger M, Ubbiali S, Wernli H (2020) Kilometer-scale climate models: prospects and challenges. Bull Am Met Soc 101:E567–E587

Scheffer M, Carpenter S, Foley JA, Folke C, Walker B (2001) Catastrophic shifts in ecosystems. Nature 413:591–596

Sellers WD (1969) A global climatic model based on the energy balance of the earth-atmosphere system. J Appl Met 8:392–400

Sherwood SC, Webb MJ, Annan JD, Armour KC, Forster PM, Hargreaves JC, Hegerl G, Klein SA, Marvel KD, Rohling EJ, Watanabe M, Andrews T, Braconnot P, Bretherton CS, Foster GL, Hausfather Z, Heydt AS, Knutti R, Mauritsen T, Norris JR, Proistosescu C, Rugenstein M, Schmidt GA, Tokarska KB, Zelinka MD (2020) An assessment of earth's climate sensitivity using multiple lines of evidence. Rev Geophys 58:e2019RG000678

Slingo J, Bates P, Bauer P, Belcher S, Palmer T, Stephens G, Stevens B, Stocker T, Teutsch G (2022) Ambitious partnership needed for reliable climate prediction. Nat Clim Change 12:499–503

Steffen W, Rockstrom J, Richardson K, Lenton TM, Folke C, Liverman D, Summerhayes CP, Barnosky AD, Cornell SE, Crucifix M, Donges JF, Fetzer I, Lade SJ, Scheffer M, Winkelmann R, Schellnhuber HJ (2018) Trajectories of the earth system in the anthropocene. Proc Natl Acad Sci USA 115:8252–8259

Stocker TF (2013) The closing door of climate targets. Science 339:280–282

Stocker TF, Johnsen SJ (2003) A minimum thermodynamic model for the bipolar seesaw. Paleoceanography 18:1087

Stocker TF, Plattner G-K (2016) Making use of the IPCC's powerful communication tool. Nat Clim Change 6:637–638

Stocker TF, Schmittner A (1997) Influence of CO_2 emission rates on the stability of the thermohaline circulation. Nature 388:862–865

Stocker TF, Wright DG (1991) Rapid transitions of the ocean's deep circulation induced by changes in surface water fluxes. Nature 351:729–732

Stocker TF, Wright DG, Broecker WS (1992) The influence of high-latitude surface forcing on the global thermohaline circulation. Paleoceanography 7:529–541
Stommel H (1961) Thermohaline convection with two stable regimes of flow. Tellus 13:224–230
Stouffer RJ, Manabe S (2003) Equilibrium response of thermohaline circulation to large changes in atmospheric CO_2 concentration. Clim Dyn 20:759–773
Stouffer RJ, Manabe S, Bryan K (1989) Interhemispheric asymmetry in climate response to a gradual increase of atmospheric CO_2. Nature 342:660–662
Stouffer RJ, Yin J, Gregory JM, Dixon KW, Spelman MJ, Hurlin W, Weaver AJ, Eby M, Flato GM, Hasumi H, Hu A, Jungclaus JH, Kamenkovich IV, Levermann A, Montoya M, Murakami S, Nawrath S, Oka A, Peltier WR, Robitaille DY, Sokolov A, Vettoretti G, Weber SL (2006) Investigating the causes of the response of the thermohaline circulation to past and future climate changes. J Clim 19:1365–1387
Tàbara JD, Frantzeskaki N, Hölscher K, Pedde S, Kok K, Lamperti F, Christensen JH, Jäger J, Berry P (2018) Positive tipping points in a rapidly warming world. Curr Opin Environ Sustain 31:120–129
Thompson V, Kennedy-Asser AT, Vosper E, Lo YTE, Huntingford C, Andrews O, Collins M, Hegerl GC, Mitchell D (2022) The western North America heat wave among the most extreme events ever recorded globally. Sci Adv 8:eabm6860
Tokarska KB, Zickfeld K, Rogelj J (2019) Path independence of carbon budgets when meeting a stringent global mean temperature target after an overshoot. Earth's Future 7:1283–1295
Vecellio DJ, Kong QQ, Kenney WL, Huber M (2023) Greatly enhanced risk to humans as a consequence of empirically determined lower moist heat stress tolerance. Proc Nat Acad Sci 120:e2305427120
Vettoretti G, Ditlevsen P, Jochum M, Rasmussen SO (2022) Atmospheric CO_2 control of spontaneous millennial-scale ice age climate oscillations. Nat Geosci 15:300–305
Wang SV, Foster A, Lenz EA, Kessler JD, Stroeve JC, Anderson LO, Turetsky MR, Betts R, Zou S, Liu W, Boos WR, Hausfather Z (2023) Mechanisms and impacts of earth system tipping elements. Rev Geophys 61:e2021RG000757
Weijer W, Cheng W, Drijfhout SS, Fedorov AV, Hu A, Jackson LC, Liu W, McDonagh EL, Mecking JV, Zhang J (2019) stability of the Atlantic meridional overturning circulation: a review and synthesis. J Geophys Res 124:5336–5375
Weitzman ML (2011) Fat-tailed uncertainty in the economics of catastrophic climate change. Rev Environ Econ Policy 5:275–292
Willcock S, Cooper GS, Addy J, Dearing JA (2023) Earlier collapse of anthropocene ecosystems driven by multiple faster and noisier drivers. Nat Sustain 6:1331–1342
Wood RA, Crucifix M, Lenton TM, Mach KJ, Moore C, New M, Sharpe S, Stocker TF, Sutton RT (2023) A climate science toolkit for high impact, low likelihood climate risks. Earth's Future 11:e2022EF003369
Worthington EL, Moat BI, Smeed DA, Mecking JV, Marsh R, McCarthy GD (2021) A 30-year reconstruction of the Atlantic meridional overturning circulation shows no decline. Ocean Sci 17:285–299
Wunderling N, Donges JF, Kurths J, Winkelmann R (2021) Interacting tipping elements increase risk of climate domino effects under global warming. Earth Syst Dyn 12:601–619
Wunderling W, Winkelmann NR, Rockstrom J, Loriani S, McKay DIA, Ritchie PDL, Sakschewski B, Donges JF (2023) Global warming overshoots increase risks of climate tipping cascades in a network model. Nature Clim Chang 13:75–82
Wunderling N, von der Heydt A, Aksenov Y, Barker S, Bastiaansen R, Brovkin V, Brunetti M, Couplet V, Kleinen T, Lear C, Lohmann J, Roman-Cuesta R, Sinet S, Swingedouw D, Winkelmann R, Anand P, Barichivich J, Bathiany S, Baudena M, Bruun J, Chiessi C, Coxall H, Docquier D, Donges JF, Falkena S, Klose AK, Obura D, Rocha J, Rynders S, Steinert N, Willeit M (2024) Climate tipping point interactions and cascades: A review. Earth Syst Dyn. https://doi.org/10.5194/esd-15-41-2024
Wunsch C (2006) Abrupt climate change: an alternative view. Quat Res 65:191–203
Zebiak SE, Cane MA (1987) A model El Niño-southern oscillation. Mon Wea Rev 115:2262–2278